The *J*-Matrix Method

Abdulaziz D. Alhaidari · Eric J. Heller ·
Hashim A. Yamani · Mohamed S. Abdelmonem
Editors

The *J*-Matrix Method

Developments and Applications

Foreword by Hashim A. Yamani and Eric J. Heller

Springer

Editors
Abdulaziz D. Alhaidari
Shura Council
Riyadh 11212
Saudi Arabia
haidari@mailaps.org

Eric J. Heller
Harvard University
Dept. of Chemistry & Physics
17 Oxford street
Cambridge MA 02138-2901
USA
heller@physics.harvard.edu

Hashim A. Yamani
Ministry of Commerce &
Industry
Riyadh 11127
Saudi Arabia
haydara@sbm.net.sa

Mohamed S. Abdelmonem
King Fahd University
of Petroleum & Minerals
Dept. of Physics
Dhahran 31261
Saudi Arabia
msmonem@kfupm.edu.sa

ISBN: 978-1-4020-6072-4 e-ISBN: 978-1-4020-6073-1

Library of Congress Control Number: 2008921100

Printed on acid-free paper

9 8 7 6 5 4 3 2 1

springer.com

Foreword

Although introduced 30 years ago, the J-matrix method has witnessed a resurgence of interest in the last few years. In fact, the interest never ceased, as some authors have found in this method an effective way of handling the continuous spectrum of scattering operators, in addition to other operators. The motivation behind the introduction of the J-matrix method will be presented in brief.

The introduction of fast computing machines enabled theorists to perform calculations, although approximate, in a conveniently short period of time. This made it possible to study varied scenarios and models, and the effects that different possible parameters have on the final results of such calculations. The first area of research that benefited from this opportunity was the structural calculation of atomic and nuclear systems. The Hamiltonian element of the system was set up as a matrix in a convenient, finite, bound-state-like basis. A matrix of larger size resulted in a better configuration interaction matrix that was subsequently diagonalized. The discrete energy eigenvalues thus obtained approximated the spectrum of the system, while the eigenfunctions approximated the wave function of the resulting discrete state. Structural theorists were delighted because they were able to obtain very accurate values for the lowest energy states of interest.

Of course, the result of diagonalization also gives information on the remaining discrete states, including those that lie in the energy continuum. The fact that the approximation yields 'discrete' scattering states could not be helped, since the Hamiltonian is represented by a finite matrix. The situation is worsened by the fact that the eigenfunctions of these "discrete" continuum states are bound-state-like, as all members of the basis set used in constructing the Hamiltonian have this property. This was deemed unnatural, and led theorists to believe that this data did not constitute information that was useful for the calculation of scattering. This belief almost put a stop to the application of basis-set techniques for the solving of scattering problems. However, a major turnaround occurred as a result of the work by Hazi and Taylor [1].

Hazi and Taylor took the natural step of asking whether a use could be made of the bound-state-like basis to describe resonances, which resemble bound states even though they are actually scattering states. This led to the "Stabilization method": real discrete energy eigenvalues closest to the resonance energy become stable as the parameters of the calculation are varied. This development rekindled confidence among theorists for the belief that the discrete energy eigenvalues and eigenfunctions may indeed contain useful information about the continuous spectrum of the Hamiltonian.

The Reinhardt group at Harvard was in the meantime developing a computational scheme to extract scattering information from the Fredholm determinant [2]. For a given short range potential V, the Fredholm determinant $\mathcal{D}(z)$ is defined as:

$$\mathcal{D}(z) = \det\left(\frac{z-H}{z-H_0}\right),$$

with the phase of $\mathcal{D}(z)$ being the negative of the phase shift $\delta(E)$ in the limit $z \to E + i\varepsilon$. It also satisfies the dispersion relation

$$\mathcal{D}(z) = 1 + \int_0^\infty \frac{A(E)}{z-E} dE$$

where $\mathcal{D}(E+i\varepsilon) = d(E) - i\pi A(E)$, and the functions $d(E)$ and $A(E)$ are both real and analytic. Encouraged by the stabilization results, the group worked with the matrix representation of H and H_0 in a finite L^2-basis set, $\{\phi_n\}_{n=0}^{N-1}$. Therefore, the Fredholm determinant has the approximate value

$$\mathcal{D}^{\text{approx}}(z) = \prod_{i=0}^{N-1}\left(\frac{z-E_i}{z-E_i^0}\right),$$

where $\{E_i\}_{i=0}^{N-1}$ and $\left\{E_i^0\right\}_{i=0}^{N-1}$ are the eigenvalues of the finite matrices $\tilde{H}$ and $\tilde{H}_0$, respectively. The above equation can be cast in the form

$$\mathcal{D}^{\text{approx}}(z) = 1 + \sum_{i=0}^{N-1}\frac{\Gamma_i}{z-E_i^0},$$

which resembles a quadrature approximation to the above dispersion relation with the set $\left\{E_i^0\right\}_{i=0}^{N-1}$ containing the abscissas. Furthermore, Schwartz observed that if the basis $\{\phi_n\}_{n=0}^{N-1}$ is a certain Laguerre functions, then the abscissas fall as the transformed zeros of an orthogonal polynomial with known properties [3]. These facts enabled the group to calculate $\mathcal{D}^{\text{approx}}(E+i\varepsilon)$ and obtain accurate scattering results.

This development added some analytical tools to the predominant numerical tools available for use in the L^2-Fredholm method. For instance, the discrete eigenfunction of the finite $N \times N$ Hamiltonian H_0 may now be written as a finite sum of L^2-basis, with the orthogonal polynomials as Fourier-like expansion coefficients:

$$|\psi_N(E)\rangle = B_N(E) \sum_{n=0}^{N-1} P_n(E)\,|\phi_n\rangle$$

where $P_n(E)$ is the orthogonal polynomial such that $P_N(E_i^0) = 0$.

Heller of the Reinhardt group, who was also studying ways of enhancing the R-matrix method of scattering, proposed a way to improve the accuracy of the L^2-Fredholm calculation. He first proposed that the exact sine-like solution of the reference Hamiltonian H_0 be expanded in the complete L^2-basis. This could formally be done since the Fourier-like expansion coefficients were obtainable, even analytically. Similar to the R-matrix method, but unlike the L^2 Fredholm method, this approach made it possible to take full account of the reference Hamiltonian H_0 by giving a full solution of the reference problem in the complete basis. Furthermore, in the spirit of the R-matrix method, Heller and Yamani proposed that the potential representation be limited to a finite subset of the complete basis. This is the basic idea behind the J-matrix method [4], so called because the matrix representation of the operator $J = H_0 - E$ in certain Laguerre or oscillator basis functions is tridiagonal (Jacobi).

The analogy with the R-matrix method [5] is strong. The language of the R-matrix in configuration space is translated for the J-matrix into the language of function space. The full model Hamiltonian and wavefunction are written for the R-matrix method and J-matrix method as follows, in their respective languages:

$$\tilde{H} = \begin{cases} H_0 + V(r) & r \leq R \\ H_0 & r > R \end{cases} \quad \rightarrow \quad \tilde{H}_{nm} = \begin{cases} (H_0)_{nm} + V_{nm} & n, m \leq N-1 \\ (H_0)_{nm} & n, m > N-1 \end{cases}$$

whereas in a pictorial representation it looks as follows

$$\tilde{H} - E \sim \left(\begin{array}{ccccc|cccccc}
\times & \times & \times & \cdots & \times & 0 & & & & & \\
\times & \times & \times & \cdots & \times & 0 & & 0 & & & \\
\vdots & \vdots & \vdots & & \vdots & \vdots & & & & & \\
\times & \times & \times & \cdots & \times & J_{N-1,N} & 0 & 0 & 0 & \cdots & \\
\hline
0 & 0 & 0 & \cdots & J_{N,N-1} & J_{N,N} & J_{N+1,N+1} & & & & \\
 & & & & 0 & J_{N+1,N} & J_{N+1,N+1} & J_{N+1,N+2} & & 0 & \\
 & & 0 & & 0 & & J_{N+2,N+1} & J_{N+2,N+2} & J_{N+2,N+3} & & \\
 & & & & \vdots & & & \cdots & \cdots & \cdots & \\
 & & & & \vdots & & 0 & & \cdots & \cdots &
\end{array}\right)$$

If we expand the solution $\psi(E, r)$ in the complete basis as $\psi(E, r) = \sum_{n=0}^{\infty} f_n(E)\, \phi_n(r)$, then

$$\psi(E,r) = \begin{cases} \psi(E,r) & r \leq R \\ \frac{1}{\sqrt{k}}\left[e^{-ikr} - S(E)e^{ikr}\right] & r > R \end{cases} \qquad \rightarrow$$

$$f_n = \begin{cases} f_n & n,m \leq N-1 \\ \frac{1}{\sqrt{k}}\left[(c_n - is_n) - S(E)(c_n + is_n)\right] & n,m > N-1 \end{cases}$$

where s_n and c_n are the expansion coefficients of the sine-like and cosine-like solutions of the reference Hamiltonian in the complete L^2 basis. In both methods, the resulting scattering matrix $S(E)$ is exact for the approximate model Hamiltonian $\tilde{H}$. This contrasts with the use of variational schemes to solve the scattering problem, which basically seek an approximate solution to the exact Hamiltonian H. The latter methods sometimes lead to the existence of anomalous pseudo-resonance behavior of the calculated cross sections. This phenomenon is not present in J-matrix calculations [6].

An example of a closed form solution provided by the J-matrix method is the explicit result for the exact S-matrix of a truncated potential:

$$S(E) = \frac{(c_{N-1} - is_{N-1}) + g_{N-1,N-1}(E)J_{N-1,N}(E)(c_N - is_N)}{(c_{N-1} + is_{N-1}) + g_{N-1,N-1}(E)J_{N-1,N}(E)(c_N + is_N)}$$

Here $g_{N-1,N-1}(E)$ is the $(N-1,N-1)$ element of the inverse of the matrix representation of $(H - E)$ in the finite basis, and $J_{nm}(E) \equiv (H_0 - E)_{nm}$. Results show that $S(E)$ is a highly accurate approximation of the S-matrix of the exact Hamiltonian. Most notably, $S(E)$ has the following two desirable properties:

(a) $S(E)$ is a smooth function of E, even as it assumes the values of the eigenvalues $\{E_i\}_{i=0}^{N-1}$ or $\{E_i^0\}_{i=0}^{N-1}$. In fact, at the points $\{E_i\}_{i=0}^{N-1}$, $S(E)$ has the special value

$$S(E_i) = \left.\frac{c_N - is_N}{c_N + is_N}\right|_{E=E_i}$$

This is a noteworthy result, which states that the S-matrix (cross section, or phase shift) for the approximate Hamiltonian $\tilde{H}$ could be calculated exactly, at positive discrete energies, by knowing only the coefficients of the wavefunction of the reference Hamiltonian H_0 in the basis. Yamani and Abdelmonem took the set $\{S(E_i)\}_{i=0}^{N-1}$, representing the value of $S(E)$ on the real (scattering) energy axis, and analytically continued it to the lower complex energy plane to search for shallow resonances [7]. These are characterized by poles of $S(z)$ in the second sheet of the complex energy plane. The same authors showed that a similar, and slightly more complicated result, holds in the multi-channel case [7].

(b) The diagonalization of H in the finite L^2 basis needs to be done only once to enable the calculation of $S(z)$ over the entire continuum range of energy.

The first application of the J-matrix method was in the Tempkin-Poet model of an S-wave e-H scattering analysis [8]. The target and incoming projectile are described on the same basis, resulting in a finite number of target channels. A six-state calculation was able to reproduce previous results using different theoretical methods, and the calculation improved when the finite basis was enlarged. Implicit in this model is the approximation of the target by a finite number of channels; some, if any, represent the bound states, while the rest represent the continuum of scattering states. This naturally leads to anomalous behavior of the scattering cross section when the scattering energy is increased, so that more discrete channels become "open." Of course, this should not be the case. The correction of this anomalous behavior is still an outstanding problem for the J-matrix method and for similar methods that approximate the target ionization region by a finite number of channels. Bray and Stelbovics showed that the anomalous behavior tends to disappear as the number of target channels is increased [9].

Another early application of the J-matrix method of scattering was carried out by Broad and Reinhardt, who calculated e-H$^-$ scattering cross-sections and accurately reproduced resonance positions [10].

Further developments in the J-matrix method ensued:

(1) The matrix elements of Green's function associated with the reference Hamiltonian H_0 were evaluated by Heller, showing the analogy to the configuration space result [6]:

$$G^0(r,r') = \tfrac{2}{k}\sin(kr_<)e^{ikr_>} \quad\rightarrow\quad G^0_{nm} = \langle\phi_n|\, G\,|\phi_m\rangle = \tfrac{2}{k}s_{n_<}(c_{n_>} + is_{n_>})$$

where $n_<$ ($n_>$) is the smaller (larger) of n and m. Heller also showed that if the regular and irregular J-matrix solutions are written, respectively, as

$$|\psi_R(E)\rangle = \sum_n R_n(E)\,|\phi_n\rangle, \qquad |\psi_I(E)\rangle = \sum_n I_n(E)\,|\phi_n\rangle,$$

Green's function is then given as

$$G_{nm}(E) = \tfrac{2}{k}R_{n_<}(I_{n_>} + i\,R_{n_>})$$

(2) Yamani and Abdelmonem generalized these results to the multi-channel case. However, they adopted a different approach that cast the results in terms of the finite Green's function $g_{N-1,N-1}(E)$ and only considered quantities associated with $G^0(E)$ [11]. In so doing, they showed that the Lippman-Schwinger equation $T = V - VGV$ could be solved over a continuous range of energies, without re-diagonalization of matrices every time the energy is varied.

(3) Horodecki achieved a relativistic generalization of the J-matrix method [12]. This generalization was refined by Alhaidari et al. [13].

(4) Yamani et al. made an important generalization of the J-matrix method to any convenient L^2-basis, without any significant loss of the advantages provided by the method [14].
(5) Vanroose et al. [15] enhanced the method, especially in the treatment of long range potentials, by introducing additional terms in the three-term recursion relation that takes into account the asymptotic behavior of the potential.
(6) Alhaidari et al. [16] present an alternative, but equivalent, approach to the regularization of the reference problem in the J-matrix method. It is an integration approach, which was found to be more direct and transparent than the classical differential approach. They also developed the relativistic J-matrix method of scattering for spin $\frac{1}{2}$ Dirac projectile with position-dependent mass [17].

The contributions in the present volume are aimed at giving a brief account of recent developments, and some selected applications, of the method in atomic and nuclear physics:

In Part II, convergence issues are revisited by Igashove who makes a comprehensive study of the convergence of the Fourier expansion of the wavefunction in the oscillator basis. After investigating the effects of the regularization procedure on convergence, Broeckhove et al. propose an alternative regularization approach in the J-matrix method and demonstrate the resulting improvements. On the other hand, a method for the accurate evaluation of the S-matrix for multi-channel analytic and non-analytic potentials in complex L^2 bases is presented in the same Section by Yamani and Abdelmonem. Using the tools of the relativistic J-matrix, Alhaidari obtains analytic expressions for the scattering phase shift of separable potentials with Laguerre-type form factors. Shirokov and Zaytsev study an interesting problem that is best addressed in the language of the J-matrix. They show that non-local interaction models could result in "isolated bound states" embedded in the continuum with positive as well as negative energy.

In Part III, Knyr et al. use the J-matrix method as a universal approach for the description of the process involving the ionization of atoms. They succeed in addressing the difficult problem of correctly describing the continuous spectrum eigenfunctions in the scattering of three charged particles using the J-matrix method. Papp exploits the J-matrix structure of the Coulomb Green's matrix to solve the Faddeev-type integral equations of the three-body Coulomb problem. In the article by Yamani and Abdelmonem, three approximation methods are proposed that endow the complex scaling method with the ability to compute resonance partial widths for multi-channel problems.

In Part IV, Lurie and Shirokov give a review of their recent work on the three-body loosely-bound nuclear systems within the J-matrix approach with an extended oscillator basis. They apply their investigation to ^{11}Li and ^{6}He nuclei. Furthermore, Shirokov et al. construct the nucleon–nucleon interaction by means of the J-matrix version pertaining to inverse scattering theory, where they eliminate the problem of ambiguity of the interaction by postulating tridiagonal and quasi-tridiagonal forms of the potential matrix. In the same Section, Arickx et al. use their proposal of modifying the J-matrix method (to account for coupling to long range interactions) to cluster descriptions of light nuclei. The method is applied to ^{6}He and ^{6}Be nuclei for scattering and reaction problems. Finally, in Part V, Johnson and Holder present a generalization of the density functional theory, which is widely used in chemical physics applications, using a theoretical framework whose structure parallels that of the J-matrix method.

It is the hope of the editors that this volume will provide the interested reader with enough material for him or her to appreciate the advantages of the J-matrix method, and to encourage its use and development into a viable method for theoretical calculation of nuclear and atomic systems.

Riyadh — Hashim A. Yamani
Cambridge, Massachusetts — Eric J. Heller

References

1. A. U. Hazi and H. S. Taylor, Phys. Rev. A **2**, 1109 (1970)
2. W. P. Reinhardt, D. W. Oxtoby, and T. N. Rescigno, Phys. Rev. Lett. **28**, 401 (1972); T. S. Murtaugh and W. P. Reinhardt, J. Chem. Phys. **57**, 2129 (1972)
3. C. Schwartz, Ann. Phys. (NY) **16**, 36 (1960)
4. E. J. Heller and H. A. Yamani, Phys. Rev. A **9**, 1201 (1974); H. A. Yamani and L. Fishman, J. Math. Phys. **16**, 410 (1975)
5. A. M. Lane and R. G. Thomas, Rev. Mod. Phys. **30**, 257 (1958); A. M. Lane and D. Robson, Phys. Rev. **178**, 1715 (1969)
6. E. J. Heller, Phys. Rev. A **12**, 1222 (1975)
7. H. A. Yamani and M. S. Abdelmonem, J. Phys. A **26**, L1183 (1993); **27**, 5345 (1994); **28**, 2709 (1996); **29**, 6991 (1996)
8. E. J. Heller and H. A. Yamani, Phys. Rev. A **9**, 1209 (1974)
9. I. Bray and A. T. Stelbovics, Phys. Rev. Lett. **69**, 53 (1992)
10. J. T. Broad and W. P. Reinhardt, J. Phys. B **9**, 1491 (1976)
11. H. A. Yamani and M. S. Abdelmonem, J. Phys. B, **30**, 1633 (1997); **30**, 3743 (1997)
12. P. Horodecki, Phys. Rev. A **62**, 052716 (2000)
13. A. D. Alhaidari, H. A. Yamani, and M. S. Abdelmonem, Phys. Rev. A **63**, 062708 (2001)
14. H. A. Yamani, A. D. Alhaidari, and M. S. Abdelmonem, Phys. Rev. A **64,** 042703 (2001)
15. W. Vanroose, J. Broeckhove, and F. Arickx, Phys. Rev. Lett. **88**, 010404 (2002)
16. A. D. Alhaidari, H. Bahlouli, M. S. Abdelmonem, F. Al-Ameen, and T. Al-Abdulaal, Phys. Lett. A **364**, 372 (2007)
17. A. D. Alhaidari, H. Bahlouli, A. Al-Hasan, and M. S. Abdelmonem, Phys. Rev. A **75**, 062711 (2007)

Contents

Contributors

M.S. Abdelmonem
Physics Department, King Fahd University of Petroleum and Minerals, Dhahran 31261, Saudi Arabia
msmonem@kfupm.edu.sa

A.D. Alhaidari
Shura Council, Riyadh 11212, Saudi Arabia
haidari@mailaps.org

F. Arickx
University of Antwerp, Group Computational Modeling and Programming, Antwerp, Belgium,
Frans.Arickx@ua.ac.be

J. Broeckhove
University of Antwerp, Group Computational Modeling and Programming, Antwerp, Belgium
Jan.Broeckhove@ua.ac.be

L. Fishman
Department of Chemistry, Harvard University, Cambridge, MA 02138, USA
Shidi53@aol.com

Eric J. Heller
Department of Chemistry and Physics, Harvard University, Cambridge, MA 02138, USA
heller@physics.harvard.edu

Dale A. Holder
Quantum Simulations, Inc., 5275 Sardis Road, Murrysville, PA 15668, USA
holder@quantumsimulations.com

S.Yu. Igashov
Moscow Engineering Physics Institute, Kashirskoe sh. 31, Moscow 115409, Russia
igashov@theor.mephi.ru

Benny G. Johnson
Quantum Simulations, Inc., 5275 Sardis Road, Murrysville, PA 15668, USA
johnson@quantumsimulations.com

V.A. Knyr
Pacific National University, Khabarovsk, Russia
knyr@fizika.khstu.ru

A. Lahmam-Bennani
Lab. des Collisions Atomiques et Moléculaires, Université de Paris-Sud XI, 91405 Orsay Cedex, France
azzedine.bennani@u-psud.fr

Yu.A. Lurie
The College of Judea and Samaria, Ariel 44837, Israel
ylurie@ycariel.yosh.ac.il

A.I. Mazur
Physics Department, Pacific National University, Tikhookeanskaya 136, Khabarovsk 680035, Russia
mazur@hpicnit.khstu.ru

A. Nesterov
Bogolyubov Institute for Theoretical Physics, Kiev, Ukraine
nesterov@bitp.kiev.ua

Z. Papp
Department of Physics and Astronomy, California State University, Long Beach, CA 90840, USA
zpapp@csulb.edu

Yu.V. Popov
Nuclear Physics Institute, Moscow State University, Moscow, Russia
popov@srd.sinp.msu.ru

A.M. Shirokov
Skobeltsyn Institute of Nuclear Physics, Moscow State University, Moscow 119992, Russia
shirokov@nucl-th.sinp.msu.ru

A.M. Sytcheva
University of Antwerp, Group Computational Modeling and Programming, Antwerp, Belgium
a.sytcheva@fz-rossendorf.de

W. Vanroose
University of Antwerp, Group Computational Modeling and Programming, Antwerp, Belgium
wim.vanroose@ua.ac.be

J.P. Vary
Department of Physics and Astronomy, Iowa State University, Ames, IA 50011-3160, USA
jvary@iastate.edu

V.S. Vasilevsky
Bogolyubov Institute for Theoretical Physics, Kiev, Ukraine
vasilev@zeos.net

T.A. Weber
Department of Physics and Astronomy, Iowa State University, Ames, IA 50011-3160, USA
taweber@iastate.edu

H.A. Yamani
Ministry of Commerce & Industry, P.O. Box 5729, Riyadh 11127, Saudi Arabia
haydara@sbm.net.sa

S.A. Zaytsev
Physics Department, Pacific National University, Tikhookeanskaya 136, Khabarovsk 680035, Russia
zaytsev@mail.khb.ru, zaytsev@fizika.khstu.ru

Part I
Two of the Original Papers

New L^2 Approach to Quantum Scattering: Theory

Eric J. Heller and Hashim A. Yamani*

Abstract By exploiting the soluble infinite tridiagonal (Jacobi)-matrix problem generated by evaluating a zero-order scattering Hamiltonian H_0 in a certain L^2 basis set, we obtain phase shift, wave functions, etc., which are exact for a full Hamiltonian H in which only the potential V is approximated. Only bound–bound (L^2) matrix elements of the Hamiltonian and finite matrix manipulations are needed. The method is worked out here for s-wave scattering using Laguerre basis functions. Kato improvement of the results and necessary generalizations to many channels are treated.

1 Introduction

In atomic and nuclear scattering, it is often desirable to use Slater (Laguerre) or oscillator (Hermite) basis functions. This chapter is the first of several in which we present a new method for performing scattering calculations entirely with square-integrable (L^2) functions. We develop techniques in which we attempt to take full advantage of the analytic properties of a given Hamiltonian and also of the L^2 basis which is used to describe the wave function. Specifically, in what follows, we develop the basic theory using Laguerre-type basis functions appropriate for s-wave scattering. In the following chapter [1], we will apply the method to electron-hydrogen elastic s-wave scattering below the n = 2 threshold, and to inelastic radial-limit scattering calculations above and below the ionization threshold.

Our basic approach is to treat an uncoupled Hamiltonian H_0 exactly in the space spanned by the complete L^2 basis. The remaining part of the Hamiltonian (i.e., the

E.J. Heller
Department of Chemistry and Physics, Harvard Univeristy, Cambridge, MA 02138, USA
e-mail: heller@physics.harvard.edu

*Supported by a fellowship from the College of Petroleum and Minerals, Dhahran, Saudi Arabia.

Reprinted with permission from:
E.J. Heller and H.A. Yamani, New L2 Approach to Quantum Scattering: Theory, Phys. Rev. A 9, 1201–1208, (1974). Copyright (1974) by the American Physical Society.
http://link.aps.org/abstract/PRA/v9/p1201

potential) is approximated to some desired degree of accuracy, V^{approx}, such that the resulting Hamiltonian $H_0 + V^{\text{approx}}$ is also exactly soluble in the complete L^2 space. Phase shifts and cross sections can then be extracted from the resulting wave function ψ_E. This wave function has the desirable property of being an exact solution to a well-defined scattering Hamiltonian. If V^{approx} is a good representation of the exact potential, and if second-order accuracy is desired, then ψ_E may be considered as a trial wave function in the standard variational formulas.

By an exact solution χ_E to the Hamiltonian $H_0 + V$, we mean of course,

$$(H_0 + V - E)\,|\chi_E\rangle = 0. \tag{1}$$

In a space of complete L^2 functions $\{\phi_n\}$, where χ_E is expanded as $\chi_E = \sum_0^\infty b_n\phi_n$, Eq. (1) is equivalent to

$$\langle\phi_m|\,(H_0 + V - E)\,|\chi_E\rangle = 0 \tag{2}$$

for all $m = 0, 1, 2, \ldots \infty$. For most potentials considered in scattering theory, it will not be possible to satisfy Eq. (2).

However, consider the basis set $\{\phi_m\}_{m=0}^\infty$ such that

$$\phi_m(r) = (\lambda r)e^{-\lambda r/2}L_m^1(\lambda r), \tag{3}$$

where λ is a scaling parameter. In Section 2.1, we show that by writing $\chi_E^0 = \sum_0^\infty b_n^0\phi_n$, the similar equation

$$\langle\phi_m|\,(H_0 - E)\,|\chi_E^0\rangle = 0, \qquad m = 0, 1, \ldots, \infty \tag{4}$$

where $H_0 = -\frac{1}{2}d^2/dr^2$, leads to a soluble Jacobi-matrix problem for the b_n^0's. The properties of the Jacobi matrix representation of H_0 in the L^2 basis play a central role in our method. For this reason we call our approach the Jacobi (or J-) matrix method.

In Section 2.2, we construct a solution ψ_E for

$$\langle\phi_m|\left(H_0 + V^N - E\right)|\psi_E\rangle = 0, \qquad m = 0, 1, \ldots, \infty \tag{5}$$

where V^N is an $N \times N$ matrix representation of V in the set $\{\phi_n\}$, thus achieving the goal of obtaining an exact solution to the Hamiltonian with an approximating potential. In Section 3, we employ this wave function as a trial function in the Kato variational formula [2]. In Section 4, the necessary extension to multi-channel scattering is developed. In Section 5, a brief discussion is presented.

2 Potential Scattering

2.1 The Unperturbed Hamiltonian $\mathbf{H_0}$

Our task in this section is to determine the coefficients b_n^0 of the expansion of χ_E^0 in Eq. (4) in terms of our basis set $\{\phi_n\}$. Substituting the expansion for χ_E^0 in Eq. (4) results in an infinite matrix problem for the set $\{b_n^0\}$:

$$J \cdot b^0 = 0, \tag{6}$$

where J is the matrix

$$\begin{aligned} J_{nm} &= \int_0^\infty \phi_n(r) \left(-\frac{1}{2}\frac{d^2}{dr^2} - E \right) \phi_m(r)\, dr \\ &= -\frac{1}{\lambda}(n+1)(m+1)\left(E + \frac{\lambda^2}{8} \right) \left(2x\delta_{n,m} - \delta_{n,m-1} - \delta_{n,m+1} \right), \end{aligned} \tag{7}$$

where $x = \left(E - \frac{\lambda^2}{8}\right) \Big/ \left(E + \frac{\lambda^2}{8}\right)$. Note that J is an infinite tridiagonal (Jacobi) matrix. Equation (6) is thus a three-term recursion relation for the s_n's [where $b_n^0 = s_n \big/ (n+1)$] of the form

$$2xs_n - s_{n-1} - s_{n+1} = 0, \quad \text{for } n \geq 1, \tag{8a}$$

with the initial relation

$$2xs_0 - s_1 = 0, \quad \text{for } n = 0. \tag{8b}$$

Equation (8a), being a second-order difference equation, naturally has two linearly independent solutions. However, Eq. (8b) provides a boundary condition and thereby completely determines the s_n's. Equation (8a) is the recursion relation satisfied by the Chebyschev polynomials [3], and Eq. (8b) gives us those polynomials of the second kind. Therefore, we may write

$$s_n = \sin(n+1)\theta, \tag{9}$$

where $\cos\theta = x = \left(E - \frac{\lambda^2}{8}\right) \Big/ \left(E + \frac{\lambda^2}{8}\right)$. $s_n \big/ \sin\theta$ is then an nth-order polynomial in x. A similar analysis of H_0 in the basis $\{\phi_n\}$ has been provided by Schwartz [4]. The expression for χ_E^0 then becomes

$$\langle r \mid \chi_E^0 \rangle = \sum_{n=0}^{\infty} \frac{s_n}{n+1} \phi_n(r) = \sum_{n=0}^{\infty} \frac{\sin(n+1)\theta}{n+1} \phi_n(r). \tag{10}$$

Since we have now solved Eq. (4) exactly in the basis set, it is not surprising that the s_n's are simply the expansion coefficients of sinkr (with $E = \frac{1}{2}k^2$) in terms of the ϕ_n's [5]. Note that although we have used a discrete (L^2) basis, H_0 nonetheless has a continuous spectrum. This stems from the fact that the set $\{\phi_n\}$ is infinite and complete in r on $[0,\infty]$.

For the purpose of Section 2.2, we will require an independent solution of the three-term recursion relation (8a). Specifically, we wish to find a set $\{c_n\}$ such that the function

$$\tilde{C}(r) = \sum_{n=0}^{\infty} \frac{c_n}{n+1}\phi_n(r)$$

behaves as coskr when $r \to \infty$. Since the c_n's form an independent solution to Eq. (8a), they satisfy the following equation:

$$2xc_n - c_{n-1} - c_{n+1} = 0, \quad n \geq 1 \tag{8c}$$

with the boundary condition

$$2xc_0 - c_1 = \beta \neq 0. \tag{8d}$$

It is easy to verify, because β is non-vanishing, that the differential equation satisfied by $\tilde{C}(r)$ is

$$(H_0 - E)\,\tilde{C}(r) = -\beta\left(E + \frac{\lambda^2}{8}\right)e^{-\lambda r/2}. \tag{11}$$

By employing the Green's function $g(r, r') = (2\sin kr_< \cos kr_>)/k$ it can be readily shown that the solution to Eq. (11) is

$$\tilde{C}(r) = -\frac{2\beta}{k}\left(E + \frac{\lambda^2}{8}\right) \times \left[\left(\int_0^r \sin kr' e^{-\lambda r'/2} dr'\right)\cos kr + \left(\int_r^\infty \cos kr' e^{-\lambda r'/2} dr'\right)\sin kr\right]$$

which upon carrying out the integration, reduces to

$$\tilde{C}(r) = -\beta\left(\cos kr - e^{-\lambda r/2}\right). \tag{12}$$

The requirement that $\tilde{C}(r) \to \cos(kr)$ as $r \to \infty$ means that $\beta = -1$. With this value for β, it is easily verified that $c_n = -\cos(n+1)\theta$ satisfies Eqs. (8c) and (8d). Therefore, $\tilde{C}(r)$ now reads

$$\tilde{C}(r) = \cos kr - e^{-\lambda r/2} = -\sum_{n=0}^{\infty} \frac{\cos(n+1)\theta}{n+1} \phi_n(r). \tag{13}$$

Note the interesting property that $\tilde{C}(r)$ behaves regularly at the origin.

2.2 *Adding an Approximating Potential*

One way to introduce an approximation to V is to truncate the representation of V in the basis $\{\phi_n\}$ to an $N \times N$ matrix; we call this new potential V^N:

$$\begin{aligned} V_{nm}^N &= \int_0^\infty \phi_n(r)V(r)\phi_m(r)dr, && n, m \le N-1 \\ &= 0, && \text{otherwise.} \end{aligned} \tag{14}$$

Our task is to solve

$$\left\langle \phi_m \right| \left(H_0 + V^N - E\right) \left|\psi_E\right\rangle = 0, \qquad m = 0, 1, \ldots, \infty \tag{15}$$

Where $\psi_E = \sum_{n=0}^{\infty} d_n \phi_n(r)$. Schematically, these equations look like

$$\begin{array}{c} \\ 0\\ 1\\ \bullet\\ \bullet\\ N-1\\ \\ N\\ N+1\\ \bullet\\ \bullet\\ \\ \\ \end{array}
\begin{array}{c} \begin{array}{ccccccccc} 0 & 1 & \bullet & \bullet & N-1 & N & N+1 & \bullet & \bullet \end{array} \\
\left[\begin{array}{ccccc|ccccc}
X & X & X & X & X & & & & & \\
X & X & X & X & X & & & & & \\
X & X & X & X & X & & & 0 & & \\
X & X & X & X & X & & & & & \\
X & X & X & X & X & X & & & & \\
\hline
 & & & & X & X & X & & & \\
 & & & & & X & X & X & 0 & \\
 & & 0 & & & & X & X & X & \\
 & & & & & & & \bullet & & \\
 & & & & & & 0 & & \bullet & \\
 & & & & & & & & & \bullet
\end{array}\right] \end{array}
\left[\begin{array}{c} d_0\\ d_1\\ \bullet\\ \bullet\\ d_{N-1}\\ \hline d_N\\ d_{N+1}\\ \bullet\\ \bullet\\ \\ \\ \end{array}\right]
=
\left[\begin{array}{c} 0\\ 0\\ \bullet\\ \bullet\\ 0\\ \hline 0\\ 0\\ \bullet\\ \bullet\\ \\ \\ \end{array}\right] \tag{16}$$

These equations can be solved in a number of ways. For example, one may use a matrix partitioning technique similar to Feshbach's method [6], treating the infinite Jacobi "tail" of the matrix by folding it in as an optical potential. However, we approach the problem from a different viewpoint, noting that V^N couples only the first N functions ϕ_m, $m = 0, 1, \ldots, N-1$, to each other. Thus outside the space spanned by these N basis functions, we expect the sine-like and the cosine-like solutions, derived in Section 2.1, to be valid. Therefore we write our solution as

$$\psi_E = \tilde{\Phi} + \tilde{S} + t\tilde{C}, \tag{17}$$

where $\tilde{\Phi} = \sum_{n=0}^{N-1} \tilde{a}_n \phi_n$, $\tilde{S}$ is the sine-like expansion χ_E^0 of Eq. (10), and $\tilde{C}$ is the cosine-like solution of Eq. (13). The unknown coefficient t, then, corresponds to the tangent of the phase shift caused by V^N. Since the $\tilde{a}_n$'s are yet to be determined, we can absorb the first N terms in the expansion of $\tilde{S}$ and $\tilde{C}$ into $\tilde{a}_n$'s, writing

$$\psi_E = \Phi + S + tC, \tag{18}$$

where

$$\Phi(r) = \sum_{n=0}^{N-1} a_n \phi_n(r), \tag{19a}$$

$$S(r) = \sum_{n=N}^{\infty} \frac{s_n}{n+1} \phi_n(r) = \sum_{n=N}^{\infty} \frac{\sin(n+1)\theta}{n+1} \phi_n(r), \tag{19b}$$

$$C(r) = \sum_{n=N}^{\infty} \frac{c_n}{n+1} \phi_n(r) = -\sum_{n=N}^{\infty} \frac{\cos(n+1)\theta}{n+1} \phi_n(r). \tag{19c}$$

The two forms (17) and (18) for ψ_E are, of course, equivalent, but Eq. (18) is more convenient.

We now proceed to verify that the $(N+1)$ unknowns $\{a_n, t\}$ are sufficient to determine an exact solution the Hamiltonian $H_0 + V^N$. Equation (15) imposes a restriction on ψ_E for each m, $m = 0, 1, \ldots, \infty$. We group these restrictions into four cases: first, the $N - 1$ conditions arising from $m = 0, 1, \ldots, N - 2$; second, the case for $m = N - 1$; third, the condition for $m = N$; and last, the remaining set of conditions arising from $m = N + 1, N + 2, \ldots, \infty$.

The first case leads to the $N - 1$ equations

$$\sum_{n=0}^{N-1} \left(J_{mn} + V_{mn}^N\right) a_n = 0, \quad m = 0, 1, \ldots, N-2. \tag{20a}$$

In case two, we have the equation

$$\sum_{n=0}^{N-1} \left(J_{N-1,n} + V_{N-1,n}^N\right) a_n + \left(J_{N-1,N} \frac{c_N}{N+1}\right) t = -J_{N-1,N} \frac{s_N}{N+1}. \tag{20b}$$

In case three, V^N is no longer operative. We consequently get

$$J_{N,N-1} a_{N-1} + \left(J_{N,N} \frac{c_N}{N+1} + J_{N,N+1} \frac{c_{N+1}}{N+2}\right) t = -\left(J_{N,N} \frac{s_N}{N+1} + J_{N,N+1} \frac{s_{N+1}}{N+2}\right)$$

which, upon using the three-term recursion relation satisfied by both c_n and s_n, reduces to

$$J_{N,N-1} a_{N-1} - J_{N,N-1} \left(c_{N-1}/N\right) t = J_{N,N-1} \left(s_{N-1}/N\right). \tag{20c}$$

So far we have $(N + 1)$ equations in $(N + 1)$ unknowns. It would seem that we are left with an infinite number of equations arising from case four with no corresponding unknowns. Therefore, if we claim that $\psi_E = \Phi + S + tC$ is an exact solution for $H_0 + V^N$, then the remaining case-four equations, for $m = N + 1, N + 2, \ldots, \infty$, must be automatically satisfied. Fortunately, this is the case, because

$$\langle \phi_m | \left(H_0 + V^N - E \right) | \tilde{\Phi} + \tilde{S} + t\tilde{C} \rangle = \langle \phi_m | (H_0 - E) | S + tC \rangle \tag{21}$$

if $m \geq N + 1$. Equation (21) follows from the fact that V^N is defined to be zero in this region of Hilbert space, and because $(H_0 - E)$ is tridiagonal in the basis $\{\phi_n\}$, and therefore does not connect the N terms in the expansion of $\tilde{\Phi}$ or the first N terms in the expansion of $\tilde{S}$ and $\tilde{C}$ with ϕ_m for $m \geq N + 1$. Furthermore, for each $m \geq N + 1$ the right-hand side of Eq. (21) leads to the three-term recursion relation (8a) and (8c) for the coefficients c_n and s_n. Therefore the right-hand side of Eq. (21) vanishes identically. Thus, we now have exactly $(N + 1)$ equations to determine the $(N + 1)$ unknowns $\{a_n, t\}$. Hence the form (18) for ψ_E is indeed capable of giving an exact solution to Eq. (15).

Equations (20) can be written in matrix form as

$$\left[\begin{array}{cccc|c}
(J+V)_{0,0} & \cdots & (J+V)_{0,N-2} & (J+V)_{0,N-1} & 0 \\
(J+V)_{1,0} & \cdots & (J+V)_{1,N-2} & (J+V)_{1,N-1} & 0 \\
\cdot & & \cdot & \cdot & \cdot \\
\cdot & & \cdot & \cdot & \cdot \\
\cdot & & \cdot & \cdot & \cdot \\
(J+V)_{N-2,0} & \cdots & (J+V)_{N-2,N-2} & (J+V)_{N-2,N-1} & 0 \\
(J+V)_{N-1,0} & \cdots & (J+V)_{N-1,N-2} & (J+V)_{N-1,N-1} & J_{N-1,N}c_N/(N+1) \\
\hline
0 & \cdots & 0 & J_{N,N-1} & -J_{N,N-1}c_{N-1}/N
\end{array}\right]$$

$$\times \left[\begin{array}{c} a_0 \\ a_1 \\ \cdot \\ \cdot \\ \cdot \\ a_{N-2} \\ a_{N-1} \\ \hline t \end{array}\right] = \left[\begin{array}{c} 0 \\ 0 \\ \cdot \\ \cdot \\ \cdot \\ 0 \\ -J_{N-1,N}s_N/(N+1) \\ \hline J_{N,N-1}s_{N-1}/N \end{array}\right] \tag{22}$$

Notice that the large $N \times N$ block of the coefficient matrix is composed of the matrix elements of $\left(H_0 + V^N - E\right)$ in the first N Laguerre basis functions. To perform a calculation, we need merely augment this $N \times N$ matrix with the extra row and column shown and with the right-hand side driving term. Equation (22) can be immediately solved for t by standard techniques. An illuminating formula for $\tan\delta = t$ can be obtained by pre-diagonalizing the inner $N \times N$ matrix $\left(H_0 + V^N - E\right)_{nm}$

with the energy-independent transformation Γ, where

$$\left[\tilde{\Gamma}\left(H_0 + V^N - E\right)\Gamma\right]_{nm} = (E_n - E)\,\delta_{nm}. \tag{23}$$

Augmenting Γ to be the $(N + 1) \times (N + 1)$ matrix

$$\Gamma_A = \begin{pmatrix} \Gamma & 0 \\ 0 & 1 \end{pmatrix} \tag{24}$$

and applying it to Eq. (22), we obtain

$$t = \tan\delta = \frac{\left(\sin N\theta/N\right) + \nu(E) J_{N,N-1}\left[\sin(N+1)\theta/(N+1)\right]}{\left(\cos N\theta/N\right) + \nu(E) J_{N,N-1}\left[\cos(N+1)\theta/(N+1)\right]}, \tag{25}$$

where $\nu(E) = \sum_{m=0}^{N-1} \Gamma_{N-1,m}^2 \Big/ (E_m - E)$. In arriving at Eq. (25), we have used the fact that $s_n = \sin(n+1)\theta$ and $c_n = -\cos(n+1)\theta$. Note that the entire energy dependence of the phase shift is given analytically by Eq. (25). It is interesting that at the N Harris eigenvalues [7] E_m, $\tan\delta$ becomes simply

$$\tan\delta(E_m) = \tan(N+1)\theta(E_m) \tag{26a}$$

Also at the $N - 1$ points E_μ where $\nu(E_\mu) = 0$, we have

$$\tan\delta(E_\mu) = \tan N\theta(E_\mu). \tag{26b}$$

3 Kato Correction

The results of Section 2 are sufficient for obtaining the exact solution ψ_E and the exact $\tan\delta = t$ for the Hamiltonian $H_0 + V^N$ at energy E. Compared to the wave function and the phase shift for the exact Hamiltonian $H_0 + V$, ψ_E and t in general contain first-order errors. However, we may reduce these errors to second order by employing ψ_E as a trial function in the Kato [2] formula. If we write ψ_t for ψ_E of Eq. (17) and $\tan\delta_t$ for $\tan\delta$ of Eq. (25), then the Kato formula reads

$$\tan\delta_s = \tan\delta_t - \frac{2}{k}\int_0^\infty \psi_t(E - H)\psi_t dr, \tag{27}$$

where $\tan\delta_s$, is the stationary result.

Since $\left(H_0 + V^N - E\right)\psi_t = 0$, we can write the last equation as

$$\tan\delta_s = \tan\delta_t + \frac{2}{k}\langle\psi_t|\left(V - V^N\right)|\psi_t\rangle = \tan\delta_t + \frac{2}{k}\langle\psi_t|\,V^R\,|\psi_t\rangle. \tag{28}$$

Equation (28) is just the distorted-wave Born formula, where V^N is the distorting potential and V^R is the perturbation which has been excluded from the calculation of $\tan \delta_t$. In order to perform the integral in Eq. (28), bound–free and free–free matrix elements of the potential are required. An approximation to the Kato correction (28) which involves only bound-bound matrix elements of V is considered in the next chapter [1].

4 Many-Channel Scattering

In this section, we extend the previous potential scattering formulas to allow collision with targets possessing internal states. Basically we will be treating the close-coupling equations, employing an s-wave Laguerre set to describe the projectile wave function in each channel. As in the close-coupling formalism, we can treat exchange by the addition of a nonlocal potential.

Assuming the target possesses coordinates which we collectively call ρ, the Schrödinger equation for the many-particle wave function Θ reads

$$\left[H_T(\rho) - \frac{1}{2} \frac{\partial^2}{\partial r^2} + V(\rho, r) - E \right] \Theta = 0 \tag{29}$$

In the above equation, H_T is the given target Hamiltonian, r is the projectile coordinate, and $V(\rho, r)$ is the interaction with the target constituents. We assume that the target Hamiltonian posses a discrete set of L^2 eigenfunctions such that

$$H_T \chi_\alpha = E_\alpha \chi_\alpha, \qquad \alpha = 1, 2, \ldots, \infty. \tag{30}$$

If the target has a dense or continuous spectrum, the method of pseudo-target states may be employed [8]. This is done in the following chapter for the case of electron-hydrogen scattering.

As in the one-particle case, we will not be able to solve Eq. (29) for Θ exactly in the Hilbert space which is spanned by the set

$$\left\{ |\chi_\alpha\rangle \, |\phi_n^{(\alpha)}\rangle \right\}, \qquad \alpha = 1, 2, \ldots, \infty; n = 0, 1, 2, \ldots, \infty. \tag{31}$$

The functions $\left\{ \phi_n^{(\alpha)} \right\}$ are the s-wave Laguerre set

$$\phi_n^{(\alpha)}(r) = (\lambda_\alpha r) e^{-\lambda_\alpha r/2} L_n^1(\lambda_\alpha r). \tag{32}$$

Here we have allowed the projectile basis to be channel-dependent through the scaling parameter λ_α. We again truncate V by defining an approximate potential $\tilde{V}$ which has the matrix elements

$$\tilde{V}_{nn'}^{\alpha\alpha'} \equiv \left\langle \chi_\alpha \phi_n^{(\alpha)} \middle| V(\rho, r) \middle| \chi_{\alpha'} \phi_{n'}^{(\alpha')} \right\rangle \tag{33}$$

for $\alpha, \alpha' \le N_c$ and $n \le N_\alpha - 1$, $n' \le N_{\alpha'} - 1$. We define $\tilde{V}_{nn'}^{\alpha\alpha'} \equiv 0$ otherwise. The number N_α is the truncation limit in the channel α and N_c is the total number of channels which are allowed to couple.

To determine an S matrix, we will need N_0 independent solutions Θ_α of Eq. (29), where N_0 is the number of open channels. In the same spirit as in the single-channel case, we expand Θ_α as

$$\Theta_\alpha = \Phi_\alpha + \frac{\chi_\alpha S_\alpha}{\sqrt{k_\alpha}} + \sum_{\alpha'=1}^{N_c} \frac{R_{\alpha\alpha'} \chi_{\alpha'} C_{\alpha'}}{\sqrt{k_{\alpha'}}}, \qquad \alpha = 1, 2, \ldots, N_0. \tag{34}$$

The quantities k_α are the channel momenta $k_\alpha = \sqrt{2\,|E_\alpha - E|}$ where E_α is the channel energy appearing in Eq. (30). Analogous to the one-channel case, the S_α's and C_α's for open channels (see below for closed channels) are

$$\begin{aligned} S_\alpha &= \sum_{n=N_\alpha}^{\infty} \frac{\sin(n+1)\theta_\alpha}{n+1} \phi_n^{(\alpha)}(r) \\ C_\alpha &= -\sum_{n=N_\alpha}^{\infty} \frac{\cos(n+1)\theta_\alpha}{n+1} \phi_n^{(\alpha)}(r), \\ \theta_\alpha &= \cos^{-1}\left[\left(k_\alpha^2 - \frac{\lambda_\alpha^2}{4}\right) \Big/ \left(k_\alpha^2 + \frac{\lambda_\alpha^2}{4}\right)\right]. \end{aligned} \tag{35}$$

The internal function Φ_α is given by

$$\Phi_\alpha = \sum_{\alpha'=1}^{N_c} \sum_{n=0}^{N_{\alpha'}-1} a_{\alpha' n}^{\alpha} \chi_{\alpha'} \phi_n^{(\alpha)}. \tag{36}$$

The $a_{\alpha' n}^{\alpha}$'s are the expansion coefficients to be determined. The number of these coefficients is $\sum_{\alpha'=1}^{N_c} N_{\alpha'}$. The remaining unknowns $R_{\alpha\alpha'}$ are of course the elements of a reactance matrix which may be used to determine the S matrix as

$$S = (1 + i[R])(1 - i[R])^{-1}, \tag{37}$$

where [R] is the $N_0 \times N_0$ open channel part of R.

The sum in Eq. (34) should formally be extended to infinity, but since $\tilde{V} \equiv 0$ for any channel $\alpha > N_c$, $R_{\alpha\alpha'}$ vanishes for α or $\alpha' > N_c$. Actually the sum need only be over open channels. But if we do things this way, we place the burden of describing the exponentially decaying closed-channel asymptotic behavior on the internal function Φ_α. Near the threshold of channel $\alpha' = N_0 + 1$, however, the decay will be so slow that Φ_α will be incapable of properly describing the asymptotic behavior. Fortunately, we can allow the sum to include the closed channels up to N_c because we can find functions $C_\alpha(r)$ that can describe the asymptotic behavior in the closed channels in the same way that $C_\alpha(r)$ and $S_\alpha(r)$ do for open channels. The proper asymptotic form in a closed channel α is $e^{-|k_\alpha|r}$. A function which has this asymptotic form can be obtained by combining the expressions for $\tilde{S}_\alpha$ and $\tilde{C}_\alpha$ for imaginary $ik_\alpha = i\sqrt{2|E_\alpha - E|}$.

The resulting function and its expansion in terms of the basis set is given by

$$\tilde{C}_\alpha(r) = e^{-k_\alpha r} - e^{-\lambda_\alpha r/2} = \sum_{n=0}^{\infty} (-1)^n \frac{e^{-(n+1)\eta_\alpha}}{n+1} \phi_n^{(\alpha)} \tag{38a}$$

for $k_\alpha \leq \lambda_\alpha/2$ and where

$$\eta_\alpha = \cosh^{-1}\left|\frac{k_\alpha^2 + \lambda_\alpha^2/4}{k_\alpha^2 - \lambda_\alpha^2/4}\right|.$$

For $k_\alpha > \lambda_\alpha/2$ the formula reads

$$\tilde{C}_\alpha(r) = \sum_{n=0}^{\infty} \frac{e^{-(n+1)\eta_\alpha}}{n+1} \phi_n^{(\alpha)}. \tag{38b}$$

The form (38b) is really not needed, since for $k_\alpha > \lambda_\alpha/2$ the decay is rapid enough to be described by Φ_α.

We wish to show that Θ of Eq. (34) is capable of describing the exact solution for the problem

$$\left(H_T - \frac{1}{2}\frac{\partial^2}{\partial r^2} + \tilde{V} - E\right)\Theta_\alpha = 0 \tag{39}$$

We demand that all projections by $\left\langle \chi_\beta \phi_n^{(\beta)} \right|$ from the left vanish:

$$\left\langle \chi_\beta \phi_n^{(\beta)} \right| \left(H_T - \frac{1}{2}\frac{\partial^2}{\partial r^2} + \tilde{V} - E\right) \left|\Theta_\alpha\right\rangle = 0. \tag{40}$$

We need not consider the χ_β's for $\beta > N_c$, for in this case it is easy to show that the left-hand side of Eq. (40) is identically zero. Consider now a typical projection for $\beta < N_c$. It can readily be written as

$$\begin{aligned} &\left\langle \phi_m^{(\beta)} \right| -\frac{1}{2}\frac{\partial^2}{\partial r^2} - (E - E_\beta) \left| \sum_{n=0}^{N_\beta - 1} a_{\beta n}^{\alpha} \phi_n^{(\beta)} + \frac{S_\alpha \delta_{\alpha\beta}}{\sqrt{k_\alpha}} + \frac{R_{\alpha\beta} C_\beta}{\sqrt{k_\beta}} \right\rangle \\ &\quad + \sum_{\alpha'=1}^{N_c} \left\langle \phi_m^{(\beta)} \right| V^{\beta\alpha'} \left| \sum_{n=0}^{N_{\alpha'}-1} a_{\alpha' n}^{\alpha} \phi_n^{(\alpha')} \right\rangle = 0 \end{aligned} \tag{41}$$

We will not consider the four cases for m as is done in the single-channel case, but show instead that for $m > N_\beta$ the left-hand side of Eq. (41) also vanishes identically. First, the potential term vanishes by the definition of $\tilde{V}$ for $m > N_\beta$. Second, $\left[-\frac{1}{2}\left(\partial^2/\partial r^2\right) - \left(E - E_\beta\right)\right]$ is tridiagonal in the $\left\{\phi^{(\beta)}\right\}$. Thus, there will be no overlap between $\phi_m^{(\beta)}$ and the first N_β functions $\phi_n^{(\beta)}$. Then Eq. (41) becomes

$$\left\langle \phi_m^{(\beta)} \left| -\frac{1}{2}\frac{\partial^2}{\partial r^2} - (E - E_\beta) \right| \frac{S_\alpha \delta_{\alpha\beta}}{\sqrt{k_\alpha}} + \frac{R_{\alpha\beta} C_\beta}{\sqrt{k_\beta}} \right\rangle = 0. \tag{42}$$

But this is automatically true for $m > N_\beta$ in direct analogy with the single-channel case, because of the three-term recursion relation satisfied by the coefficients of S_β and C_β. The remainder of the equations; i.e., Eq. (41) for $m \le N_\beta$, lead to the same number of conditions as there are unknowns. The totality of these equations can be organized in a similar fashion as done in the form (22), with the L^2 matrix elements of $\tilde{V}$ appearing in a large inner block. One extra row and column are added to this block for each channel $\alpha \le N_c$. The right-hand side driving term and the solution "vector" containing the $a_{\beta n}^{\alpha}$'s and $R_{\alpha\alpha'}$'s have as many columns as open channels. The R matrix can then be obtained by solving the resulting linear equations. As before, the calculation may be facilitated by a pre-diagonalization of the inner block using an energy-independent transformation.

5 Discussion

The comparison between the approach taken in this chapter and the R-matrix method is considered first. In R-matrix theory [9], Hilbert space is divided into two parts, an inner coordinate-space up to a radius A and the remaining space from A to infinity. In the present work and in the spirit of Feshbach's generalization of R-matrix theory [6], we have divided the Hilbert space into two function spaces. We have an inner space consisting of those functions coupled by V^N ($\phi_n, n = 0, 1, 2, \ldots, N-1$), and an outer space in which the Hamiltonian is already solved and consists of the remaining functions (ϕ_n, n = N, ..., ∞). It is interesting to compare the Wigner R matrix which has the form

$$R(E) = \frac{1}{2}\sum_n \frac{|\psi_n(A)|^2}{E_n - E} = \sum_n \frac{\gamma_n^2}{E_n - E}$$

with what we have designated $\nu(E)$ in Eq. (25):

$$\nu(E) = \sum_n \frac{\Gamma_{N-1,n}^2}{E_n - E}$$

If the set $\{\phi_n\}$ were orthogonal, $\Gamma_{N-1,n}$ would just be

$$\Gamma_{N-1,n} = \langle \phi_{N-1} \mid \psi_n \rangle$$

Note that $\Gamma_{N-1,n}$ and γ_n are both the components of the wave function ψ_n at the boundary of their respective inner spaces.

Recent work seems to indicate that the R-matrix method works best using eigenfunctions of the scattering H_0 as a basis [10]. In this basis, H_0 is of course diagonal, and may be treated exactly by the addition of the Buttle correction. In our basis set, H_0 is tridiagonal and is also treated exactly. Other types of basis sets can also be used in the R-matrix method. In general, however, the ability to account for H_0, exactly is lost. The same is true in the present method, in the analogous situation when other basis functions not belonging to the tridiagonal set $\{\phi_n\}$ are used.

As in the R-matrix approach, we expect to find no Kohn-type pseudo-resonances [11] appearing in computed cross sections. Because in some sense V^N uniformly approximates V, we expect ψ^N to uniformly approximate ψ as $N \to \infty$. A series of ever denser, but narrower pseudo-resonances as found by Schwartz in this limit seems rather unlikely. This conjecture has been borne out by extensive calculations, some of which appear in the next chapter [1].

It is interesting to note that our truncated potential V^N leads to a separable kernel [12] in the T-matrix equation

$$T = V + VG^0T$$

It is easy to see that T possesses non-vanishing matrix elements only in the same L^2 subspace as that of V^N. This means that we can solve the finite matrix problem

$$T_{nn'} = V^N_{nn'} + \sum_{m,l=0}^{N-1} V^N_{nl} G^0_{lm} T_{mn'}; \quad n = 0, 1, \ldots, N-1, n' = 0, 1, \ldots, N-1. \quad (43a)$$

Then, the on-shell T matrix is obtained as

$$\sum_{n,n'=0}^{N-1} \langle E_0 \mid n \rangle\, T_{nn'} \left\langle n' \mid E_0 \right\rangle, \quad (43b)$$

where $|E_0\rangle$ is the properly normalized continuum eigenfunction of H_0. Compared to the approach taken in Sections 2–4, the separable approach leads to a less convenient algorithm. The kernel of Eq. (43a) is energy dependent and must be regenerated, and the linear equations must be resolved, at each new value of the total energy.

Finally, we compare the Jacobi-matrix approach to the recently developed L^2 Fredholm techniques [13–15]. Both approaches enjoy the advantage of requiring only L^2 matrix elements of the potential. The L^2 Fredholm method employs the devices of analytic continuation [13], dispersion correction [14], and contour rotation [15]. These techniques can be viewed as supplying, in an approximate fashion, information about the continuous spectrum of H_0 which is not explicitly contained in a finite L^2 matrix representation. Unfortunately, the amount of information concerning H_0 that can be extracted decreases with the number of basis functions per channel, and this can cause difficulties when basis size is a restriction. On the other hand, approximate treatment of H_0 can be advantageous when for example channel threshold details are of no interest or are unwanted artifacts of particular models. In the J-matrix method, H_0 is accounted for exactly independent of basis size. Thus we start with a large part of the problem "diagonalized" and the full analytic structure of the S matrix is built into the problem, raising the hope that quite small basis sets will be sufficient for many problems. The analytic nature of the solutions allows variational corrections to be made and provides a solid footing for further theoretical work.

We now summarize the steps necessary to perform a calculation with the J-matrix method. First, the potential V^N (or $\tilde{V}$) is evaluated in the Laguerre basis set; and is then added to the $N \times N$ tridiagonal representation of $H_0 - E$. To this inner matrix we add one extra row and column, for each asymptotic channel, containing matrix elements of H_0 and the $\cos(n+1)\theta$ terms. The right-hand side "driving" terms are similarly constructed with the $\sin(n+1)\theta$ terms. The resulting linear equations can be solved efficiently if a pre-diagonalizing transformation Γ is applied to the inner matrix as in Section 2. If desired, the matrix elements of $H_0 + V^N - E$ can be evaluated in the Slater set $(\lambda r)^n e^{-\lambda r/2}$, $n = 1, 2, \ldots, N$, since these are just transformed Laguerres. Then a different transformation Γ' will be necessary to pre-diagonalize the inner matrix.

In the following chapter we apply the method presented here to s-wave electron-hydrogen scattering model. The generalization of the method to all partial waves for both Laguerre and Hermite basis sets has been derived and will be the subject of a future publication. The case where H_0, contains the term α/r (i.e., the Coulomb case) is also worked out for Laguerre sets.

Acknowledgments We are grateful for helpful discussions with Professor William P. Reinhardt and his support of this work. We have also benefited greatly from conversations with L. Fishman, A. Hazi, T. Murtaugh, and T. Rescigno. This work was supported by a grant from the National Science Foundation.

References

1. E.J. Heller and H.A. Yamani, following paper, Phys. Rev. A 9, 1209 (1974).
2. T. Kato, Progr. Theor. Phys. 6, 394 (1951).
3. M. Abramowitz and I.A. Stegun, *Handbook of Mathematical Functions* (Dover, New York, 1965).
4. C. Schwartz, Ann. Phys. (N.Y.) 16, 36 (1961).
5. E.J. Heller, W.P. Reinhardt, and Hashim A. Yamani, J. Comp. Phys. **13**, 536 (1973).
6. H. Feshbach, Ann. Phys. (N.Y.) 19, 287 (1962).
7. F.E. Harris, Phys. Rev. Lett. 19, 173 (1967).
8. P.G. Burke, D.F. Gallaher, and S. Geltman, J. Phys. B 2, 1142 (1962).
9. See for example, A.M. Lane and R.G. Thomas, Rev. Mod. Phys. 30, 257 (1958); A. M. Lane and D. Robson, Phys. Rev. 178, 1715 (1969).

10. P.J.A. Buttle, Phys. Rev. 160, 719 (1967); P.G. Burke and W.D. Robb, J. Phys. B 5, 44 (1972); E.J. Heller, Chem. Phys. Lett. 23, 102 (1973).
11. C. Schwartz, Phys. Rev. 124, 1468 (1961). For a comprehensive review of the algebraic methods for scattering, including a discussion of pseudo-resonances, see D.G. Truhlar, J. Abdallah, and R.L. Smith, *Advances in Chemical Physics* (to be published).
12. R.G. Newton, *Scattering Theory of Waves and Particles* (McGraw-Hill, New York, 1966), p. 274.
13. W.P. Reinhardt, D.W. Oxtoby, and T.N. Rescigno, Phys. Rev. Lett. 28, 401 (1972); T.S. Murtaugh and W.P. Reinhardt, J. Chem. Phys. 57, 2129 (1972).
14. E.J. Heller, W.P. Reinhardt, and H.A. Yamani, J. Comp. Phys. **13**, 536 (1973); E.J. Heller, T.N. Rescigno, and W.P. Reinhardt, Phys. Rev. A 8, 2946 (1973); H.A. Yamani and W.P. Reinhardt Phys. Rev. A**11**, 1144 (1975).
15. T.N. Rescigno and W.P. Reinhardt, Phys. Rev. A 8, 2828 (1973); J. Nuttall and H. L. Cohen, Phys. Rev. 188, 1542 (1969).

J-Matrix Method: Extensions to Arbitrary Angular Momentum and to Coulomb Scattering

Hashim A. Yamani* **and Louis Fishman**

Abstract The J-matrix method introduced previously for s-wave scattering is extended to treat the ℓth partial wave kinetic energy and Coulomb Hamiltonians within the context of square integrable (L^2), Laguerre (Slater), and oscillator (Gaussian) basis sets. The determination of the expansion coefficients of the continuum eigenfunctions in terms of the L^2 basis set is shown to be equivalent to the solution of a linear second order differential equation with appropriate boundary conditions, and complete solutions are presented. Physical scattering problems are approximated by a well-defined model which is then solved exactly. In this manner, the generalization presented here treats the scattering of particles by neutral and charged systems. The appropriate formalism for treating many channel problems where target states of differing angular momentum are coupled is spelled out in detail. The method involves the evaluation of only L^2 matrix elements and finite matrix operations, yielding elastic and inelastic scattering information over a continuous range of energies.

1 Introduction

In two previous publications [1,2] (referred to as I and II) the J-matrix (Jacobi matrix) method was introduced as a new approach for solution of quantum scattering problems. As discussed in I, the principal characteristics of the method are its use

---

H.A. Yamani
Ministry of Commerce & Industry, P.O. Box 5729, Riyadh 11127, Saudi Arabia
e-mail: haydara@sbm.net.sa

*Supported by a fellowship from the College of Petroleum and Minerals, Dhahran, Saudi Arabia.

Reprinted with permission from: J-matrix Method: Extension to Arbitrary Angular Momentum and to Coulomb Scattering, H.A. Yamani and L. Fishman, Journal of Mathematical Physics 16, 410–420 (1975). Copyright 1975, American Institute of Physics.

A.D. Alhaidari et al. (eds.), *The J-Matrix Method*, 19–46.

of only square integrable (L^2) basis functions and its ability to yield an exact solution to a model scattering Hamiltonian, which, in a well-defined and systematically improvable manner, approximates the actual scattering Hamiltonian. The method is numerically highly efficient as scattering information is obtained over a continuous range of energies from a single matrix diagonalization.

The development of the J-matrix method as presented in I is based primarily upon the observation that the s wave kinetic energy,

$$H_0 = -\frac{1}{2}\frac{d^2}{dr^2} \tag{1}$$

can be analytically diagonalized in the Laguerre(Slater) basis:

$$\phi_n(\lambda r) = (\lambda r)e^{-\lambda r/2}L_n^1(\lambda r), \quad n = 0, 1, \ldots, \infty, \tag{2}$$

where λ is a scaling parameter. This follows from the fact that the infinite matrix representation of $(H_0 - k^2/2)$ in the above basis is tridiagonal (i.e., J or Jacobi matrix) and that the resulting three-term recursion scheme can be analytically solved yielding the expansion coefficients of both a "sine-like" $\tilde{S}(r)$ and a "cosine like" $\tilde{C}(r)$ function. The J-matrix solutions $\tilde{S}(r)$ and $\tilde{C}(r)$ are used to obtain the exact solution of the model scattering problem defined by approximating the potential V by its projection V^N onto the finite subspace spanned by the first N basis functions. That is, the exact solution Ψ of the scattering problem,

$$(H_0 + V^N - k^2/2)\Psi = 0, \tag{3}$$

is obtained by determining its expansion coefficients in terms of the basis set $\{\phi_n\}$ subject to the asymptotic boundary condition,

$$\Psi \to \tilde{S}(r) + \tan\delta\tilde{C}(r), \tag{4}$$

where δ is the phase shift due to the potential V^N.

This chapter is intended to generalize the formalism developed in I in three areas. First, the results of I are extended to all partial waves, in which case the uncoupled Hamiltonian becomes the ℓth partial wave kinetic energy operator,

$$H_0 = -\frac{1}{2}\frac{d^2}{dr^2} + \frac{\ell(\ell+1)}{2r^2}, \tag{5}$$

which has a Jacobi representation in the Laguerre basis,

$$\phi_n(\lambda r) = (\lambda r)^{\ell+1}e^{-\lambda r/2}L_n^{2\ell+1}(\lambda r), \quad n = 0, 1, \ldots, \infty, \tag{6}$$

Secondly, a similar analysis of the Hamiltonian of Eq. (5) is presented within the context of the oscillator(Gaussian) basis,

$$\phi_n(\lambda r) = (\lambda r)^{\ell+1} e^{-\lambda^2 r^2/2} L_n^{\ell+1/2}(\lambda^2 r^2), \quad n = 0, 1, \ldots, \infty, \tag{7}$$

which preserves the Jacobi representation and is also analytically soluble. The third generalization involves the analysis of the ℓth partial wave Coulomb Hamiltonian,

$$H_0 = -\frac{1}{2}\frac{d^2}{dr^2} + \frac{\ell(\ell+1)}{2r^2} + \frac{z}{r} \tag{8}$$

in the Laguerre function space of Eq. (6), which again yields a Jacobi form and is subsequently analytically soluble. It is noted that the analysis of the Coulomb Hamiltonian in the oscillator set of Eq. (7) does not lead to a Jacobi form.

For the solution of these problems, a general technique is developed which reduces the solution of the infinite recurrence problem for the asymptotically "sine-like" J-matrix eigenfunction to the solution of a linear second order differential equation with appropriate boundary conditions. An asymptotically "cosine-like" solution which obeys the same differential equation with different boundary conditions is then constructed. The fact that both J-matrix solutions obey the same recurrence scheme is essential to the success of the method as an efficient technique for solving scattering problems [1].

The program of the chapter is as follows: In Section 2.1, the generalized H_0 problem is considered and a general procedure for obtaining the expansion coefficients of the sine-like and cosine-like functions in terms of the basis sets is outlined. In Section 2.2, the general method is illustrated in detail for the case of the radial kinetic energy in a Laguerre basis. The analogous results for the oscillator basis and for the Coulomb problem are outlined in Sections 2.3 and 2.4, respectively. The details of the Coulomb derivation are given in the Appendix. Section 3 contains the application of the results thus obtained to potential scattering problems. This section presents a formula which allows for the computation of phase shifts. Section 4 presents the natural generalization of the J-matrix method to multi-channel scattering. Finally, Section 5 contains a brief discussion of the over all results and suggestions for applications and areas of further theoretical interest.

2 The H_0 Problem

The problem examined in this section is the "solution" of the equation,

$$\left(H_0 - k^2/2\right)\Psi = 0 \tag{9}$$

within the framework of the L^2 function space $\{\Phi_n\}$ in such a manner as to obtain both an asymptotically sine-like and asymptotically cosine-like function. The two J-matrix solutions, $\tilde{S}(r)$ and $\tilde{C}(r)$, form the basis for the asymptotic representation of the scattering wavefunction associated with the full problem. It will also be required that the expansion coefficients of both $\tilde{S}(r)$ and $\tilde{C}(r)$ satisfy the same three-term recursion scheme,

2.1 Generalized $\mathbf{H_0}$ *Problem*

The basic differential equation

$$\left(H_0 - k^2/2\right)\Psi^0 = 0 \tag{10}$$

possesses both a regular and an irregular solution which behave near the origin as

$$\Psi^0_{\text{reg}} \underset{r\to 0}{\sim} r^{\ell+1}, \tag{11a}$$

$$\Psi^0_{\text{irreg}} \underset{r\to 0}{\sim} r^{-\ell} \tag{11b}$$

and asymptotically as

$$\Psi^0_{\text{reg}} \underset{r\to \infty}{\sim} \sin\xi, \tag{12a}$$

$$\Psi^0_{\text{irreg}} \underset{r\to \infty}{\sim} \cos\xi, \tag{12b}$$

where $\xi = kr - \pi\ell/2$ in the free particle case and $\xi = kr + t\ln(2kr) - \pi\ell/2 + \sigma_\ell$ in the Coulomb case. In the above, the definitions $t = -z/k$ and $\sigma_\ell = \arg\Gamma(\ell+1-it)$ have been used [3].

Since the basis set $\{\phi_n\}$ is complete for functions regular at the origin, Ψ^0_{reg} can be expanded as

$$\Psi^0_{\text{reg}} \equiv \tilde{S}(r) = \sum_{n=0}^{\infty} s_n \phi_n(\lambda r) \tag{13}$$

with the expansion coefficients s_n being formally given by [4]

$$s_n = \left\langle \bar{\phi}_n(\lambda r) \mid \tilde{S}(r) \right\rangle \tag{14}$$

with $\bar{\phi}_n$ satisfying $\left\langle \bar{\phi}_n \mid \phi_m \right\rangle = \delta_{nm}$. A differential equation satisfied by the set of coefficients $\{s_n\}$ can be constructed in the following manner. Since the basis set $\{\phi_n\}$ tridiagonalizes the operator $\left(H_0 - k^2/2\right)$, the $\{s_n\}$ satisfy a three-term recursion relation of the form,

$$[a_{1n} + a_{2n}g(\eta)]\, s_n + a_{3n}s_{n-1} + a_{4n}s_{n+1} = 0, \quad n > 0, \tag{15a}$$

$$[a_{10} + a_{20}g(\eta)]\, s_0 + a_{40}s_1 = 0, \quad n = 0, \tag{15b}$$

where η is the energy variable defined by $\eta = k/\lambda$ and $g(\eta)$ is a function dependent upon the particular choice of $\{\phi_n\}$ and H_0. Differentiating Eq. (14) with respect to x, where x is a function of the energy variable η appropriate to the particular case, leads to a differential difference equation of the form

$$\eta \frac{dx}{d\eta} \frac{ds_n}{dx} = b_{1n} s_{n+1} + b_{2n} s_n + b_{3n} s_{n-1}, \quad n > 0 \tag{16a}$$

$$\eta \frac{dx}{d\eta} \frac{ds_0}{dx} = b_{10} s_1 + b_{20} s_0, \quad n = 0. \tag{16b}$$

For the case of the Laguerre function space, $b_{2n} = 0$, while for the oscillator function space the general form of Eq. (16) is appropriate. Combining Eqs. (15) and (16) yields a linear second order differential equation of the form

$$A(x) \frac{d^2 s_n}{dx^2} + B(x) \frac{ds_n}{dx} + D(x) s_n = 0 \tag{17}$$

with two linearly independent solutions χ_1 and χ_2 and a general solution of the form

$$s_n = \alpha_1 \chi_1 + \alpha_2 \chi_2. \tag{18}$$

Equation (15b) determines s_n to within a normalization constant. The advantage of the differential equation approach is that a cosine-like solution $\tilde{C}(r)$, whose expansion coefficients will also satisfy the differential equation (17), can be readily constructed.

The cosine-like *J*-matrix solution,

$$\tilde{C}(r) = \sum_{n=0}^{\infty} c_n \phi_n(\lambda r), \tag{19}$$

is constructed to be (1) regular at the origin like Ψ^0_{reg} so as to be expandable in the basis set $\{\phi_n\}$, (2) behave asymptotically as Ψ^0_{irreg}, and (3) to have its expansion coefficients $\{c_n\}$ satisfy Eq. (15b). This immediately means that $\tilde{C}(r)$ cannot satisfy the homogeneous differential equation (10). By choosing $\tilde{C}(r)$ to satisfy the inhomogeneous differential equation

$$\left(H_0 - k^2/2\right) \tilde{C}(r) = \beta \bar{\phi}_0(\lambda r), \tag{20}$$

the Green's function [5]

$$G(r, r') = 2\Psi^0_{\text{reg}}(r_<)\Psi^0_{\text{irreg}}(r_>) \Big/ W\left(\Psi^0_{\text{reg}}, \Psi^0_{\text{irreg}}\right) \tag{21}$$

may be used to obtain the solution [5]

$$\tilde{C}(r) = -\frac{2\beta}{W}\left[\Psi^0_{\text{irreg}}(r) \int_0^r dr' \Psi^0_{\text{reg}}(r')\bar{\phi}_0(\lambda r') + \Psi^0_{\text{reg}}(r) \int_r^\infty dr' \Psi^0_{\text{irreg}}(r')\bar{\phi}_0(\lambda r')\right], \tag{22}$$

where $W\left(\Psi^0_{\text{reg}}, \Psi^0_{\text{irreg}}\right)$ is the Wronskian of the two independent solutions Ψ^0_{reg} and Ψ^0_{irreg} and is independent of r and β is a free parameter [6]. $\tilde{C}(r)$ as given by Eq. (22) is regular at the origin and with the choice

$$\beta = -W/2s_0 \tag{23}$$

goes asymptotically as Ψ^0_{irreg}. The fact that the inhomogeneity of Eq. (20) is orthogonal to the set $\{\phi_n\}$ for $n = 1, 2, \ldots, \infty$, implies that, for $n > 0$, the $\{c_n\}$ satisfy the same three-term recursion relation as the $\{s_n\}$, Eq. (15a). The $n = 0$ case has the form

$$[a_{10} + a_{20}g(\eta)]\, c_0 + a_{40}c_1 = l(\eta) \neq 0 \tag{24}$$

where $l(\eta)$ is a function which depends upon the form of β and upon any terms that were divided out in the derivation of the homogeneous recursion relation, Eq. (15). Equation (24) is to be contrasted with the homogeneous initial condition

$$[a_{10} + a_{20}g(\eta)]\, s_0 + a_{40}s_1 = 0, \tag{15b}$$

which occurs in the sine-like J-matrix solution. It may be shown from Eqs. (22) and (23) that the set $\{c_n\}$ satisfies a differential difference equation analogous to Eq. (16), which when combined with Eqs. (15a) and (24) leads to the differential equation (17). The application of the inhomogeneous initial condition given by Eq. (24), and an additional boundary condition specific to the case being considered, determine the two integration constants γ_1 and γ_2 in the solution

$$c_n = \gamma_1\chi_1 + \gamma_2\chi_2. \tag{25}$$

2.2 Radial Kinetic Energy: Laguerre Basis

For the case of the radial kinetic energy and a Laguerre basis, the detailed construction of the J-matrix solutions is given following the general technique out lined in Section 2.1 The Hamiltonian is

$$H^0_\ell = -\frac{1}{2}\frac{d^2}{dr^2} + \frac{\ell(\ell+1)}{2r^2}, \tag{26}$$

while the L^2 expansion set is given by

$$\phi_n(\lambda r) = (\lambda r)^{\ell+1} e^{-\lambda r/2} L_n^{2\ell+1}(\lambda r), \quad n = 0, 1, \ldots, \infty, \tag{27}$$

In essence the infinite matrix problems

$$\langle \phi_m | \left(H^0_\ell - k^2/2\right) |\tilde{S}(r)\rangle = 0, \quad m = 0, 1, 2, \ldots, \infty, \tag{28a}$$

and

$$\langle \phi_m | \left(H_\ell^0 - k^2/2 \right) | \tilde{C}(r) \rangle = 0, \quad m = 1, 2, \ldots, \infty, \tag{28b}$$

are solved where

$$\tilde{S}(r) = \sum_{n=0}^{\infty} s_n \phi_n(\lambda r), \tag{13}$$

$$\tilde{C}(r) = \sum_{n=0}^{\infty} c_n \phi_n(\lambda r), \tag{19}$$

subject to the asymptotic boundary conditions

$$\tilde{S}(r) \underset{r\to\infty}{\sim} \sin(kr - \ell\pi/2) \tag{29a}$$

and

$$\tilde{C}(r) \underset{r\to\infty}{\sim} \cos(kr - \ell\pi/2). \tag{29b}$$

These solutions have the appropriate asymptotic forms to allow for the formulation of partial wave scattering problems for potentials falling off faster than $1/r^2$ at infinity, where the solution of the scattering problem will have the asymptotic form [3]

$$\Psi \underset{r\to\infty}{\sim} \sin(kr - \ell\pi/2) + \tan\delta \cos(kr - \ell\pi/2), \tag{30}$$

δ being the scattering phase shift.

From the boundary conditions of Eq. (29), $\tilde{S}(r)$ is designated the "sine-like" J-matrix solution, and $\tilde{C}(r)$ the "cosine-like" J-matrix solution. The sine-like solution is discussed in Section 2.2.1, where the recurrence relation for the coefficients $\{s_n\}$ is solved explicitly, giving closed form expressions. The discussion of $\tilde{C}(r)$ is somewhat more complex: In Section 2.2.2, a function $\tilde{C}(r)$ with the appropriate cosine-like behavior is constructed such that, for $n > 0$, the expansion coefficients $\{c_n\}$ obey the same recursion scheme as the set $\{s_n\}$; a fact that is an essential ingredient of the J-matrix method as will be seen in Sections 3 and 4 and has been discussed in I and II.

2.2.1 Sine-Like Solution

One of the linearly independent eigenfunctions of the radial kinetic energy,

$$H_\ell^0 = -\frac{1}{2}\frac{d^2}{dr^2} + \frac{\ell(\ell+1)}{2r^2}, \tag{26}$$

may be taken to be regular at $r = 0$ and sine-like asymptotically, that is,

$$\Psi_{\text{reg}}(r) \underset{r\to 0}{\sim} r^{\ell+1}, \tag{31a}$$

$$\Psi_{\text{reg}}(r) \underset{r\to \infty}{\sim} \sin(kr - \ell\pi/2) \tag{31b}$$

where the eigenfunction satisfying Eq. (31) is referred to as the regular solution [3]. A J-matrix solution,

$$\tilde{S}(r) \equiv \Psi_{\text{reg}}(r) = \sum_{n=0}^{\infty} s_n \phi_n(\lambda r), \tag{13'}$$

satisfying the boundary conditions of Eq. (31), is easily found within the context of the Laguerre set of Eq. (27).

The matrix

$$\langle\phi_n|\left(H_\ell^0 - k^2/2\right)|\phi_m\rangle = J_{nm} = \int_0^\infty dr\phi_n(\lambda r)\left[-\frac{1}{2}\frac{d^2}{dr^2} + \frac{\ell(\ell+1)}{2r^2} - \frac{k^2}{2}\right]\phi_m(\lambda r), \tag{32}$$

may, upon application of the orthogonality and recursion properties of the Laguerre functions $L_n^{2\ell+1}(\lambda r)$ [4], be reduced to the Jacobi (J-matrix) form

$$J_{nm} = -\frac{\lambda\eta}{2}\frac{\Gamma(n+2\ell+2)}{\Gamma(n+1)\sin\theta}\left[2x(n+\ell+1)\delta_{n,m} - n\delta_{n,m-1} - (n+2\ell+2)\delta_{n,m+1}\right], \tag{33}$$

where

$$x = \cos\theta = \left(\eta^2 - \frac{1}{4}\right)\Big/\left(\eta^2 + \frac{1}{4}\right) \tag{34a}$$

and

$$\eta = k/\lambda = \frac{1}{2}\cot(\theta/2) \tag{34b}$$

The expansion coefficients $\{s_n\}$ which satisfy the matrix equation $J \cdot s = 0$ may be determined by the solution of the three-term recursion relation

$$2x(n+\ell+1)u_n(x) - (n+2\ell+1)u_{n-1}(x) - (n+1)u_{n+1}(x) = 0 \tag{35a}$$

with the initial condition

$$2x(\ell-1)u_0(x) - u_1(x) = 0, \tag{35b}$$

where

$$s_n(x) = \left[\Gamma(n+1)/\Gamma(n+2\ell+2)\right]u_n(x). \tag{36}$$

Formally, of course, the $\{u_n\}$ are given by the Fourier projection [4]

$$u_n(x) = \int_0^\infty dr \Psi_{\text{reg}}(\eta r)\phi_n(r)/r \tag{37}$$

Rather than obtaining the $\{u_n\}$ by the direct evaluation of the integral, Eq. (37), a linear second order differential equation for the $\{u_n\}$ is derived. This differential equation formulation will be utilized for the construction of the cosine-like solution $\tilde{C}(r)$, where the analog of the projection of Eq. (37) does not exist.

Differentiation of Eq. (37) with respect to x, utilizing the fact that Ψ_{reg} is a function of (ηr), gives, after application of the chain rule, integration by parts, and application of the Laguerre recursion relations [4],

$$(x^2-1)\frac{du_n(x)}{dx} = \frac{n+1}{2}u_{n+1}(x) - \frac{n+2\ell+1}{2}u_{n-1}(x) \tag{38a}$$

with the initial condition

$$(x^2-1)\frac{du_0(x)}{dx} = \frac{1}{2}u_1(x). \tag{38b}$$

Combining Eqs. (35) and (38) gives the differential equation

$$(1-x^2)u_n''(x) - xu_n'(x) - \left[\ell(\ell+1)\frac{x^2}{1-x^2} - (n+\ell)^2 + \ell(\ell-1) - 2n - 1\right]u_n(x) = 0 \tag{39}$$

where the differentiation is with respect to x.

Equation (39) is easily solved. Letting $u_n(x) = \left(1-x^2\right)^{\frac{\ell+1}{2}} v_n(x)$ gives

$$(1-x^2)v_n''(x) - (2\ell+3)xv_n'(x) - n(n+2\ell+2)v_n(x) = 0 \tag{40}$$

which is the differential equation satisfied by the Gegenbauer polynomial $C_n^{\ell+1}(x)$ [6]. The general solution of Eq. (39) is then [6]

$$u_n^{\text{gen}}(x) = A_{n\ell}\chi_1(x) + B_{n\ell}\chi_2(x), \tag{41}$$

where

$$\chi_1(x) = (\sin\theta)^{\ell+1}C_n^{\ell+1}(\cos\theta), \tag{42a}$$

$$\chi_2(x) = \frac{(\cos\theta/2)^{\ell+1}}{(\sin\theta/2)^{\ell}}\,{}_2F_1\left(n+\ell+\frac{3}{2}, -n-\ell-\frac{1}{2}; \frac{1}{2}-\ell; \sin^2\frac{\theta}{2}\right), \tag{42b}$$

where again $x = \cos\theta$ and ${}_2F_1(a, b; c; z)$ is the Gauss hypergeometric function [7]. The coefficients $A_{n\ell}$ and $B_{n\ell}$ can be determined to within an ℓ dependent factor by substitution into Eq. (35a), resulting in

$$u_n^{\text{gen}}(x) = a_\ell \chi_1(x) + b_\ell \chi_2(x). \tag{43}$$

The form of $u_n^{\text{gen}}(x)$ appropriate to the initial condition

$$2x(\ell + 1)u_0(x) - u_1(x) = 0, \tag{35b}$$

may be determined from Eq. (42) as

$$u_n(x) = a_\ell \chi_1(x) \tag{44}$$

since [4]

$$2x(\ell + 1)C_0^{\ell+1}(x) - C_1^{\ell+1}(x) = 0 \tag{45}$$

while $\chi_2(x)$ does not satisfy Eq. (35b) [7]. Substitution of Eq. (44) into Eq. (13) gives

$$\tilde{S}(r) = a_\ell (\sin\theta)^{\ell+1} \sum_{n=0}^{\infty} \frac{\Gamma(n+1)}{\Gamma(n+2\ell+2)} C_n^{\ell+1}(\cos\theta)\phi_n(\lambda r), \tag{46}$$

where the requirement that

$$\lim_{\substack{r\to 0\\ k\to 0}} \left[\tilde{S}(kr) - \Psi_{\text{reg}}(kr)\right] = 0 \tag{47}$$

determines a_ℓ as $2^\ell\Gamma(\ell + 1)$ and thus that

$$s_n(x) = \left[2^\ell\Gamma(\ell + 1)\Gamma(n + 1)/\Gamma(n + 2\ell + 2)\right](\sin\theta)^{\ell+1} C_n^{\ell+1}(\cos\theta). \tag{48}$$

The coefficients $\{s_n\}$ of the regular, sine-like, eigenfunction of the radial kinetic energy have now been determined by the solution of a linear second order differential equation followed by the imposition of the appropriate boundary conditions.

2.2.2 Cosine-Like Solution

The cosine-like eigenfunction of the radial kinetic energy, which is irregular at the origin and defined by the conditions

$$\Psi_{\text{irreg}}(r) \underset{r\to 0}{\sim} r^{-\ell}, \tag{49a}$$

$$\Psi_{\text{irreg}}(r) \underset{r\to\infty}{\sim} \cos(kr - \ell\pi/2), \tag{49b}$$

will be referred to as the irregular solution [3]. For the construction of a cosine-like J-matrix solution

$$\tilde{C}(r) = \sum_{n=0}^{\infty} c_n \phi_n(\lambda r), \tag{19}$$

with the asymptotic boundary condition,

$$\tilde{C}(r) \underset{r\to\infty}{\sim} \cos(kr - \ell\pi/2) \tag{50}$$

it is seen that $\tilde{C}(r) \neq \Psi_{\text{irreg}}(r)$ since $\tilde{C}(r) \underset{r\to 0}{\sim} r^{\ell+1}$ as follows from Eq. (19). Thus, the expansion coefficients $\{c_n\}$ cannot be written as a Fourier projection of the form

$$\int_0^{\infty} dr \Psi_{\text{irreg}}(\eta r)\phi_n(r)/r \tag{51}$$

in analogy to Eq. (37). A cosine-like J-matrix solution $\tilde{C}(r)$ must thus be constructed with the following requirements: (1) $\tilde{C}(r)$ should have a cosine-like asymptotic form; (2) $\tilde{C}(r)$ should be regular at the origin; and (3) the coefficients $\{c_n\}$ should satisfy the same three-term recursion relation as the set $\{s_n\}$ for $n > 0$. Actually, from I it is seen that the most general condition in requirement (3) is $n \geq N+1$ where N is the number of functions in the subspace onto which the potential V is projected in the formulation of the model problem. For the purposes of the J-matrix method, however, it is sufficient to consider the condition $n > 0$. It is immediately seen that $\tilde{C}(r)$ will not satisfy $\langle \phi_m | \left(H_\ell^0 - k^2/2\right) | \tilde{C}(r) \rangle = 0, m = 0, 1, 2, \ldots, \infty$, as the cosine-like eigenfunction of H_ℓ^0 which is linearly independent of Ψ_{reg} is Ψ_{irreg} which is not regular at the origin.

The function $\tilde{C}(r)$ satisfying the above conditions is given by the solution of the equation,

$$\left(H_\ell^0 - k^2/2\right) \tilde{C}(r) = \beta \bar{\phi}_0(\lambda r) \tag{52}$$

subject to the boundary conditions

$$\tilde{C}(r) \underset{r\to 0}{\sim} r^{\ell+1}, \tag{53a}$$

$$\tilde{C}(r) \underset{r\to\infty}{\sim} \cos(kr - \ell\pi/2), \tag{53b}$$

where $\bar{\phi}_0(\lambda r) = \phi_0(\lambda r)/r\Gamma(2\ell + 2)$ [4]. It is noted that the particular choice of the inhomogeneity in Eq. (52) gives the infinite matrix problem

$$\langle \phi_m | \left(H_\ell^0 - k^2/2\right) | \tilde{C}(r) \rangle = \beta \delta_{m0}, \quad m = 0, 1, 2, \ldots, \infty, \tag{54}$$

immediately implying that, for $n > 0$, the $\{c_n\}$ and $\{s_n\}$ satisfy the same three-term pure recurrence relation. The parameter β is determined from a Green's function construction of the solution to Eq. (52) building in the boundary conditions of Eq. (53). Using Ψ_{reg} and Ψ_{irreg} as the two linearly independent solutions of the homogeneous equation $\left(H_\ell^0 - k^2/2\right)\tilde{C}(r) = 0$, the solution of the inhomogeneous problem of Eq. (52) may be written as [5]

$$\tilde{C}(r) = -\frac{2\beta}{W}\left[\Psi_{\text{irreg}}(r)\int_0^r dr'\Psi_{\text{reg}}(r')\bar{\phi}_0(\lambda r') + \Psi_{\text{reg}}(r)\int_r^\infty dr'\Psi_{\text{irreg}}(r')\bar{\phi}_0(\lambda r')\right], \tag{55}$$

where W is the Wronskian and is equal to $(-k)$ [3] for the radial kinetic energy case. By taking the $r \to \infty$ limit, β is determined as

$$\beta = \frac{-W}{2s_0} = \frac{2^\ell k\Gamma(\ell + 3/2)}{\sqrt{\pi}(\sin\theta)^{\ell+1}}, \tag{56}$$

where s_0 is the zeroth expansion coefficient of $\tilde{S}(r)$.

To demonstrate that the $\{c_n\}$ satisfy the same differential equation as the $\{s_n\}$ the equation

$$u_n(x) = \frac{\Gamma(n + 2\ell + 2)}{\Gamma(n+1)}c_n(x) = \int_0^\infty dr\tilde{C}(r/\lambda)\frac{\phi_n(r)}{r}, \tag{57}$$

where $\tilde{C}(r)$ is given by Eq. (55) with β given by Eq. (56), is differentiated with respect to x. Applying the same procedures that were used in going from Eq. (37) to (38) yields the differential difference equation of Eq. (38a), which, when combined with the previously derived recursion relation, Eq. (35b), gives the differential equation (39). The solution $u_n(x)$ satisfying the initial condition

$$2x(\ell + 1)u_0(x) - u_1(x) = -2^{\ell+1}\Gamma\left(\ell + \frac{3}{2}\right)\Big/\sqrt{\pi}(\sin\theta)^\ell \tag{58}$$

is given by

$$c_n(x) = a_\ell\chi_1(x) - \frac{\Gamma(\ell + 1/2)}{\sqrt{\pi}}\frac{\Gamma(n+1)}{\Gamma(n + 2\ell + 2)}\chi_2(x), \tag{59}$$

where $\chi_1(x)$ and $\chi_2(x)$ are given by Eq. (42). From the explicit expressions for $\tilde{S}(r)$ and $\tilde{C}(r)$ in terms of Ψ_{reg} and Ψ_{irreg} it follows that

$$c_n(-\theta) = (-)^\ell c_n(\theta), \tag{60a}$$

$$s_n(-\theta) = (-)^{\ell+1}s_n(\theta), \tag{60b}$$

immediately establishing that $a_\ell = 0$ and giving

$$c_n(x) = -\frac{\Gamma(\ell + 1/2)}{\sqrt{\pi}}\frac{\Gamma(n+1)}{\Gamma(n + 2\ell + 2)}\frac{(\cos\theta/2)^{\ell+1}}{(\sin\theta/2)^\ell} \times {}_2F_1\left(n + \ell + \frac{3}{2}, -n - \ell - \frac{1}{2}; \frac{1}{2} - \ell; \sin^2\frac{\theta}{2}\right). \tag{61}$$

A useful alternative form of $c_n(x)$ is given by [7]

$$c_n(x) = -2^{\ell}\frac{\Gamma(\ell+1/2)\Gamma(n+1)}{\sqrt{\pi}\Gamma(n+2\ell+2)}\frac{1}{(\sin\theta/2)^{\ell}} \times {}_2F_1\left(-n-2\ell-1, n+1; \frac{1}{2}-\ell; \sin^2\frac{\theta}{2}\right), \tag{62}$$

which is a finite polynomial in $(\sin^2\theta/2)$.

2.3 Radial Kinetic Energy: Oscillator Basis

Within the framework of the oscillator basis

$$\phi_n(\lambda r) = (\lambda r)^{\ell+1}e^{-\lambda^2 r^2/2}L_n^{\ell+1/2}(\lambda^2 r^2), \quad n = 0, 1, \ldots, \infty, \tag{63}$$

the J-matrix defined by

$$J_{nm} = \langle\phi_n|\left(H_\ell^0 - k^2/2\right)|\phi_m\rangle = \int_0^\infty dr\phi_n(\lambda r)\left[-\frac{1}{2}\frac{d^2}{dr^2}+\frac{\ell(\ell+1)}{2r^2}-\frac{k^2}{2}\right]\phi_m(\lambda r), \tag{64}$$

is a tridiagonal (Jacobi) matrix leading to the fundamental recursion relation

$$-\left(2n+\ell+\frac{3}{2}-\eta^2\right)u_n\left(\eta^2\right)+\left(n+\ell+\frac{1}{2}\right)u_{n-1}\left(\eta^2\right)+(n+1)u_{n+1}\left(\eta^2\right) = 0, \; n>0, \tag{65}$$

where again $\eta = k/\lambda$, for the solution of the infinite matrix problems

$$\langle\phi_m|\left(H_\ell^0 - k^2/2\right)|\tilde{S}(r)\rangle = 0, \quad m = 0, 1, 2, \ldots, \infty, \tag{66a}$$

$$\langle\phi_m|\left(H_\ell^0 - k^2/2\right)|\tilde{C}(r)\rangle = 0, \quad m = 0, 1, 2, \ldots, \infty, \tag{66b}$$

for the sine-like arid cosine-like J-matrix solutions.

The set $\{u_n\}$ satisfies the differential equation

$$u_n''\left(\eta^2\right)+\left[4n+2\ell+3-\eta^2-\frac{\ell(\ell+1)}{\eta^2}\right]u_n\left(\eta^2\right) = 0, \tag{67}$$

where the differentiation is with respect to η. Equation (67) has the general solution [6]

$$u_n^{\mathrm{gen}}(\eta) = A_{n\ell}\eta^{\ell+1}e^{-\eta^2/2}L_n^{\ell+1/2}\left(\eta^2\right)+B_{n\ell}\eta^{-\ell}e^{-\eta^2/2}{}_1F_1\left(-n-\ell-\frac{1}{2}, \frac{1}{2}-\ell, \eta^2\right), \tag{68}$$

where ${}_1F_1(a, c, z)$ is the confluent hypergeometric function [7]. The sine-like solution is deduced from Eq. (68) by substitution into Eq. (65), imposition of the initial condition

$$-\left(\ell + \frac{3}{2} - \eta^2\right) u_0\left(\eta^2\right) + u_1\left(\eta^2\right) = 0, \tag{69}$$

and normalization in the manner of Eq. (47), giving

$$s_n\left(\eta^2\right) = \frac{(-)^n \Gamma(n+1)}{\Gamma(n+\ell+3/2)} u_n\left(\eta^2\right) = \sqrt{2\pi} \frac{(-)^n \Gamma(n+1)}{\Gamma(n+\ell+3/2)} \eta^{\ell+1} e^{-\eta^2/2} L_n^{\ell+1/2}\left(\eta^2\right), \tag{70}$$

where

$$\tilde{S}(r) = \sum_{n=0}^{\infty} s_n\left(\eta^2\right) \phi_n(\lambda r). \tag{71}$$

Substitution into Eq. (65), imposition of the initial condition appropriate to the construction of a cosine like solution,

$$-\left(\ell + \frac{3}{2} - \eta^2\right) u_0\left(\eta^2\right) + u_1\left(\eta^2\right) = -\sqrt{2/\pi}\,\Gamma\left(\ell + \frac{3}{2}\right) \eta^{-\ell} e^{\eta^2/2}, \tag{72}$$

and use of the symmetry conditions given in Eq. (60) give

$$c_n\left(\eta^2\right) = (-)^n \sqrt{\frac{2}{\pi}} \frac{\Gamma\left(\ell + \frac{1}{2}\right) \Gamma(n+1)}{\Gamma\left(n + \ell - \frac{3}{2}\right)} \eta^{-\ell} e^{-\eta^2/2} {}_1F_1\left(-n - \ell - \frac{1}{2}, \frac{1}{2} - \ell, \eta^2\right), \tag{73}$$

where

$$\tilde{C}(r) = \sum_{n=0}^{\infty} c_n\left(\eta^2\right) \phi_n(\lambda r). \tag{74}$$

$\tilde{S}(r)$ and $\tilde{C}(r)$ are real, regular at the origin, and have sine-like and cosine-like asymptotic forms, respectively.

2.4 Radial Coulomb Hamiltonian: Laguerre Basis

Within the framework of the Laguerre basis, Coulomb J-matrix solutions $\tilde{S}(r)$ and $\tilde{C}(r)$ are constructed, which are regular at the origin and behave asymptotically as [3]

$$\tilde{S}(r) \underset{r\to\infty}{\sim} \sin\left(kr + t\,ln(2kr) - \pi\ell/2 + \sigma_\ell\right), \tag{75a}$$

$$\tilde{C}(r) \underset{r\to\infty}{\sim} \cos\left(kr + t\,ln(2kr) - \pi\ell/2 + \sigma_\ell\right). \tag{75b}$$

The Coulomb J matrix

$$\begin{aligned} J_{nm} &= \langle\phi_n|\left(H^0_{\ell,z} - k^2/2\right)|\phi_m\rangle \\ &= \int_0^\infty dr\phi_n(\lambda r)\left[-\frac{1}{2}\frac{d^2}{dr^2} + \frac{\ell(\ell+1)}{2r^2} + \frac{z}{r} - \frac{k^2}{2}\right]\phi_m(\lambda r) \end{aligned} \tag{76}$$

is a tridiagonal form, yielding the recurrence relation

$$\begin{aligned} &\left[(n+\ell+1)\left(\frac{x^2+1}{x}\right) - it\left(\frac{x^2-1}{x}\right)\right]v_n(x) \\ &\quad -(n+1)v_{n+1}(x) - (n+2\ell+1)v_{n-1}(x) = 0 \end{aligned}, \quad n>0 \tag{77}$$

where

$$x = e^{i\theta} = -\left[\left(\frac{1}{2} - i\eta\right)\Big/\left(\frac{1}{2} + i\eta\right)\right] \tag{78}$$

and

$$s_n(x) = \frac{2^\ell\Gamma(n+1)\,|\Gamma(\ell+1-it)|}{\Gamma(2\ell+2)\Gamma(n+2\ell+2)}e^{\pi t/2}v_n(x) \tag{79}$$

for the solution of the infinite matrix equations

$$\left\langle\phi_m\right|\left(H^0_{\ell,z} - k^2/2\right)\left|\tilde{S}(r)\right\rangle = 0, \qquad m = 0,1,2,\ldots,\infty, \tag{80a}$$

$$\left\langle\phi_m\right|\left(H^0_{\ell,z} - k^2/2\right)\left|\tilde{C}(r)\right\rangle = 0, \qquad m = 0,1,2,\ldots,\infty, \tag{80b}$$

for the sine-like and cosine-like J-matrix solutions.

The $\{v_n\}$ satisfy the differential equation

$$\begin{aligned} (x^2-1)v_n''(x) + \frac{x^2-1}{x}v_n'(x) - \Bigg\{&\left(\frac{x^2-1}{x^2}\right)\left[(n+\ell)^2 - \ell(\ell-1) + 2n + 1 - t^2\right] \\ &+ \left(\frac{x^2+1}{x^2}\right)\left[\left(\frac{x^2+1}{x^2-1}\right)\ell(\ell+1) - 2it(n+\ell+1)\right]\Bigg\}v_n(x) = 0 \end{aligned} \tag{81}$$

where the differentiation is with respect to $x = \exp(i\theta)$ and where $t = -z/k$ is considered to be independent of x. The derivation of this equation is discussed in the Appendix. The general solution of Eq. (81) is [7]

$$
\begin{aligned}
v_n^{\mathrm{gen}}(x) = {} & A_{n\ell t}(\sin\theta)^{\ell+1}e^{\theta t}e^{-in\theta}{}_2F_1\left(-n, \ell+1-it; 2\ell+2; 1-e^{2i\theta}\right) \\
& + B_{n\ell t}(\sin\theta)^{it}e^{-i(n+\ell+1)\theta}{}_2F_1\Big(-\ell-it, \ell+1-it; n+\ell+2-it; \\
& \times\left(1-e^{2i\theta}\right)^{-1}\Big)
\end{aligned}
\tag{82}
$$

The sine-like solution is determined from Eq. (82) by substitution into Eq. (77), imposition of the initial condition

$$
\left[(\ell+1)\left(\frac{x^2+1}{x}\right) - it\left(\frac{x^2-1}{x}\right)\right] v_0(x) - v_1(x) = 0, \tag{83}
$$

and by application of a normalization procedure discussed in the Appendix, giving

$$
s_n(\theta) = 2^{\ell}\frac{n!\,|\Gamma(\ell+1-it)|}{\Gamma(n+2\ell+2)}(\sin\theta)^{\ell+1}e^{\theta t}\exp\left[\left(\varepsilon+\frac{1}{2}\right)\pi t\right]P_n^{\ell+1}\left(\cos\theta; \frac{2z}{\lambda}; -\frac{2z}{\lambda}\right), \tag{84}
$$

where

$$
\varepsilon = \begin{cases} -1 \text{ for } \theta \in [0, +\pi] \\ +1 \text{ for } \theta \in [0, -\pi] \end{cases} \tag{85}
$$

and $P_n^{\lambda}(z; a; b)$ is the Pollaczek polynomial [4] as discussed in the Appendix. The cosine-like solution is determined by substitution into Eq. (77), imposition of the initial condition

$$
\left[(\ell+1)\left(\frac{x^2+1}{x}\right) - it\left(\frac{x^2-1}{x}\right)\right] v_0(x) - v_1(x) = \frac{-2^{-2\ell}e^{-\theta t}[\Gamma(2\ell+2)]^2}{(\sin\theta)^{\ell}\,|\Gamma(\ell+1-it)|^2} \tag{86}
$$

and application of a limiting procedure as discussed in the Appendix, giving

$$
c_n + is_n = \frac{-n!e^{i\sigma_\ell}e^{\pi t/2}e^{-\theta t}e^{-i(n+1)\theta}}{(2\sin\theta)^{\ell}\Gamma(n+\ell+2-it)}{}_2F_1(-\ell-it, n+1; n+\ell+2-it; e^{-2i\theta}). \tag{87}
$$

The functions $\tilde{S}(r)$ and $\tilde{C}(r)$ are real and reduce to the radial kinetic energy results when $z = 0$.

3 Potential Scattering

In this section, it will be assumed that the potential V does not couple angular momentum eigenstates; the generalization to the case where coupling occurs is straightforward and will be considered in the multi-channel case discussed in Section 4. Thus, the good angular momentum quantum number ℓ will be suppressed, it being implicitly assumed that a definite partial wave is under consideration. The aim of this section is then to determine the phase shift caused by the potential V with respect to the uncoupled Hamiltonian H_0 which may be taken to be the ℓth partial wave kinetic energy or Coulomb Hamiltonian.

As alluded to in the Introduction, and motivated in I, the potential V is approximated by truncating its representation in the function space $\{\phi_n\}$ to a finite, $N \times N$, representation V^N defined by

$$V_{nm}^N = \begin{cases} \int_0^\infty \phi_n(\lambda r)V(r)\phi_m(\lambda r)dr, & n, m \leq N-1, \\ 0, & \text{otherwise.} \end{cases} \tag{88}$$

The problem is then to solve

$$\langle \Phi_m | \left(H_0 + V^N - k^2/2\right) |\Psi_E\rangle = 0, \qquad m = 0, 1, 2, \ldots, \infty, \tag{89}$$

where

$$\Psi_E = \sum_{n=0}^{\infty} d_n\phi_n(\lambda r). \tag{90}$$

The form of V^N as defined in Eq. (88) is such, however, that it only couples the first N functions ϕ_m, $m = 0, 1, 2, \ldots, N-1$, in the infinite function space. Thus, outside the space spanned by these N basis functions the generalized sine-like and cosine-like solutions associated with the generalized H_0 problem discussed in Section 2 are valid. This leads to the following form for the wavefunction $\Psi_E(r)$,

$$\Psi_E(r) = \Phi(r) + S(r) + tC(r), \tag{91}$$

where

$$\Phi(r) = \sum_{n=0}^{N-1} a_n\phi_n(\lambda r), \tag{92a}$$

$$S(r) = \sum_{n=N}^{\infty} s_n\phi_n(\lambda r), \tag{92b}$$

$$C(r) = \sum_{n=N}^{\infty} c_n\phi_n(\lambda r), \tag{92c}$$

and t is the tangent of the phase shift caused by V^N with respect to H_0. Note that the first N terms in the expansions of $\tilde{S}(r)$ and $\tilde{C}(r)$ have been incorporated into the $\Phi(r)$ term, the remainder of the expansions being designated as $S(r)$ and $C(r)$, respectively. The sets of coefficients $\{s_n\}$ and $\{c_n\}$ are just those that were determined in Section 2 and are dependent upon the particular H_0 and basis set $\{\phi_n\}$ being considered. The solution (89) given in Eq. (91) contains $N+1$ unknowns, (t, $\{a_n\}$, $n = 0, 1, \ldots, N-1$).

By returning to Eq. (89), it is immediately established that it is satisfied for $m \geq N+1$: Since V^N is de fined to be zero in this region of the function space and due to the tridiagonal representation of $\left(H_0 - k^2/2\right)$,

$$\begin{aligned}
&\left\langle \phi_m \right| \left(H_0 + V^N - k^2/2\right) \left|\Psi_E\right\rangle \\
&\quad = \left\langle \phi_m \right| \left(H_0 - k^2/2\right) \left|S + tC\right\rangle && \text{(93a)}\\
&\quad = \langle \phi_m | \, J \, | S + tC \rangle && \text{(93b)}\\
&\quad = \sum_{n=0}^{\infty} J_{mn}(s_n + tc_n) && \text{(93c)}\\
&\quad = J_{m,m-1}(s_{m-1} + tc_{m-1}) + J_{m,m}(s_m + tc_m) + J_{m,m+1}(s_{m+1} + tc_{m+1}) && \text{(93d)}\\
&\quad = 0. && \text{(93e)}
\end{aligned}$$

Equation (93e) follows from the recursion relation satisfied by both s_n and c_n. The remaining $N+1$ conditions corresponding to the cases $m = 0, 1, .., N$ are sufficient to determine the $N+1$ unknowns, and by following the analysis presented in I the resulting system matrix is obtained:

$$\left[\begin{array}{cccc|c}
(J+V)_{0,0} & \cdots & (J+V)_{0,N-2} & (J+V)_{0,N-1} & 0 \\
(J+V)_{1,0} & \cdots & (J+V)_{1,N-2} & (J+V)_{1,N-1} & 0 \\
\cdot & & \cdot & \cdot & \cdot \\
\cdot & & \cdot & \cdot & \cdot \\
\cdot & & \cdot & \cdot & \cdot \\
(J+V)_{N-2,0} & \cdots & (J+V)_{N-2,N-2} & (J+V)_{N-2,N-1} & 0 \\
(J+V)_{N-1,0} & \cdots & (J+V)_{N-1,N-2} & (J+V)_{N-1,N-1} & J_{N-1,N}c_N \\
\hline
0 & \cdots & 0 & J_{N,N-1} & -J_{N,N-1}c_{N-1}
\end{array}\right]$$

$$\times \left[\begin{array}{c} a_0 \\ a_1 \\ \cdot \\ \cdot \\ \cdot \\ a_{N-2} \\ a_{N-1} \\ \hline t \end{array}\right] = \left[\begin{array}{c} 0 \\ 0 \\ \cdot \\ \cdot \\ \cdot \\ 0 \\ -J_{N-1,N}s_N \\ \hline J_{N,N-1}s_{N-1} \end{array}\right] \tag{94}$$

Equation (94) can be immediately solved for t by standard techniques. In particular, an expression for $t = \tan\delta$ can be obtained by pre-diagonalizing the inner $N \times N$ matrix $\left(H_0 + V^N - k^2/2\right)_{nm}$ with the energy independent transformation Γ defined by

$$\left[\tilde{\Gamma}\left(H_0 + V^N - k^2/2\right)\Gamma\right]_{nm} = (E_n - E)\delta_{nm}, \tag{95}$$

where $\{E_n\}$, are the Harris eigenvalues. Augmenting Γ to be the $(N+1) \times (N+1)$ matrix,

$$\Gamma_A = \begin{bmatrix} \Gamma & 0 \\ 0 & 1 \end{bmatrix}, \tag{96}$$

and applying it to Eq. (94) gives

$$t = \tan\delta = -\frac{s_{N-1} + \nu(E) J_{N,N-1} s_N}{c_{N-1} + \nu(E) J_{N,N-1} c_N}, \tag{97}$$

where

$$\nu(E) = \sum_{m=0}^{N-1} \frac{\Gamma_{N-1,m}^2}{E_m - E}. \tag{98}$$

At the positive Harris eigenvalues E_n [8], $\tan\delta$ reduces to

$$\tan\delta(E_n) = s_N(\eta_n)/c_N(\eta_n), \qquad \eta_n = k_n/\lambda. \tag{99}$$

Also, at the positive energies E_μ, where $\nu(E_\mu) = 0$, $\tan\delta$ becomes

$$\tan\delta(E_\mu) = -s_{N-1}(\eta_\mu)/c_{N-1}(\eta_\mu), \quad \eta_\mu = k_\mu/\lambda. \tag{100}$$

Equations (94), (97), (99), and (100) are the generalizations of the corresponding results obtained in I.

4 Multi-Channel Scattering

The potential scattering results of the previous section will now be generalized to include collisions with targets having internal states. Due to the generalizations developed for solving the H_0 problem, the results of which are given in Sections 2.2–2.4, collisions with charged, as well as neutral targets, can be considered, while employing the appropriate basis set (Laguerre or oscillator in the neutral case, and Laguerre in the charged case) to describe the projectile wavefunction. Since the method is formulated in terms of the close-coupling [9] equations, exchange can be treated by the inclusion of the appropriate nonlocal potential followed by its truncation in the J-matrix sense as was done in II.

The target will be described by the collective coordinate ρ, and its dynamics by $H_t(\rho)$. It is assumed that the target possesses a discrete set of L^2 eigenfunctions $R_\gamma(\rho)$ and, further, that the collective quantum number γ includes the total orbital quantum number of the target, L_t, its projection on some specified direction (the z axis), M_t, in addition to the quantum numbers μ that are needed further to completely define the target states. If the target has a dense or continuous spectrum, the method of pseudotarget states may be employed [2, 10]. It then follows that

$$H_t(\rho)R_\gamma(\rho) = E_{\mu,L_t} R_\gamma(\rho), \tag{101}$$

where $\gamma = \{\mu, L_t, M_t\}$.

The wavefunction Θ describing the projectile-target system satisfies the following Schrödinger equation:

$$[H_t(\rho) + H_0(r) + V(r,\rho) - E]\,\Theta(r,\rho) = 0, \tag{102}$$

where r is the projectile coordinate, $H_0(r) = -\frac{1}{2}\nabla_r^2 + z/r$, its Hamiltonian in atomic units, and $V(r,\rho)$, the interaction between the projectile and the target. In the neutral case z is equal to zero.

For most cases of interest, such as the scattering of electrons by light atoms, the total angular momentum of the system, L, its projection along the z axis, M, as well as the parity, Π, are conserved in the collision process. Therefore, it is more convenient to use a representation which is diagonal in these quantum numbers. Coupling the projectile with the target [9] in a picture with definite total L and total M leads to defining

$$\chi_\Gamma(\hat{r},\rho) = \sum_{m,M_t} C(\ell L_t L, m M_t M) Y_{\ell m}(\hat{r}) R_\gamma(\rho), \tag{103}$$

where $C(\ell L_t L, m M_t M)$ is the Clebsch-Gordan coefficient, Γ, the channel index $\{\mu L_t \ell L M\}$, and $Y_{\ell m}(\hat{r})$ the spherical harmonic functions. It is noted that χ_Γ is an eigenfunction of L and M, satisfying the same equation as R_γ, namely,

$$H_t(\rho)\chi_\Gamma(\hat{r},\rho) = E_{\mu,L_t}\chi_\Gamma(\hat{r},\rho), \tag{104}$$

where L satisfies the triangular relations, $|L_t - \ell| \le L \le L_t + \ell$. Since the function χ_Γ is composed of L^2 functions and is an eigenfunction of $H_t(\rho)$, the set $\{\chi_\Gamma\}$ may be taken to be orthonormal,

$$\langle \chi_\Gamma \mid \chi_{\Gamma'} \rangle = \delta_{\Gamma\Gamma'} = \delta_{\mu\mu'}\delta_{L_t L_t'}\delta_{\ell\ell'}, \tag{105}$$

where $\Gamma = \{\mu L_t \ell L M\}$ and $\Gamma' = \{\mu' L_t' \ell' L M\}$.

Within a definite *LM* picture, the Hilbert space of the system is spanned by the set

$$\left\{ |\chi_\Gamma\rangle \, |\phi_n^{(\ell,\Gamma)}\rangle \right\} \tag{106}$$

for all μ, L_t, ℓ, and for $n = 0, 1, \ldots, \infty$. For neutral targets the set $\{\phi_n^{(\ell,\Gamma)}\}$ can be either the Laguerre functions

$$\phi_n^{(\ell,\Gamma)}(\lambda_\Gamma r) = (\lambda_\Gamma r)^{\ell+1} e^{-\lambda_\Gamma r/2} L_n^{2\ell+1}(\lambda_\Gamma r), \qquad n = 0, 1, \ldots, \infty, \tag{107}$$

or the oscillator functions

$$\phi_n^{(\ell,\Gamma)}(\lambda_\Gamma r) = (\lambda_\Gamma r)^{\ell+1} e^{-\lambda_\Gamma^2 r^2/2} L_n^{\ell+1/2}(\lambda_\Gamma^2 r^2), \qquad n = 0, 1, \ldots, \infty, \tag{108}$$

while for charged targets the set $\{\phi_n^{(\ell,\Gamma)}\}$ can only be the Laguerre functions. Note that the projectile basis can be made channel dependent through the channel dependent scaling parameter λ_Γ.

In general it is not possible to solve Eq. (102) for $\Theta(r, \rho)$ exactly in the Hilbert space of Eq. (106). By following the procedure in I, the interaction potential V is approximated by $\tilde{V}$, which is defined by the following truncation scheme:

$$\tilde{V}_{nn'}^{\Gamma\Gamma'} = \begin{cases} \left\langle \chi_\Gamma \phi_n^{(\ell,\Gamma)} \middle| V(r,\rho) \middle| \chi_{\Gamma'} \phi_{n'}^{(\ell',\Gamma')} \right\rangle & \text{for } \Gamma, \Gamma' \leq N_c, n \leq N_\Gamma - 1, n' \leq N_{\Gamma'} - 1 \\ 0, & \text{otherwise} \end{cases} \tag{109}$$

where N_Γ is the truncation limit in the channel Γ and N_c is the total number of channels that are allowed to couple through the potential.

It is now proposed to solve the model equation

$$\left[H_t(\rho) + H_0(r) + \tilde{V}(r,\rho) - E\right] \Theta(r,\rho) = 0 \tag{110}$$

exactly in the Hilbert space of Eq. (106). There will be as many independent solutions Θ_Γ as there are open channels. By following the approach taken in I, Θ_Γ is expanded as

$$\Theta_\Gamma(r,\rho) = \Phi_\Gamma(r,\rho) + \frac{\chi_\Gamma(\hat{r},\rho) S_\Gamma(r)}{\sqrt{k_\Gamma}} + \sum_{\Gamma'}^{N_c} \frac{\chi_{\Gamma'}(\hat{r},\rho) C_{\Gamma'}(r) R_{\Gamma'\Gamma}}{\sqrt{k_{\Gamma'}}},$$
$$\Gamma = 1, 2, \ldots, N_0 \tag{111}$$

where N_0 is the number of open channels. The scattering matrix $\mathcal{S}^{LM}$ is related to the reactance matrix elements by the relation,

$$\mathcal{S}^{LM} = e^{i\sigma} (1 + i[R]) (1 - i[R])^{-1} e^{i\sigma}, \tag{112a}$$

where $[R]$ is the $N_0 \times N_0$ open-channel part of R and

$$\left(e^{i\sigma}\right)_{\Gamma\Gamma'} = e^{i\sigma_\ell} \delta_{\Gamma\Gamma'}. \tag{112b}$$

In Eq. (111), the quantity k_Γ is the wave number of the scattered electron and is given by

$$k_\Gamma = \sqrt{2\left|E - E_{\mu,L_t}\right|}. \tag{113}$$

Furthermore, for an open channel Γ,

$$S_\Gamma(r) = \sum_{n=N_\Gamma}^{\infty} s_n(k_\Gamma)\phi_n^{(\ell,\Gamma)}(\lambda_\Gamma r), \tag{114a}$$

$$C_\Gamma(r) = \sum_{n=N_\Gamma}^{\infty} c_n(k_\Gamma)\phi_n^{(\ell,\Gamma)}(\lambda_\Gamma r), \tag{114b}$$

where $\{s_n\}$ and $\{c_n\}$ are given by the results of Section 2, while, for a closed channel Γ, the cosine-like term $C_\Gamma(r)$ is replaced by the linear combination

$$C_\Gamma(r) + i S_\Gamma(r) \tag{115}$$

and evaluated at $k_\Gamma = i\,|k_\Gamma|$. The resulting function correctly describes the exponentially decaying closed channel asymptotic behavior. In the neutral case, the function of Eq. (115) is related to the spherical Hankel function of the first kind, $h_\ell^{(1)}(z)$, evaluated at $z = i\,|k_\Gamma|\,r$. The internal function, $\Phi_\Gamma(r,\rho)$, which describes the scattering process at close distances, is written as

$$\Phi_\Gamma(r,\rho) = \sum_{\Gamma'}^{N_c} \sum_{n=0}^{N_{\Gamma'}-1} a_n^{\Gamma'\Gamma} \chi_{\Gamma'}(\hat{r},\rho)\phi_n^{(\ell',\Gamma')}(\lambda_{\Gamma'} r). \tag{116}$$

The remainder of this section briefly demonstrates that the wavefunction Θ is capable of being an exact solution of the model Hamiltonian in the Hilbert space. This is accomplished by uniquely determining the unknowns $\{a_n^{\Gamma'\Gamma}, R_{\Gamma'\Gamma}\}$. It is required that all projections by $\left\langle \chi_{\Gamma''}\phi_{n''}^{(\ell'',\Gamma'')}\right|$ from the left-hand side of Eq. (110) vanish:

$$\left\langle \chi_{\Gamma''}\phi_{n''}^{(\ell'',\Gamma'')}\right| \left[H_t(\rho) + H_0(r) + \tilde{V}(r,\rho) - E\right] \left|\Theta(r,\rho)\right\rangle = 0. \tag{117}$$

For the case of $\Gamma = 1,2,\ldots,N_0$, $\Gamma'' = 1,2,\ldots,N_c$, and $n'' = 0,1,\ldots,N_{\Gamma''}$, $N_0\left[\sum_{\Gamma''=1}^{N_c}(N_{\Gamma''}+1)\right]$ equations are obtained which is equal to the total number of unknowns: $N_0\left(\sum_{\Gamma'=1}^{N_c} N_{\Gamma'}\right)$ of the a's and $N_0 N_c$ of the R's. It only remains now to show that for all other cases Eq. (117) is satisfied. The assertion is clear when $\Gamma'' > N_c$. Suppose now that $\Gamma'' \le N_c$, but than $n'' \ge N_{\Gamma''} + 1$. Then $\tilde{V}_{n''n'''}^{\Gamma''\Gamma'''} = 0$ by definition and Eq. (117) reduces to

$$\left\langle \phi_{n''}^{(\ell'',\Gamma'')}\right| \left[-\frac{1}{2}\frac{d^2}{dr^2} + \frac{\ell''(\ell''-1)}{2r^2} + \frac{z}{r} - \left(E - E_{\mu'',L_t''}\right)\right]$$
$$\times \left| \sum_{n=0}^{N_{\Gamma''}-1} a_n^{\Gamma''\Gamma}\phi_n^{(\ell'',\Gamma'')} + \frac{S_{\Gamma''}}{\sqrt{k_{\Gamma''}}}\delta_{\Gamma\Gamma''} + \frac{C_{\Gamma''}R_{\Gamma''\Gamma}}{\sqrt{k_{\Gamma''}}}\right\rangle = 0 \tag{118}$$

where $z = 0$ in the neutral target case. Since the operator appearing in Eq. (118) is tridiagonal in the $\{\phi_n^{(\ell'',\Gamma'')}\}$ representation, the contribution of the S and C functions vanish since their expansion coefficients $\{s_n\}$ and $\{c_n\}$ satisfy the resulting three-term recursion relation. Since $n'' \geq N_{\Gamma''} + 1$ while $n \leq N_{\Gamma''} - 1$, the contribution of the sum term in Eq. (118) vanishes, proving that Eq. (118) is identically true.

As in I, the nontrivial equations can be arranged such that the L^2 matrix elements of $\left(H_t + H_0 + \tilde{V}\right)$ appear in an inner block. Additionally, one extra row and column are added to this block for each channel $\Gamma \leq N_c$. The right-hand side driving term and the solution vector, containing the $a_n^{\Gamma\Gamma'}$ and $R_{\Gamma\Gamma'}$, have as many columns as open channels. The R matrix can then be extracted by solving the resulting linear equations by standard techniques. After having done so, the $\mathcal{S}^{LM}$ matrix is constructed via Eq. (112), from which physical quantities are then obtained: e.g., the cross section for the transition $\left(\mu_0 L_t^0 \rightarrow \mu L_t\right)$, averaged over M_t^0 states and summed over M_t states, is given by

$$\sigma\left(\mu_0 L_t^0 \rightarrow \mu L_t\right) = \frac{\pi}{\left(2L_t^0 + 1\right) k_{\Gamma^0}^2} \sum_{L_0 \ell L} (2L+1) \left|\delta_{\Gamma^0\Gamma} - \mathcal{S}_{\Gamma^0\Gamma}^{LM}\right|^2. \tag{119}$$

Other physical quantities of interest can be similarly constructed [9].

5 Discussion

The discussion in I compared the J-matrix approach with the R-matrix, separable kernel, and Fredholm methods. The remarks were of such a general nature as to apply to the extensions made in this chapter. In particular, there are two points that should be stressed. First, no Kohn-type pseudo-resonances are expected to appear in the computed cross sections. This can be demonstrated by bounding these quantities for all energies [11]. Secondly, since H_0 is solved analytically within the function space, it is expected that small basis sets would be adequate to account for the addition of the approximate potential. The resulting physical quantities contain first order errors; however, the analytic nature of the solutions allows for variational correction. This can be accomplished through the application of the Kato correction [12] as discussed in I, and results in reducing the errors to second order.

Presently, work is being done on some of the more mathematical aspects of the J- matrix method. These include the sense of convergence of the expansions for the sine-like and cosine-like functions, possible analytic approaches to the second order Kato correction, and the generality of the solution scheme for the H_0 problem. In addition, the J-matrix method is being applied to the e—He^+ scattering calculation.

Acknowledgments The authors are grateful to Professor William P. Reinhardt for encouragement and support of this work along with a critical reading of the manuscript. Additional thanks are expressed to Dr. E. J. Heller for continued interest in this work and a critical reading of the manuscript. Helpful conversations with J. Broad, T. Murtaugh, and J. Winick are acknowledged. This work was supported by grants from the National Science Foundation and the Camille and Henry Dreyfus Foundations.

Appendix: The Coulomb J-Matrix Solutions

The Coulomb J matrix of Eq. (76) leads to the fundamental recursion relation which may be cast into the form

$$2\left[\left(n+\ell+1+\frac{2z}{\lambda}\right)\cos\theta-\frac{2z}{\lambda}\right]v_n-(n+2\ell+1)v_{n-1}-(n+1)v_{n+1}=0,\ n>0, \tag{A1}$$

which may be cast into the form

$$2\left[(n+\ell+1)\cos\theta+t\sin\theta\right]v_n-(n+2\ell+1)v_{n-1}-(n+1)v_{n+1}=0,\quad n>0, \tag{A2}$$

where $t=-z/k=(-2z/\lambda)\tan\theta/2$ for $-\pi<\theta\leq\pi$. Since Ψ_{reg} for the Coulomb case has the form [3]

$$\Psi_{\text{reg}}(k,r)=\frac{1}{2}(2kr)^{\ell+1}e^{ikr}e^{\pi t/2}\frac{|\Gamma(\ell+1-it)|}{\Gamma(2\ell+2)}{}_1F_1(\ell+1-it,2\ell+2,-2ikr), \tag{A3}$$

the Fourier projection of Eq. (37) has the general form

$$v_n=\int_C^{\infty}dr\Phi[\eta r,t(k)]\phi_n(r)/r. \tag{A4}$$

Since the k dependence in Ψ_{reg} appears in both the variable and the order parameter of the confluent hypergeometric function, a differential equation for v_n, cannot immediately be constructed as in the radial kinetic energy case [4]. However, since the only requirement is to satisfy a pure recurrence relation and not any differential properties, the procedure of Section 2.1 can be applied after making one modification. Noting that Eq. (A2) can be viewed as a recursion relation, where t and θ are independent of one another, t is taken to be independent of θ (or k) in Eq. (A4), resulting in the form

$$v_n=\int_0^{\infty}dr\Phi(\eta r)\phi_n(r)/r \tag{A5}$$

and the procedure to obtain a differential difference relation can be carried through as before. Only at the end of the procedure is t set equal to $-z/k$. The differential relations derived will then no longer hold; however, Eq. (A2) will become Eq. (A1), and the original problem will be solved.

It is then straightforward to derive the differential equation

$$\begin{aligned}(x^2-1)v_n''+\frac{x^2-1}{x}v_n'-\Bigg\{\left(\frac{x^2-1}{x^2}\right)\left[(n+\ell)^2-\ell(\ell-1)+2n+1-t^2\right]\\+\left(\frac{x^2+1}{x^2}\right)\left[\left(\frac{x^2+1}{x^2-1}\right)\ell(\ell+1)-2it(n+\ell+1)\right]\Bigg\}v_n=0\end{aligned} \tag{A6}$$

which can be solved by standard techniques to give [7]

$$\begin{aligned} v_n &= A_{n\ell t}(\sin\theta)^{\ell+1} e^{\theta t} e^{-in\theta}\,{}_2F_1\left(-n, \ell+1-it; 2\ell+2; 1-e^{2i\theta}\right) \\ &\quad + B_{n\ell t}(\sin\theta)^{it} e^{-i(n+\ell+1)\theta}\,{}_2F_1\left(-\ell-it, \ell+1-it; n+\ell+2-it; \left(1-e^{2i\theta}\right)^{-1}\right) \end{aligned} \tag{A7}$$

Substitution into Eq. (A2) followed by the imposition of the homogeneous initial condition determines s_n to with in a factor dependent upon ℓ and t, $a_{\ell t}$. Due to the cut structure of the solutions, $a_{\ell t}$ must be determined in two normalization steps, one corresponding to $k \to 0$ for $\theta \in [0, \pi]$ and other to $k \to 0$ for $\theta \in [0, -\pi]$. Applying Eq. (47), first in the limit $r \to 0, k \to 0^+$, corresponding to $\theta \in [0, \pi]$, and then in the limit $r \to 0, k \to 0^-$ corresponding to $\theta \in [0, -\pi]$, finally yields

$$\begin{aligned} s_n &= 2^{\ell} \frac{|\Gamma(\ell+1-it)|}{\Gamma(2\ell+2)} (\sin\theta)^{\ell+1} e^{\theta t} e^{-in\theta} \exp\left[\left(\varepsilon + \frac{1}{2}\right)\pi t\right] \\ &\quad \times {}_2F_1\left(-n, \ell+1-it; 2\ell+2; 1-e^{2i\theta}\right) \end{aligned} \tag{A8}$$

Letting $t = -z/k$ recovers the original recursion relation and noting the definition of the Pollaczek polynomials [4],

$$P_n^{\lambda}(\cos\theta; a; b) = \frac{(2\lambda)_n}{n!} e^{-in\theta}\,{}_2F_1(-n, \lambda - i\omega; 2\lambda; 1-e^{2i\theta}), \tag{A9}$$

where $\omega = (a\cos\theta + b)/\sin\theta$, gives

$$s_n = 2^{\ell} \frac{n!\,|\Gamma(\ell+1-it)|}{\Gamma(n+2\ell+2)} (\sin\theta)^{\ell+1} e^{\theta t} \exp\left[\left(\varepsilon + \frac{1}{2}\right)\pi t\right] P_n^{\ell+1}\left(\cos\theta; \frac{2z}{\lambda}; -\frac{2z}{\lambda}\right). \tag{A10}$$

Three properties of $\{s_n\}$ will now be verified. Since Ψ_{reg} is real [3] and the basis set is real, $\{s_n\}$ must be real. Application of the linear transformation [7]

$${}_2F_1(a, b; c; z) = (1-z)^{-a}\,{}_2F_1\left(a, c-b; c; \frac{z}{z-1}\right) \tag{A11}$$

to Eq. (A8) immediately establishes the reality of $\{s_n\}$. Since Ψ_{reg} transforms as [3]

$$\Psi_{\text{reg}}(-k, r) = (-)^{\ell+1} e^{-\pi t} \Psi_{\text{reg}}(k, r), \tag{A12}$$

the set $\{s_n\}$ should transform as

$$s_n(-k) = (-)^{\ell+1} e^{-\pi t} s_n(k). \tag{A13}$$

The transformation $k \to -k$ implies $\theta \to -\theta$, $\varepsilon \to -\varepsilon$, and $t \to -t$, which when combined with the reality property immediately establishes Eq. (A13). Finally, in the limit as $z \to 0$, the radial kinetic energy in a Laguerre basis results should be recovered. Noting that [4]

$$P_n^{\lambda}(\cos\theta; 0; 0) = C_n^{\lambda}(\cos\theta) \tag{A14}$$

immediately establishes the reduction.

Substitution into Eq. (A2) followed by the imposition of the inhomogeneous initial condition, in the derivation of which the result

$$W(k) = -k \exp\left[(1+\varepsilon)\pi t\right], \tag{A15}$$

as can be derived from the results in Ref. [3], is employed, yields

$$\begin{aligned} c_n + \alpha(\ell, t)s_n = {} & \frac{-n!e^{i\sigma_\ell}e^{\pi t/2}e^{-\theta t}e^{-i(n+1)\theta}}{(2\sin\theta)^{\ell}\Gamma(n+\ell+2-it)} \\ & \times {}_2F_1\left(-\ell - it, n+1; n+\ell+2-it; e^{-2i\theta}\right). \end{aligned} \tag{A16}$$

The coefficient $\alpha(\ell, t)$ is determined in three steps. Enforcing the reality of $\{c_n\}$ through the application of the linear transformation [7]

$$\begin{aligned} {}_2F_1(a, b; c; z) = {} & \frac{\Gamma(c)\Gamma(b-c)}{\Gamma(b)\Gamma(c-a)}(-z)^{-a}{}_2F_1\left(a, 1-c+a; 1-b+a; 1/z\right) \\ & + \frac{\Gamma(c)\Gamma(a-b)}{\Gamma(a)\Gamma(c-b)}(-z)^{-b}{}_2F_1\left(b, 1-c+b; 1-a+b; 1/z\right) \end{aligned} \tag{A17}$$

results in

$$\alpha(\ell, t) = \rho(\ell, t) + i, \tag{A18}$$

where $\rho(\ell, t)$ is a real function. Enforcing the symmetry condition analogous to Eq. (A13),

$$c_n(-k) = (-)^{\ell} e^{-\pi t} c_n(k), \tag{A19}$$

results in the fact that $\rho(\ell, t)$ is an odd function of t,

$$\rho(\ell, -t) = -\rho(\ell, t). \tag{A20}$$

Since $\rho(\ell, t)$ is independent of n, it is sufficient to consider $n = 0$ in Eq. (A16), which takes the form

$$c_0 + is_0 = \frac{-e^{i\sigma_\ell}e^{\pi t/2}e^{-\theta t}e^{-i\theta}}{(2\sin\theta)^{\ell}\Gamma(\ell+2-it)}{}_2F_1\left(-\ell - it, 1; \ell+2-it; e^{-2i\theta}\right) - \rho(\ell, t)s_0. \tag{A21}$$

The lhs of Eq. (A21) corresponds for $k > 0$ to the zeroth expansion coefficient of the function

$$\tilde{C}(r) + i\tilde{S}(r) = \frac{2^{-\ell}\lambda e^{\pi t/2} e^{-\theta t}}{(\sin\theta)^{\ell+1}\,|\Gamma(\ell+1-it)|}\left[\left(\Psi_{\text{irreg}} + i\Psi_{\text{reg}}\right)\int_0^r dr'\Psi_{\text{reg}}\frac{\phi_0(\lambda r')}{\lambda r'} + \Psi_{\text{reg}}\int_r^\infty dr'\left(\Psi_{\text{irreg}} + i\Psi_{\text{reg}}\right)\frac{\phi_0(\lambda r')}{\lambda r'}\right] \tag{A22}$$

The coefficient is thus

$$c_0 + is_0 = \frac{1}{\Gamma(2\ell+2)}\int_0^\infty dr\left[\tilde{C}(\frac{r}{\lambda}, t) + i\tilde{S}(\frac{r}{\lambda}, t)\right]\phi_0(r)/r \tag{A23}$$

Now, letting $k \to ik$ in Eqs. (A22) and (A23), and correspondingly letting $\eta \to i\eta$ in Eq. (A21), and taking the limit $\eta \to \frac{1}{2} - \delta$, gives for Eq. (A21) two terms of order,

$$O(\delta^{\ell+1-it}) - \rho(\ell, t)O(\delta^{-\ell-1+it}) \tag{A24}$$

As $\delta \to 0$, for $(\ell+1+Imt) > 0$, the $\rho(\ell, t)$ term is the dominant term in Eq. (A24). The oddness property of $\rho(\ell, t)$, Eq. (A20), removes the restriction imposed by the inequality. By referring to the expansions given in Eqs. (11) and (12), it is seen that the integral in Eq. (A23) is convergent in the limit taken above. Thus, in that limit, Eq. (A23) goes as $O(\delta^{\ell+1-it})$ implying that

$$\rho(\ell, t) = 0 \tag{A25}$$

and thus that

$$c_n + is_n = \frac{-n!e^{i\sigma_\ell}e^{\pi t/2}e^{-\theta t}e^{-i(n+1)\theta}}{(2\sin\theta)^\ell\Gamma(n+\ell+2-it)}{}_2F_1\left(-\ell-it, n+1; n+\ell+2-it; e^{-2i\theta}\right). \tag{A26}$$

That Eq. (A26) reduces properly in the limit of $z \to 0$ can be established from the theory of Legendre functions [7]. Taking the limit $z \to 0$ in Eq. (A26) and applying formula 3.2 (30) of Ref. [7] establishes the reduction to the radial kinetic energy case.

References

1. E. J. Heller and H. A. Yamani, Phys. Rev. A **9**, 1201 (1974).
2. E. J. Heller and H. A. Yamani, Phys. Rev. A **9**, 1209 (1974).
3. A. Messiah, *Quantum Mechanics*, Vol. I (North-Holland, Amsterdam, 1965).
4. A. Erdélyi, Ed., *Higher Transcendental Functions*, Vol. II (McGraw-Hill, New York, 1953).

5. R. Courant and D. Hilbert, *Methods of Mathematical Physics*, Vol. I (Interscience, New York, 1966).
6. P. M. Morse and H. Feschbach, *Methods of Theoretical Physics*, Vol. I (McGraw-Hill, New York 1953).
7. A. Erdélyi, Ed., *Higher Transcendental Functions*, Vol. I (McGraw-Hill, New York, 1953).
8. F. E. Harris, Phys. Rev. Lett. **19**, 173 (1967).
9. Standard reference: S. Geltman, *Topics in Atomic Collision Theory* (Academic, New York, 1969).
10. P. G. Burke, D.F. Gallaber, and S. Geltman, J. Phys. B **2**, 1142 (1962).
11. E. J. Heller, unpublished work.
12. G. Kato, Prog. Theor. Phys. **6**, 394 (1951).

Part II
Theoretical and Mathematical Considerations

Oscillator Basis in the J-Matrix Method: Convergence of Expansions, Asymptotics of Expansion Coefficients and Boundary Conditions

S.Yu. Igashov

Abstract Important mathematical aspects of the J-matrix method are considered in the case of the oscillator basis. The asymptotic form of the Fourier coefficients for the expansions over the oscillator basis is found by the use of the asymptotic approximations for the basis functions. These results are applied to investigation of the pointwise convergence of the expansions.

1 Introduction

The Chapter [1] and developed there J-matrix method give rise to the further wide utilization of the expansions over the square-integrable basises as an effective tool for solving Schrodinger equation. The J-matrix method is grounded on the expansion of solution over the basis functions. One of these basises, offered in [1] is the basis of the oscillator eigenfunctions. The expansions over the oscillator basis are of particular interest and have numerous applications in physics due to the unique properties of the oscillator eigenfunctions. First, the many-particle oscillator eigenfunctions admit separation of the common center-of-mass motion. Second, these functions have rather simple transformation properties under the transformations between different sets of Jacobi coordinates. Third, the overlapping integrals of many-particle oscillator functions can be calculated analytically. Also should be mentioned, that rather effective methods were developed for calculation of matrix elements between the many-particle oscillator wave functions [2–5]. Thus, the J-matrix approach with the oscillator functions occurs a unique and very effective tool in many physical problems, especially in physics of light nuclei. In subsequent chapters, methods for solving Schrodinger equation using expansions in the oscillator basis were further developed and extended. In particular, the extension of the oscillator basic J-matrix formalism on the case of true A-body scattering is discussed in [6]. An analysis of the algebraic equations for the scattering problem in

S.Yu. Igashov
Moscow Engineering Physics Institute, Kashirskoe sh. 31, Moscow 115409, Russia
e-mail: igashov@theor.mephi.ru

A.D. Alhaidari et al. (eds.), *The J-Matrix Method*, 49–66.

the oscillator representation is given and a new method for solving these equations is suggested in [7, 8]. This modification [8] of the J-matrix approach for scattering allows to improve convergence of the method and reduce the computational cost. Inverse problem applications of the J-matrix method are developed in [9–13].

The asymptotic behavior of expansion coefficients plays the key role in the J-matrix approach. The point is that the boundary conditions in the J-matrix method are formulated in terms of the asymptotics of expansion coefficients. Hence, it is necessary to investigate precisely in rather general case the asymptotic behavior of the Fourier coefficients. Also restrictions on the functions expanded and estimates of the remainder term in the asymptotic formulas should be considered. Here we present the comprehensive study of these problems, basing on the asymptotic approximations for the oscillator functions. Such approach is most relevant for our purposes.

Another important problem is the investigation of convergence of the expansions. Study of convergence of expansions of the continuos spectra wave functions is connected with certain difficulties. They are caused with the different nature of the oscillator functions and the wave functions of continuous spectra: the latter functions do not belong to the class of square integrable functions. The convergence of the expansion series can be proved using the asymptotic behavior of the expansion coefficients. But it is not enough for applications. It is necessary to prove the convergence to the function expanded. Here we apply for this purpose the approach, based on the Christoffel-Darboux formula for the Laguerre polynomials.

Below we consider more general case of expansion in the set of the oscillator type functions. Let us turn to calculation of expansion coefficients $C(n)$ of the function $f(r)$:

$$C(n) = \int_0^\infty f(r)\,\chi_n(r)\,dr, \tag{1}$$

where

$$\chi_n(r) = r^{\alpha+1/2} e^{-r^2/2} L_n^{(\alpha)}\left(r^2\right) \tag{2}$$

are the oscillator type functions (for simplicity, a trivial normalization factor is omitted here). These functions satisfy the equation

$$\chi'' + \left(k^2 - r^2 + \frac{1/4 - \alpha^2}{r^2}\right)\chi = 0, \tag{3}$$

where $k^2 = 4n+2\alpha+2$, $L_n^{(\alpha)}$ is the Laguerre polynomial [14–16]. Below we assume that $\alpha \geq 0$. In quantum mechanics, at integer values of $l = \alpha - 1/2$, the functions (2) are the radial wave functions of the three-dimensional oscillator corresponding to the radial quantum number n and the orbital angular momentum l. Integral (1) can be calculated analytically in a limited number of cases. Some of these cases are considered here as illustrations of the general asymptotic formula for the coefficients $C(n)$.

We split the integral (1) into three integrals: over the segments $[0, \beta k]$ and $[\beta k, k]$ and interval $[k, +\infty)$, $0 < \beta < 1$. We will calculate below these integrals by the use of appropriate asymptotic approximations for functions (2). Then, we consider the convergence of expansions in the set of the oscillator type functions $\chi_n(r)$.

2 Integration over the Segment $[0, \beta k]$

Neglecting the term r^2 in (3) at small r, we obtain the simplified equation that has the two linearly independent solutions

$$\Phi_1(r) = \sqrt{r}\, J_\alpha(kr), \qquad \Phi_2(r) = \sqrt{r}\, Y_\alpha(kr), \tag{4}$$

where J_α and Y_α are the Bessel functions [14, 15]. The exact solution of (3) is approximated by linear combination of functions (4). An accuracy of this approximation is determined by the Liouville's method [16].

The WKB approximation [17] is efficient in the regions far from the turning points. Neglecting the term $1/r^2$ in (3), we can obtain the approximate WKB solutions

$$\Psi_1(r) = \left(k^2 - r^2\right)^{-1/4} sin\left(\xi(r)\right), \qquad \Psi_2(r) = \left(k^2 - r^2\right)^{-1/4} cos\left(\xi(r)\right), \tag{5}$$

where

$$\xi(r) = \int_0^r \left(k^2 - x^2\right)^{1/2} dx. \tag{6}$$

Consider the functions

$$F_{1,2}(r) = \left(k^2 - r^2\right)^{-1/4} \Phi_{1,2}\left(\frac{\xi(r)}{k}\right). \tag{7}$$

Functions (7) behave asymptotically as functions (4) or linear combinations of functions (5) in respective regions. This suggests the assumption that functions (7) provide more general approximation than functions (4) and (5). The accuracy estimation can be based on the Liouville's method [16]. Taking into account the differential equation for the Bessel functions, we can easily find that functions (7) satisfy the second-order differential equation [18]

$$F'' + \left(k^2 - r^2 - \frac{\alpha^2 - 1/4}{r^2} - Q(r)\right) F = 0, \tag{8}$$

where

$$Q(r) = \frac{1}{2\left(k^2 - r^2\right)} + \frac{5r^2}{4\left(k^2 - r^2\right)^2} + \left(\alpha^2 - \frac{1}{4}\right) h(r), \tag{9}$$

$$h(r) = \left(1 - \frac{r^2}{k^2}\right)\left(\frac{k}{\xi(r)}\right)^2 - \frac{1}{r^2}. \tag{10}$$

Equation (3) can be rewritten in the form similar to (8)

$$\chi'' + \left(k^2 - r^2 - \frac{\alpha^2 - 1/4}{r^2} - Q(r)\right)\chi = -Q(r)\chi. \tag{11}$$

We seek the solutions of (11) in the form

$$\chi(r) = C_1^I(r) F_1(r) + C_2^I(r) F_2(r) \tag{12}$$

with the additional condition

$$C_1^{I\prime}(r) F_1(r) + C_2^{I\prime}(r) F_2(r) = 0. \tag{13}$$

The functions $C_{1,2}^{I\prime}(r)$ are determined by substituting (12) into (11) with regard for condition (13). Obtained equations for $C_{1,2}^{I\prime}(r)$ should be formally integrated over the segment $[\Lambda, r]$, then multiplied by $F_{1,2}(r)$, respectively, and add together. As a result, we obtain the Volterra-type integral equation

$$\begin{aligned}\chi(r) &= C_1^I(\Lambda) F_1(r) + C_2^I(\Lambda) F_2(r)\\ &+ \frac{\pi k}{2}\int_{\Lambda}^{r} Q(y)\left(F_1(r) F_2(y) - F_1(y) F_2(r)\right)\chi(y)\, dy.\end{aligned} \tag{14}$$

The values $C_{1,2}^I(\Lambda)$ are obtained with the use of (12), (13) and the values of $\chi(\Lambda)$ and $\chi'(\Lambda)$. Using the asymptotic formulas for the Bessel functions in the limit $\Lambda \to 0$, we obtain [18]

$$C_1^I(0) = 2^{\alpha} k^{1/2-\alpha} \Gamma(\alpha + 1) L_n^{(\alpha)}(0); \quad C_2^I(0) = 0 \tag{15}$$

Let us consider the function $X(r) = \left(k^2 - r^2\right)^{1/4} \chi(r)$. Then (14) takes the form

$$X(r) = X_0(r) + \frac{\pi}{2}\int_0^r \frac{Q(y)}{\left(k^2 - y^2\right)^{1/2}} G(r, y) X(y)\, dy, \tag{16}$$

where the notation $X_0(r)$ and $G(x, y)$ is clear from comparison with (14). Equation (16) we will solve by the method of successive approximations:

$$X(r) = \sum_{m=0}^{\infty} X_m(r), \tag{17}$$

$$X_{m+1}(r) = \frac{\pi}{2} \int_0^r \frac{Q(y)}{\left(k^2 - y^2\right)^{1/2}} G(r, y) X_m(y) dy. \tag{18}$$

Convergence of series (17) can be analyzed with the use of majorizing estimates of its terms. For this purpose we consider the functions entering into the integrand in (18). It follows from expression (9) that

$$\frac{Q(y)}{\left(k^2 - y^2\right)^{1/2}} < \frac{D}{k^3}, \quad \text{if } y \in [0, \beta k], \tag{19}$$

where D is nonnegative and independent of k. The estimates for $G(x, y)$ can be obtained from the following obvious inequalities for the Bessel functions: $|J_\alpha(z)| < A_1 z^\alpha, \alpha \geq 0, |Y_\alpha(z)| < A_2 z^{-\alpha}, \alpha > 0$, if $0 < z \leq \delta$ and $|J_\alpha(z)|, |Y_\alpha(z)| < A_3 z^{-1/2}$ if $z \geq \delta$. The case of $\alpha = 0$ should be considered separately with the use of the relation [14, 15]

$$\frac{\pi}{2} Y_0(z) = \left[\gamma + ln\frac{z}{2}\right] J_0(z) - \sum_{m=1}^{\infty} (-1)^m \frac{(z/2)^{2m}}{(m!)^2} \left(1 + \frac{1}{2} + \ldots + \frac{1}{m}\right). \tag{20}$$

The mathematical induction and the above estimates enable us to obtain the inequality

$$|X_m(r)| < E_m \left(\frac{r}{k^3}\right)^m, \quad \text{if } r \in [0, \beta k], \tag{21}$$

where the sequence $\{E_m\}$ is majorized by a sequence $ab^m/m!$ (with a and b independent of m). The dependence of E_m on k is asymptotically estimated as $E_m = E_0\, O(1)$ at $k \to \infty$. Thus the absolute convergence of series (17) follows from inequality (21). It should be mentioned that the estimate for each term in series (17) has the additional parameter r/k^3 as compared to the estimate for the preceding term. It means that (21) allows to obtain the asymptotic approximation for the function χ_n. The similar approximations (for the Laguerre polynomials) were obtained in [19] by the use of other equations. We use below the approach based on (14) that is more appropriate for our calculation.

Now, let us turn to calculation of the contribution

$$I_1 = \int_0^{\beta k} f(r)\,\chi_n(r)\,dr$$
$$= \int_0^{\beta k} f(r)\left(k^2 - r^2\right)^{-1/4}\left(X_0(r) + X_1(r) + \ldots\right)dr \equiv I_1^0 + I_1^1 + \ldots \quad (22)$$

for the expansion coefficients from the first domain $\left[0, \beta k\right]$ at large k values. First, the asymptotic of the integral

$$\tilde{I}_1 = \int_0^{\beta k} g(z)\sqrt{kz}\,J_\alpha(kz)\,dz \quad (23)$$

should be considered. For this aim, we transform (23) to the form

$$\tilde{I}_1 = \int_0^{\beta k} g(z)\left[\sqrt{kz}\,J_\alpha(kz) - \sqrt{\frac{2}{\pi}}cos\left(kz - \frac{\alpha\pi}{2} - \frac{\pi}{4}\right)\right]dz$$
$$+ \sqrt{\frac{2}{\pi}}\int_0^{\beta k} g(z)\,cos\left(kz - \frac{\alpha\pi}{2} - \frac{\pi}{4}\right)dz. \quad (24)$$

Let us suppose that the function $g(z)$ and its first to third derivatives are continuous and bounded within the interval $[0, +\infty)$. Performing three integrations by parts in (24), we obtain

$$\tilde{I}_1 = \frac{g(0)}{k}\left(\lambda_\alpha(0) + \sqrt{\frac{2}{\pi}}sin\left(\frac{\alpha\pi}{2} + \frac{\pi}{4}\right)\right)$$
$$+ \frac{g(\beta k)}{k}\sqrt{\frac{2}{\pi}}sin\left(\beta k^2 - \frac{\alpha\pi}{2} - \frac{\pi}{4}\right) + O\left(\frac{1}{k^2}\right), \quad (25)$$

where

$$\lambda_\alpha(x) = \int_x^{\infty}\left[\sqrt{z}\,J_\alpha(z) - \sqrt{\frac{2}{\pi}}cos\left(z - \frac{\alpha\pi}{2} - \frac{\pi}{4}\right)\right]dz. \quad (26)$$

The value

$$\lambda_\alpha(0) = \frac{2^{1/2}\Gamma((\alpha + 3/2)/2)}{\Gamma((\alpha - 1/2)/2)} - \sqrt{\frac{2}{\pi}}sin\left(\frac{\alpha\pi}{2} + \frac{\pi}{4}\right) \quad (27)$$

involved in (25) is easy to find from (26) with the use of the recursion relation

$$\sqrt{z}J_{\alpha}(z) = \left[\sqrt{z}J_{\alpha+1}(z)\right]' + \frac{\alpha+1/2}{\sqrt{z}}J_{\alpha+1}(z). \tag{28}$$

Introducing new variable $z = \xi(r)/k$ and taking into account (25), for the term I_1^0 we obtain the asymptotic expression

$$\begin{aligned} I_1^0 = & \frac{k^{\alpha-3/2}\Gamma((\alpha+3/2)/2)}{2^{\alpha-1/2}\Gamma((\alpha+1/2)/2)} f(0)\left(1+O\left(\frac{1}{k}\right)\right)+ \\ & + \frac{k^{\alpha-3/2} f(\beta k)}{2^{\alpha-1/2}\pi^{1/2}\left(1-\beta^2\right)^{3/4}}\left(sin\left(\xi(\beta k)-\frac{\alpha\pi}{2}-\frac{\pi}{4}\right)+O\left(\frac{1}{k}\right)\right). \end{aligned} \tag{29}$$

It remains to estimate the terms I_1^m, $m > 0$. First, consider the case $m = 1$. We split the integral in I_1^1 into two – over the segments $[0, \delta]$ and $\left[\delta, \beta k\right]$, where δ is a constant, independent of k. The estimate for the first integral over the first segment $[0, \delta]$ follows directly from (21). Consider the second integral

$$\int\limits_{\delta}^{\beta k} dr \frac{f(r)}{\left(k^2-r^2\right)^{1/4}}\left(\int\limits_{0}^{\delta_1/k} + \int\limits_{\delta_1/k}^{\delta} + \int\limits_{\delta}^{r}\right) dy \frac{Q(y)}{\left(k^2-y^2\right)^{1/2}} G(r,y)\sqrt{\xi(y)}J_{\alpha}(\xi(y)). \tag{30}$$

The integrals over the segments $[0, \delta_1/k]$ and $[\delta_1/k, \delta]$ are estimated with the use of inequality (19) and the mentioned above estimates for the Bessel functions. The third integral in (30) (over the segment $[\delta, r]$) can be estimated by the use of the asymptotic relation

$$\begin{aligned} G(r,y)\sqrt{\xi(y)}J_{\alpha}(\xi(y)) = & -\frac{2^{1/2}}{\pi^{3/2}}\left(sin\left(\xi(r)-2\xi(y)+\frac{\alpha\pi}{2}+\frac{\pi}{4}\right)+\right. \\ & \left. + sin\left(\xi(r)-\frac{\alpha\pi}{2}-\frac{\pi}{4}\right)\right)+O\left(\frac{1}{k}\right), \quad \text{if } r, y > \delta. \end{aligned} \tag{31}$$

Integrating by parts, we obtain the estimation for this case. The estimates for remainder term are evident. The relevant estimate of the integral with respect to r is straightforward. As a result we obtain $I_1^1 = O\left(k^{\alpha-5/2}\right)$. The estimates for I_1^m, $m \geq 2$ follow directly from (21). These estimates do not exceed the estimate for I_1^1. These conclude the consideration of the contribution of the first domain $\left[0, \beta k\right]$.

3 Integration over the Interval $[k, +\infty)$

Let us consider asymptotic approximations of the function $\chi_n(r)$ in the interval $[k, +\infty)$ and its contribution to the expansion coefficient $C(n)$. The approximation of the exact solution in the vicinity of the turning point $r = k$ by a linear combination of the Airy functions is rather efficient. The Airy functions $A_i\left(-k^{4/3}t\right)$ and $B_i\left(-k^{4/3}t\right)$ are solutions of the differential equation

$$\frac{d^2W}{dt^2}+k^4tW=0. \tag{32}$$

The substitution $t=\phi(x)$, $W=\left[-\phi'(x)\right]^{1/2}\Xi(x)$ reduces (32) to the form

$$\Xi''(x)+\left[k^4\phi(x)\left(\phi'(x)\right)^2+\frac{1}{2}\{\phi,x\}\right]\Xi(x)=0, \tag{33}$$

where

$$\{\phi,z\}=\frac{\phi'''}{\phi'}-\frac{3}{2}\left(\frac{\phi''}{\phi'}\right)^2 \tag{34}$$

is the Schwarzian derivative. By comparing (3) and (33), we can choose

$$\phi\left(\phi'\right)^2=1-x^2, \tag{35}$$

where $x=r/k$. It follows from (35) and from the boundedness constraint for the term $\{\phi,x\}$ in (33) at $x=1$, that [19]

$$\phi(x)=-\left[\frac{3}{2}\int\limits_1^x\left(z^2-1\right)^{1/2}dz\right]^{2/3}. \tag{36}$$

Thus, the functions

$$\Xi_1(x)=\left[-\phi'(x)\right]^{-1/2}A_i\left(-k^{4/3}\phi(x)\right),\quad \Xi_2(x)=\left[-\phi'(x)\right]^{-1/2}B_i\left(-k^{4/3}\phi(x)\right) \tag{37}$$

are solutions of (33), which is rather close to (3). Thus we can expect that appropriate linear combination of the functions (37) approximate solutions of (3). The integral equation

$$\chi(kx)=C_1^{III}(\Lambda)\Xi_1(x)+C_2^{III}(\Lambda)\Xi_2(x)+\frac{\pi}{k^{4/3}}\int\limits_x^\Lambda\left(\frac{\alpha^2-1/4}{y^2}+\frac{1}{2}\{\phi,y\}\right)\times$$
$$\times\left(\Xi_1(x)\Xi_2(y)-\Xi_1(y)\Xi_2(x)\right)\chi(ky)\,dy \tag{38}$$

is established similar to (14). For large Λ, the values $C_{1,2}^{III}(\Lambda)$ are determined with the use of asymptotic formulas for the Airy functions [14, 15, 17] and $\chi(r)$. Introducing the function $Z(x)=\left[-\phi'(x)\right]^{1/2}\chi(kx)$, we obtain

$$Z(x) = Z_0(x) + \frac{\pi}{k^{4/3}} \int_x^\infty \left[-\phi'(y)\right]^{-1} \left(\frac{\alpha^2 - 1/4}{y^2} + \frac{1}{2}\{\phi, y\}\right) \Theta(x, y) Z(y) dy, \tag{39}$$

where

$$Z_0(x) = C_1^{III}(\infty) A_i\left(-k^{4/3}\phi(x)\right), \tag{40}$$

$$C_1^{III}(\infty) = \frac{(-1)^n 2\pi^{1/2}}{n! k^{1/6}} \left(\frac{k}{2}\right)^{k^2/2} e^{-k^2/4}, \qquad C_2^{III}(\infty) = 0. \tag{41}$$

We construct the solution of (39) by the method of successive approximations. Using the mathematical induction and the majorizing estimates for the Airy functions, we obtain

$$|Z_m(x)| < \frac{\tilde{D}^m}{k^{2m}} |Z_0(x)|. \tag{42}$$

Convergence of the iteration series follows from (42) for sufficiently large values of k.

Let us apply the obtained results to calculation of the integral

$$\begin{aligned} I_3 &= \int_k^\infty f(r) \chi_n(r) dr \\ &= \int_k^\infty \frac{f(r)}{[-\phi'(r/k)]^{1/2}} \left(Z_0\left(\frac{r}{k}\right) + Z_1\left(\frac{r}{k}\right) + \ldots\right) dr \equiv I_3^0 + I_3^1 + \ldots \end{aligned} \tag{43}$$

Assume that the function $f(r)$ and its first six derivatives are continuous and bounded within the interval $\left[\beta k, k\right)$. Performing six integrations by parts, we obtain

$$I_3^0 = C_1^{III}(\infty) \left(\sum_{m=1}^{6} \frac{h_m}{k^{m/3}} + \frac{j_6}{k^2}\right), \tag{44}$$

where

$$h_m = -a_m\left(-k^{4/3}\phi\left(\frac{z}{k}\right)\right) \left(\left[-\phi'\left(\frac{z}{k}\right)\right]^{-1} \frac{d}{dz}\right)^{m-1} \left[-\phi'\left(\frac{z}{k}\right)\right]^{-3/2} f(z)\Bigg|_{z=k}^{z=\infty}, \tag{45}$$

$$j_m = \int_k^\infty dz\, a_m\left(-k^{4/3}\phi\left(\frac{z}{k}\right)\right) \left(\frac{d}{dz}\left[-\phi'\left(\frac{z}{k}\right)\right]^{-1}\right)^m \left[-\phi'\left(\frac{z}{k}\right)\right]^{-1/2} f(z), \tag{46}$$

$$a_{-1}(x) = -A_i'(x), \qquad a_{m+1}(x) = \int_x^\infty a_m(x)\,dx \tag{47}$$

Relation (45) yields

$$\begin{aligned}
h_1 &= 2^{-1/2} a_1(0)\, f(k), \\
h_2 &= 2^{-5/6} a_2(0) \left[f^{(1)}(k) - \frac{3}{10}\frac{f(k)}{k} \right], \\
h_3 &= 2^{-7/6} a_3(0) \left[f^{(2)}(k) - \frac{4}{5}\frac{f^{(1)}(k)}{k} \right] + O\left(\frac{1}{k^2}\right), \\
h_4 &= 2^{-3/2} a_4(0)\, f^{(3)}(k) + O\left(\frac{1}{k}\right), \\
h_5 &= 2^{-11/6} a_5(0)\, f^{(4)}(k) + O\left(\frac{1}{k}\right), \\
h_6 &= 2^{-13/6} a_6(0)\, f^{(5)}(k) + O\left(\frac{1}{k}\right).
\end{aligned} \tag{48}$$

Here $f^{(m)}(z)$ denotes the mth-order derivative of the function $f(z)$. The values

$$a_m(0) = \frac{1}{3^{(m+2)/3}\Gamma((m+2)/3)} \tag{49}$$

can be found with the recurrence relation

$$a_m(x) = -\frac{x}{m-1} a_{m-1}(x) + \frac{1}{m-1} a_{m-3}(x). \tag{50}$$

It is easily established by induction and integration by parts.

It remains to estimate j_m. Consider the function

$$\left(\frac{d}{dz} \left[-\phi'\left(\frac{z}{k}\right) \right]^{-1} \right)^m \left[-\phi'\left(\frac{z}{k}\right) \right]^{-1/2} f(z). \tag{51}$$

It is easy to observe that this function is bounded. As a result we obtain from (46) $j_m = O\left(k^{-1/3}\right)$. The estimates for the terms I_3^M $(M \geq 1)$ follow directly from (42) and do not exceed the estimates for the remainder term in I_3^0.

4 Integration over the Segment $[\beta k, k]$

The asymptotic approximations in the case $r \in [\beta k, k]$ is deduced similarly to the case considered in the preceding section. We introduce the functions $\Xi_{1,2}(x)$ by the relations (37) in which the function

$$\phi(x) = \left[\frac{3}{2} \int_{x}^{1} \left(1 - z^2\right)^{1/2} dz \right]^{2/3}. \tag{52}$$

It can be verified that the function $Z(x)$ satisfies the equation

$$Z(x) = Z_0(x) + \frac{\pi}{k^{4/3}} \int_{x}^{1} \left[-\phi'(y)\right]^{-1} \left(\frac{\alpha^2 - 1/4}{y^2} + \frac{1}{2}\{\phi, y\} \right) \times$$

$$\times \left(A_i\left(-k^{4/3}\phi(x)\right) B_i\left(-k^{4/3}\phi(y)\right) - A_i\left(-k^{4/3}\phi(y)\right) B_i\left(-k^{4/3}\phi(x)\right)\right) Z(y)\, dy, \tag{53}$$

where

$$Z_0(x) = C_1^{II}(1)\, A_i\left(-k^{4/3}\phi(x)\right) + C_2^{II}(1)\, B_i\left(-k^{4/3}\phi(x)\right), \tag{54}$$

$$C_1^{II}(1) = C_1^{III}(\infty)\left(1 + O\left(1/k^2\right)\right), \qquad C_2^{II}(1) = C_1^{III}(\infty)\, O\left(1/k^2\right). \tag{55}$$

The values $C_{1,2}^{II}(1)$ are determined by the use of the values of $\chi(k)$ and $\chi'(k)$ and relations similar to (12) and (13). The asymptotic value of $\chi(k)$ is directly obtained from (40) and (42). The value of $\chi'(k)$ can be determined from the relations $C_1^{III}(1) = C_1^{III}(\infty)\left(1 + O\left(1/k^2\right)\right)$ and $C_2^{III}(1) = C_1^{III}(\infty)\, O\left(1/k^2\right)$. These relations follow from the integral equations for the functions $C_{1,2}^{III}(x)$ (see the argumentation before (14)) and the asymptotic approximations for $\chi(r)$ (see (40) and (42)). We solve (53) by the method of successive approximations. The estimates

$$|Z_m(x)| < \frac{\tilde{\tilde{D}}^m}{k^{2m}} \left|C_1^{II}(1)\right| \left(\left|A_i\left(-k^{4/3}\phi(x)\right)\right| + \left|B_i\left(-k^{4/3}\phi(x)\right)\right|\right), \quad x \in [\beta, 1] \tag{56}$$

for the mth iteration can be deduced by the mathematical induction. The convergence of the iteration series follows from (56) for sufficiently large k.

Let us consider the asymptotic behavior of the following integral at large k

$$I_2 = \int_{\beta k}^{k} f(r)\, \chi_n(r)\, dr$$

$$= \int_{\beta k}^{k} \frac{f(r)}{\left[-\phi'(r/k)\right]^{1/2}} \left(Z_0\left(\frac{r}{k}\right) + Z_1\left(\frac{r}{k}\right) + \ldots\right) dr \equiv I_2^0 + I_2^1 + \ldots \tag{57}$$

We perform six integrations by parts in the term I_2^0 and obtain

$$I_2^0 = C_1^{II}(1)\left(\sum_{m=1}^{6} \frac{(-1)^{m+1} s_m}{k^{m/3}} + \frac{i_6}{k^2}\right)$$

$$+ C_2^{II}(1) \int_{\beta k}^{k} f(r) \left[-\phi'\left(\frac{r}{k}\right)\right]^{-1/2} B_i\left(-k^{4/3}\phi\left(\frac{r}{k}\right)\right) dr, \tag{58}$$

where

$$s_m = A_m\left(-k^{4/3}\phi\left(\frac{z}{k}\right)\right)\left(\left[-\phi'\left(\frac{z}{k}\right)\right]^{-1}\frac{d}{dz}\right)^{m-1}\left[-\phi'\left(\frac{z}{k}\right)\right]^{-3/2} f(z)\Big|_{z=\beta k}^{z=k}, \tag{59}$$

$$i_m = \int_{\beta k}^{k} dz\, A_m\left(-k^{4/3}\phi\left(\frac{z}{k}\right)\right)\left(\frac{d}{dz}\left[-\phi'\left(\frac{z}{k}\right)\right]^{-1}\right)^{m}\left[-\phi'\left(\frac{z}{k}\right)\right]^{-1/2} f(z), \tag{60}$$

$$A_1(x) = \int_{-\infty}^{x} A_i(x)\, dx, \qquad A_{m+1} = \int_{-\infty}^{x} A_m(x) dx. \tag{61}$$

Equation (59) allows us to determine the asymptotic values of s_m. With the aid of the asymptotic relations

$$A_1(-\nu) = \frac{-1}{\pi^{1/2}\nu^{3/4}}\left(sin\left(\frac{2}{3}\nu^{3/2} - \frac{\pi}{4}\right) + O\left(\frac{1}{\nu^{3/2}}\right)\right), \qquad \nu \to +\infty \tag{62}$$

$$A_m(-\nu) = O\left(\nu^{-(2m+1)/4}\right), \qquad \nu \to +\infty \tag{63}$$

we obtain

$$s_1 = 2^{-1/2} A_1(0) f(k) - \left[-\phi'(\beta)\right]^{-3/2} A_1\left(-k^{4/3}\phi(\beta)\right) f(\beta k), \tag{64}$$

$$s_2 = 2^{-5/6} A_2(0)\left[f^{(1)}(k) - \frac{3}{10}\frac{f(k)}{k}\right] + O\left(\frac{1}{k^{5/3}}\right). \tag{65}$$

The asymptotic expressions for the other $s_m, m = 3, \ldots 6$ are similar to corresponding expressions (48) for h_m (where the values $a_m(0)$ are replaced by the values $A_m(0) = 2cos((m-1)\pi/3)\, a_m(0)$ which were found in [17]). The estimate $i_m = O\left(k^{-1/3}\right)$, $m \geq 2$, is obtained similarly to the one for j_m in the preceding section. The estimate $O\left(k^{-1/3}\right)$ for the latter term in (58) is obtained by the integrating by parts.

It remains to consider the terms I_2^m for $m \geq 1$. Inequality (56) does not sufficient to estimate I_2^1. We use an explicit expression for the first order approximation Z_1. Substituting Z_1 into I_2^1, we obtain

$$I_2^1 = \frac{\pi}{k^{4/3}} (\vartheta_1 - \vartheta_2), \tag{66}$$

$$\vartheta_1 = \int\limits_{\beta k}^{k} dz\, f(z) \left[-\phi'\left(\frac{z}{k}\right)\right]^{-1/2} A_i\left(-k^{4/3}\phi\left(\frac{z}{k}\right)\right) \int\limits_{z/k}^{1} dy \left(\frac{\alpha^2 - 1/4}{y^2} + \frac{1}{2}\{\phi, y\}\right) \times$$
$$\times \left[-\phi'(y)\right]^{-1} B_i\left(-k^{4/3}\phi(y)\right) Z_0(y). \tag{67}$$

The expression for ϑ_2 differs from the expression for ϑ_1 by replacing A_i and B_i by each other. Integrating by parts in ϑ_1, we obtain

$$\vartheta_1 = \frac{-f(\beta k)}{k^{1/3}\left[-\phi'(\beta)\right]^{3/2}} A_1\left(-k^{4/3}\phi(\beta)\right) \int\limits_{\beta}^{1} dy \left(\frac{\alpha^2 - 1/4}{y^2} + \frac{1}{2}\{\phi, y\}\right) \times$$
$$\times \left[-\phi'(y)\right]^{-1} B_i\left(-k^{4/3}\phi(y)\right) Z_0(y) - \tilde{\vartheta}_1. \tag{68}$$

The integral with respect to z in $\tilde{\vartheta}_1$ should be represented as a sum of integrals over the segments $\left[\beta k, k - \Delta k^{-1/3}\right]$ and $\left[k - \Delta k^{-1/3}, k\right]$ (where Δ is a constant, $\Delta > 0$). Then, the estimate for ϑ_1 follows from the estimates for the Airy functions and the function A_1. The estimate for ϑ_2 is obtained in the similar way. As a result, we obtain $I_2^1 = C_1^{III}(\infty)\, O\left(k^{-7/3}\right)$. Thus, the contribution of the first-order approximation does not exceed the estimate for the remainder term in I_2^0. The contributions of the higher-order approximations can be estimated by means of inequality (56). These contributions also do not exceed the estimates for the remainder term in I_2^0.

5 The General Asymptotic Form of the Expansion Coefficients Specific Examples

Adding together I_1, I_2, and I_3, we obtain the general asymptotic expression for $C(n)$. Now we can formulate the following result.

Theorem 1. *Assume that function $f(r)$ satisfies the following conditions: (i) function $f(r)$ and its first three derivatives are continuous and bounded at $r \geq 0$, (ii) there exists $r_0 > 0$ such that the derivatives of function $f(r)$ up to the six order inclusive are continuous and bounded at $r > r_0$. Then, the following asymptotic formula is valid:*

$$C(n) = \frac{k^{\alpha-1/2}}{2^\alpha} \left\{(-1)^n \left[f(k) + \frac{1}{6k} f^{(3)}(k)\right]\right.$$
$$\left. + \frac{2^{1/2}\Gamma((\alpha+3/2)/2)}{k\Gamma((\alpha+1/2)/2)} f(0) + O\left(\frac{1}{k^2}\right)\right\}. \tag{69}$$

It should be mentioned that for the normalized functions $\tilde{\chi}_n(r) = \chi_n(r) / \|\chi_n\|$, where $\|\chi_n\|^2 = \int_0^\infty \chi_n^2(r)\,dr$, the factor $(2/k)^{1/2}$ appears in (69) instead of $k^{\alpha-1/2}/2^\alpha$.

Let us illustrate general result (69).

Example 1. Consider the function $f(r) = \Omega(pr)$, where $\Omega(z) = z\, j_l(z)$ is the Riccati-Bessel function [15] and $\alpha = l + 1/2$ (in quantum mechanics this corresponds to the free motion with the orbital angular momentum l). In this case, an explicit expression for the coefficient $C(n)$ has the form [20].

$$C(n) = (-1)^n \sqrt{\frac{\pi}{2}} p^{l+1} e^{-p^2/2} L_n^{(l+1/2)}\left(p^2\right). \tag{70}$$

The asymptotic formula for $C(n)$ follows directly from (70) and (17), (21):

$$C(n) = (-1)^n \frac{k^l}{2^{l+1/2}} \left(\Omega(\xi(p)) + O\left(\frac{1}{k^2}\right)\right). \tag{71}$$

This expression can be transformed to the form similar to (69). Using the asymptotic relation $\xi(p) = kp - p^3/6k + O\left(k^{-3}\right)$ and the Taylor series for the function $\Omega(z)$ at the point $z = kp$, we obtain $\Omega(\xi(p)) = \Omega(kp) - \Omega'(kp)\, p^3/6k + O\left(k^{-2}\right)$. Here the first derivative can be related to the third derivative by the use of the equation

$$\Omega'' + \left(1 + \frac{1 - \alpha^2/4}{z^2}\right)\Omega = 0. \tag{72}$$

Differentiating this equation, we obtain $\Omega'(kp) = -\Omega'''(kp) + O\left(k^{-2}\right)$. The final result has the form

$$C(n) = (-1)^n \frac{k^l}{2^{l+1/2}} \left(f(k) + \frac{1}{6k} f^{(3)}(k) + O\left(\frac{1}{k^2}\right)\right). \tag{73}$$

This example demonstrates that, in the general case, the estimate for the remainder term in asymptotic formula (69) can not be improved.

Let us consider another example.

Example 2. $f(r) \equiv 1$, $\alpha = 1/2$. In this case, an explicit expression for the coefficient $C(n)$ has the form [20]

$$C(n) = \sum_{m=0}^{n} \frac{(-1)^{n-m}\,\Gamma(m + 1/2)}{\Gamma(1/2)\, m!}. \tag{74}$$

The asymptotic value of the coefficients $C(n)$ can be obtained directly from (74). We represent the sum with respect to m as the difference $\sum_{m=0}^{\infty} - \sum_{m=n+1}^{\infty}$. The value of the former sum is known [21]. In the latter sum we add in pairs the terms with the indices $m = 2N$ and $2N + 1$. Thus, the alternating series is reduced to the series with positive terms

$$\sum_{m=(n+1)/2}^{\infty} \frac{\Gamma(2m+1/2)}{\Gamma(2m+2)}, \quad \text{if } n \text{ is odd} \tag{75}$$

$$-\sum_{m=n/2}^{\infty} \frac{\Gamma(2m+3/2)}{\Gamma(2m+3)}, \quad \text{if } n \text{ is even} \tag{76}$$

Substituting the asymptotic value [15] of the ratio of the Γ-functions into these series, we obtain

$$C(n) = \frac{1}{2^{1/2}} \left\{ (-1)^n + \frac{2^{1/2}}{k\,\Gamma(1/2)} + O\left(\frac{1}{k^3}\right) \right\}. \tag{77}$$

Concluding this section, we consider the application of the general formula (69) to the scattering problem. The asymptotic behavior of the expansion coefficients of the channel radial wave function is described by the following improved formula:

$$\tilde{C}_i(n) = C^-_{l_i n_i} \delta_{i i_0} - S_{i i_0} C^+_{l_i n_i} + O\left(\frac{1}{n^{5/4}}\right), \tag{78}$$

where

$$C^{\pm}_{ln} = \frac{e^{\mp i\sigma_l}}{\sqrt{k r_0 p}} \left\{ G_l(\eta, k r_0 p) \pm i F_l(\eta, k r_0 p) - \frac{p^3 r_0^3}{6k} \left[G'_l(\eta, k r_0 p) \pm i F'_l(\eta, k r_0 p) \right] \right\}. \tag{79}$$

Here F_l and G_l are the regular and irregular Coulomb wave functions [15], η is the Coulomb parameter, p is the channel wave number, r_0 is the oscillator radius (in the preceding formulas we put $r_0 = 1$), σ_l is the Coulomb phase [15], l is the orbital momentum, $S_{i i_0}$ is the S-matrix element, i is the channel index (i_0 is the entrance channel). The improvement in the asymptotic formula (79) is connected with account (in accordance with (69)) of the term with third derivative of the function expanded. Here we express the third derivative in terms of the first in the similar way as in Example 1. The formula (78) means the boundary condition in the J-matrix method.

6 Convergence Properties of the Expansion Series

The approach developed above for deriving the asymptotic form of expansion coefficients enables us to analyze convergence of the Fourier series $\sum_{m=0}^{\infty} \tilde{C}(m)\,\tilde{\chi}_m$, where $\tilde{C}(m) = C(m)/\|\chi_m\|$. From the Darboux-Christoffel identity [14–16] for the Laguerre polynomials it follows that

$$\mathrm{K}_n(r,t) = \sum_{m=0}^{n} \tilde{\chi}_m(r)\,\tilde{\chi}_m(t)$$
$$= -(n+1)\left(1+\frac{\alpha}{r+1}\right)^{1/2} \frac{\tilde{\chi}_{n+1}(r)\,\tilde{\chi}_n(t) - \tilde{\chi}_n(r)\,\tilde{\chi}_{n+1}(t)}{r^2-t^2}. \tag{80}$$

Denote the partial sum of the Fourier series for function $f(r)$ as $S_n(r)$, then

$$f(r) - S_n(r) = \int_0^\infty \left(f(r)(t/r)^{\alpha+1/2} e^{-(t^2-r^2)/2} - f(t)\right)\mathrm{K}_n(r,t)\,dt. \tag{81}$$

Thus, the convergence of the sequence of the partial sums $S_n(r)$ is related to the behavior of the expansion coefficients of the function

$$T_r(t) = \frac{f(r)(t/r)^{\alpha+1/2} e^{-(t^2-r^2)/2} - f(t)}{r^2-t^2}. \tag{82}$$

This function and its first- and second-order derivatives are continuous and bounded since the function $f(r)$ satisfies the condition (i) of Theorem 1 (at the point $t = r$ the function $T_r(t)$ is defined by its limit value, then the function $T_r(t)$ and its first and second order derivatives are continuous at the point $t = r$). The function $T_r(t)$ and its first- and second-order derivatives decrease (as $1/t^2$) at large t (r is fixed). These properties are sufficient to find the asymptotic form of the expansion coefficient of the function $T_r(t)$. Let us apply the approach developed in Sections 2–5. Calculation of I_1, I_2 and I_3 is much simplified due to rapid decreasing of the function $T_r(t)$ and its derivatives. It is sufficient to perform two integrations by parts in expressions for I_1^0, I_2^0 and I_3^0. Omitting the details, we present the final result:

$$\tilde{C}(n) = \frac{2\Gamma((\alpha+3/2)/2)}{k^{3/2}\Gamma((\alpha+1/2)/2)} T_r(0) + O\left(\frac{1}{k^{5/2}}\right). \tag{83}$$

The asymptotic expression for the function $\tilde{\chi}_n(r)$ at fixed $r \neq 0$ follows from (17) and (21):

$$\tilde{\chi}_n(r) = \frac{2}{\pi^{1/2}k^{1/2}}\left[\cos\left(kr - \frac{\alpha\pi}{2} - \frac{\pi}{4}\right) + O\left(\frac{1}{k}\right)\right]. \tag{84}$$

Substituting (80) into (81) and using (83) and (84), we obtain the estimate for the remainder term

$$R_n(r) = f(r) - S_n(r) = O\left(\frac{1}{k}\right). \tag{85}$$

Thus, we state the convergence theorem:

Theorem 2. *If a function $f(r)$ satisfies the condition (i) of the Theorem 1, the Fourier series of $f(r)$ converges pointwisely to the function $f(r)$ at any $r > 0$. For the remainder term the estimate (85) is valid.*

The case $r = 0$ is trivial because $\chi_n(0) = 0$ and all the terms of the series vanish. We illustrate these results for the simplest case $f(r) \equiv 1$. The behavior of some partial sums is represented in Figs. 1, 2.

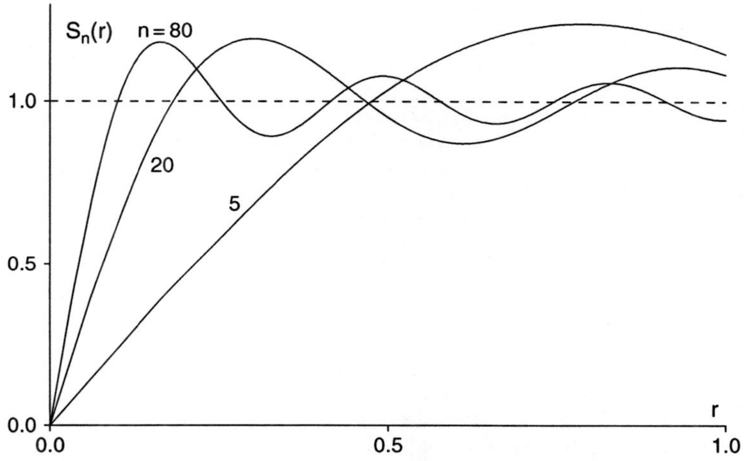

Fig. 1 Partial sums $S_n(r)$, $n = 5; 20; 80$ for the case $f(r) \equiv 1$: behavior in the vicinity of $r = 0$

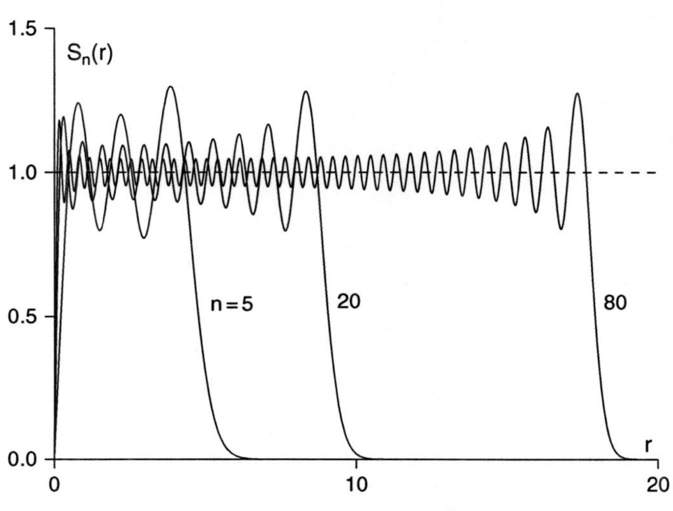

Fig. 2 Partial sums $S_n(r)$, $n = 5; 20; 80$ for the case $f(r) \equiv 1$

It should be noted that the condition (i) for the second- and the third-order derivative of the function $f(r)$ can be weakened. It is sufficient to assume that the derivatives $f''(r)$ and $f'''(r)$ are bounded and continuous with the exception of a finite set of points where these derivatives have finite jumps. This remark enables us to apply J-matrix method in the case of discontinuous potentials.

References

1. H.A. Jamani and L. Fishman, J. Math. Phys., **16**, 410 (1975).
2. V.S. Vasilevsky, G.F. Filippov, and L.L. Chopovsky, Sov. J. Part. Nucl., **16**, 153 (1985); ibid., **15**, 1338 (1984).
3. I.P. Okhrimenko, Nucl. Phys. A, **424**, 121 (1984).
4. S.Yu. Igashov and Yu.M. Tchuvil'sky, Bull. Russ. Acad. Sci., Phys., **66**(3), 385 (2002).
5. S.Yu. Igashov and A.M. Shirokov, Bull. Russ. Acad. Sci., Phys., **71**(6), 797 (2007).
6. Y.A. Lurie and A.M. Shirokov, Ann. Phys., **312**(2), 284 (2004).
7. V.S. Vasilevsky and F. Arickx, Phys. Rev. A, **55**(1), 265 (1997).
8. J. Broeckhove, F. Arickx, W. Vanroose and V.S. Vasilevsky, J. Phys. A, **37**(31), 7769 (2004).
9. S.A. Zaytsev, Theor. Math. Phys., **140**(1), 918 (2004).
10. A.M. Shirokov, A.I. Mazur, S.A. Zaytsev, J.P. Vary, and T.A. Weber, Phys. Rev. C, **70**(4), 044005 (2004).
11. S.A. Zaytsev, Inverse Probl., **21**(3), 1061 (2005).
12. A.M. Shirokov, J.P. Vary, A.I. Mazur, S.A. Zaytsev, and T.A. Weber, J. Phys. G, **31**(8), S1283 (2005).
13. N. Barnea, W. Leidemann, and G. Orlandini, Phys. Rev. C, **74**(3), 034003 (2006).
14. A. Erdelyi, Ed., Higher Transcendental Functions, Vol. II (Mc Graw-Hill, New York, 1953).
15. M. Abramowitz and I.A. Stegun, Eds., Handbook of Mathematical Functions, (Dover, New York, 1965).
16. G. Szego, Orthogonal Polynomials, Am. Math. Soc. Colloq. Publ., Vol. 23, 4th ed. (AMS, Providence, RI, 1975).
17. F.W.J. Olver, Asymptotics and Special Functions (Academic, London, 1974).
18. S.Yu. Igashov, Int. Transf. Spec. Funct., **8**, 209 (1999).
19. A. Erdelyi, J. Indian Math. Soc., **24**, 235 (1960).
20. A.P. Prudnikov, Yu.A. Brychkov, and O.I. Marichev, Integrals and Series, Vol. 2 (Gordon and Breach, New York, 1986).
21. A.P. Prudnikov, Yu.A. Brychkov, and O.I. Marichev, Integrals and Series, Vol. 1 (Gordon and Breach, New York, 1986).

Scattering Phase Shift for Relativistic Separable Potentials with Laguerre-Type Form Factors

A.D. Alhaidari

Abstract Using the tools of the J-matrix method, we obtain analytic expression for the relativistic scattering phase shift of M-term separable potential with Laguerre-type form factors for $M = 1, 2$, and 3. The Dirac Hamiltonian with spherical symmetry is taken as the reference Hamiltonian. For higher order separable potentials ($M \geq 4$) an exact numerical evaluation could be obtained in a simple and straightforward way.

Motivated by the diversity of applications of separable potentials to the solution of various models in nuclear and condensed matter physics, we considered in an earlier publication [1] relativistic scattering for separable potentials with exponential-type form factors. Taking the Dirac Hamiltonian as the reference Hamiltonian and maintaining manifest relativistic invariance gave that work added significance. We studied two classes of spherically symmetric separable potentials. One of them has the form factor $r^{\nu-1}e^{-\lambda r/2}$ while the other one is of the form $r^{2\nu}e^{-\lambda^2 r^2/2}$, where λ is the potential range parameter and ν is a non-negative integer. It turned out that these potentials are $(\nu+1)$-term separable. That is, their exact and full matrix representations in a suitable L^2 spinor basis are of dimension $\nu+1$. The tools of the relativistic J-matrix method of scattering [2–4] were used to obtain analytic expressions for the phase shift for $\nu = 0$, 1, and 2. For higher order potentials ($\nu \geq 3$), we found that it was sufficient and more practical to calculate the *exact* phase shift numerically using the relativistic J-matrix method. The resulting structure of the phase shift as a function of energy turned out to be very rich and highly interesting as demonstrated graphically in Ref. [1]. In this article, we extend that work significantly and in two directions. On the one hand, the results obtained here are for all values of the angular momentum, which is in contrast to the previous work where we could only obtain the phase shift for S-wave scattering. On the other hand, the form factors in the relativistic separable potential will be Laguerre-type of the form $(\lambda r)^{\ell+\alpha}e^{-x/2}L_n^{\mu}(x)$, where ℓ is the angular momentum, $L_n^{\mu}(x)$ is the associated Laguerre polynomial [5], and either:

A.D. Alhaidari
Shura Council, Riyadh 11212, Saudi Arabia
e-mail: haidari@mailaps.org

A.D. Alhaidari et al. (eds.), *The J-Matrix Method*, 67–81.

$$1)\quad x=\lambda r,\ \mu=2\ell+\beta,\ \alpha=0\text{ or }1,\ \text{ and }\beta=0\text{ or }2. \tag{1a}$$

$$2)\quad x=(\lambda r)^2,\ \mu=\ell+\beta,\ \alpha\in\{0,1,2,3\},\ \text{ and }\beta=-\frac{1}{2}\text{ or }\frac{3}{2}. \tag{1b}$$

Next, we start by setting up the problem and review briefly our findings in Ref. [1].

In the atomic units ($m=\hbar=1$) and writing the speed of light $c=\lambda\!\!\!^{-}{}^{-1}$, the radial component of the free Dirac equation reads

$$\begin{pmatrix} 1-\varepsilon & \lambda\!\!\!^{-}\left(\frac{\kappa}{r}-\frac{d}{dr}\right) \\ \lambda\!\!\!^{-}\left(\frac{\kappa}{r}+\frac{d}{dr}\right) & -1-\varepsilon \end{pmatrix}\begin{pmatrix} \sum_n d_n(\varepsilon)\phi_n(r) \\ \sum_n d_n(\varepsilon)\theta_n(r) \end{pmatrix}=0, \tag{2}$$

where $\lambda\!\!\!^{-}$ is the Compton wavelength $\hbar/mc$, ε is the relativistic energy in units of $mc^2=\lambda\!\!\!^{-}{}^{-2}$, and κ is the spin-orbit quantum number defined by $\kappa=\pm\left(j+\frac{1}{2}\right)$ for $\ell=j\pm\frac{1}{2}$. $\{d_n(\varepsilon)\}_{n=0}^{\infty}$ are the expansion coefficients of the radial component of the two-spinor wavefunction $\Phi(\vec{r})$. Introducing nonlocal interaction in (2) via the potential $V(\vec{r},\vec{r}')$ gives

$$\begin{pmatrix} 1-\varepsilon & \lambda\!\!\!^{-}\left(\frac{\kappa}{r}-\frac{d}{dr}\right) \\ \lambda\!\!\!^{-}\left(\frac{\kappa}{r}+\frac{d}{dr}\right) & -1-\varepsilon \end{pmatrix}\begin{pmatrix} \sum_n d_n(\varepsilon)\phi_n(r) \\ \sum_n d_n(\varepsilon)\theta_n(r) \end{pmatrix}+\lambda\!\!\!^{-}{}^2\int V(\vec{r},\vec{r}')\Phi(\vec{r}')d^3\vec{r}'=0. \tag{3}$$

For spherically symmetric separable potentials, the relativistic kernel $V(\vec{r},\vec{r}')$ was taken in [1] as the following 2×2 general matrix

$$V(\vec{r},\vec{r}')=\begin{pmatrix} V_+U(r)U(r') & V_0U(r)W(r') \\ V_0W(r)U(r') & V_-W(r)W(r') \end{pmatrix}, \tag{4}$$

where $U(r)$ and $W(r)$ are real radial functions – the form factors. $V_\pm$ and V_0 are real coupling parameters. It is obvious that this representation satisfies unitarity, namely $V(\vec{r},\vec{r}')^\dagger=V(\vec{r}',\vec{r})$. The potential matrix elements are

$$V_{nm}^{I,I'}=\lambda\!\!\!^{-}{}^2\iint \chi_n^I(\vec{r})^\dagger V(\vec{r},\vec{r}')\chi_m^{I'}(\vec{r}')d^3\vec{r}d^3\vec{r}', \tag{5}$$

where $\chi_n^I(\vec{r})$ is an element of the spinor basis with radial components $\phi_n(r)$ and $\theta_n(r)$ and spin indices I. In [1], we found that the integral (5) is tractable only when $\ell=0$ (i.e., $\kappa=-1$).[1] Therefore, that problem was limited to S-wave scattering only. The angular contribution to the integral is a factor of 4π. Hence,

$$V_{nm}=4\pi\lambda\!\!\!^{-}{}^2\left[V_+I_nI_m+V_-J_nJ_m+V_0\left(I_nJ_m+I_mJ_n\right)\right], \tag{6}$$

[1] The inadmissible assignment $\kappa=0$ was made in several places in Ref. [1]. It is an error that should be corrected everywhere to read $\ell=0$ (equivalently, $\kappa=-1$).

where we've defined the following integrals

$$I_n = \int rU(r)\phi_n(r)dr, \text{ and} \tag{7a}$$

$$J_n = \int rW(r)\theta_n(r)dr. \tag{7b}$$

The relativistic two-component spinor bases that are compatible with this problem were given in Appendix A of [1]. The Laguerre basis functions given by (A.7) in [1] is the natural basis for the solution space of the problem with separable potential whose form factor is $r^{\nu-1}e^{-\lambda r/2}$. For $\ell = 0$, the separable potential (4) with $U(r) = W(r) = r^{\nu-1}e^{-\lambda r/2}$ results in a finite number of terms for both integrals in (7a) and (7b) simultaneously. That is, by substituting elements of the Laguerre spinor basis, given in Appendix A of [1] with $\ell = 0$, in the integrals (7) we obtain

$$I_n = \frac{1}{\lambda^{\nu+1/2}\sqrt{n+1}} \int x^{\nu+1}e^{-x}L_n^1(x)dx, \text{ and} \tag{8a}$$

$$J_n = \frac{\sqrt{n+1}}{2}\frac{\lambda C}{\lambda^{\nu+1/2}} \int x^{\nu}e^{-x}\left[L_n^0(x) + L_{n+1}^0(x)\right]dx, \tag{8b}$$

where $x = \lambda r$ and C is the strength parameter of the small spinor component. Using the three-term recursion relation of the Laguerre polynomial [5] repeatedly on the term $x^{\nu+\mu}L_m^\mu(x)$ inside the integral results in a sum of Laguerre polynomials multiplying x^μ. Then, inserting $L_0^\mu(x)$, which is equal to one, inside these integrals and using the orthogonality relation of the Laguerre polynomials [5] one obtains $\nu + 1$ terms. For example, we obtain the following potential elements for $\nu = 0$

$$V_{nm} = 4\pi\frac{\tilde{\lambda}^2}{\lambda}\left[V_+ + (\lambda C/2)^2 V_- + \lambda C V_0\right]\delta_{n0}\delta_{m0}, \tag{9}$$

whereas, for $\nu = 1$

$$V_{nm} = 8\pi\frac{\tilde{\lambda}^2}{\lambda^3}\begin{pmatrix} 2V_+ & -\sqrt{2}\left(V_+ + \frac{\lambda C}{2}V_0\right) \\ -\sqrt{2}\left(V_+ + \frac{\lambda C}{2}V_0\right) & V_+ + \left(\frac{\lambda C}{2}\right)^2 V_- + \lambda C V_0 \end{pmatrix}_{nm}. \tag{10}$$

Now, the spinor basis is chosen such that the matrix representation of the Dirac operator in (2) is tridiagonal. Therefore, the solution of the problem defined by (3) comes down to solving a three-term recursion relation with the $(\nu + 1) \times (\nu + 1)$ symmetric matrix, coming from the separable potential, acting as a source term. In other words, the potential terms create new initial relations for the three-term recursion relation. The details of how to solve this recursion relation, which is found in Ref. [1], will be summarized below when we investigate our current problem.

The same procedure outlined above could be implemented in the case of the second type of separable potential with the Gaussian form factors $r^{2\nu}e^{-\lambda^2r^2/2}$. However, the natural basis in this case is the oscillator spinor basis given by (A.15) in [1]. Moreover, for $\ell = 0$, the two integrals in (7a) and (7b) produce finite $(\nu + 1)$ terms only if we take $U(r) = \lambda r W(r) = r^{2\nu} e^{-\lambda^2 r^2/2}$. In this case, we obtain the following integrals

$$I_n = \frac{1/\sqrt{2}}{\lambda^{2\nu+3/2}} \sqrt{\frac{\Gamma(n+1)}{\Gamma(n+3/2)}} \int x^{\nu+1/2} e^{-x} L_n^{1/2}(x) dx, \text{ and} \tag{11a}$$

$$J_n = \frac{\lambda C/\sqrt{2}}{\lambda^{2\nu+3/2}} \sqrt{\frac{\Gamma(n+1)}{\Gamma(n+3/2)}} \int x^{\nu-1/2} e^{-x} \left[(n+1/2) L_n^{-1/2}(x) + (n+1) L_{n+1}^{-1/2}(x) \right] dx, \tag{11b}$$

where $x = \lambda^2 r^2$. Evaluation of these integrals gives the $(\nu + 1)$-term separable potential (6) which acts as a source term (i.e., initial relations) for the three-term recursion relation that results in algebraic solution of (3). With this background, we now turn attention to our present problem.

Let $\psi_n(r)$ stand for the two-component radial spinor basis whose upper component is $\phi_n(r)$ and lower component $\theta_n(r)$ and consider the following alternative representation of the M-term separable potential

$$\tilde{V} = \tilde{\lambda}^2 \sum_{n,m=0}^{M-1} \left| \bar{\psi}_n \right\rangle V_{nm} \left\langle \bar{\psi}_m \right|, \tag{12}$$

where $\bar{\psi}_n$ is an element of the conjugate L^2 space (i.e., $\langle \bar{\psi}_n \mid \psi_m \rangle = \langle \psi_n \mid \bar{\psi}_m \rangle = \delta_{nm}$) and V_{nm} are elements of an $M \times M$ real symmetric matrix. They are the real separable potential parameters with a dimension of inverse squared length. The Laguerre and oscillator spinor bases and their conjugates that are suitable for the current problem were constructed recently in [6] and are given in Table 1. These were chosen to support a tridiagonal matrix representation of the Dirac operator of (2). In configuration space, the separable potential (12) could be written as

$$\tilde{V}(r, r') \equiv \langle r | \tilde{V} \left| r' \right\rangle = \begin{pmatrix} V^+(r, r') & V^0(r, r') \\ V^0(r', r) & V^-(r, r') \end{pmatrix}, \tag{13}$$

where

$$V^+(r, r') = \tilde{\lambda}^2 \sum_{n,m=0}^{M-1} V_{nm} \bar{\phi}_n(r) \bar{\phi}_m(r') \tag{14a}$$

$$V^-(r, r') = \tilde{\lambda}^2 \sum_{n,m=0}^{M-1} V_{nm} \bar{\theta}_n(r) \bar{\theta}_m(r') \tag{14b}$$

$$V^0(r, r') = \tilde{\lambda}^2 \sum_{n,m=0}^{M-1} V_{nm} \bar{\phi}_n(r) \bar{\theta}_m(r') \tag{14c}$$

Table 1 The two-component Laguerre and oscillator spinor bases and their conjugates for positive and negative κ. $x = \omega r$ and ω is a length scale parameter, which is a measure of the effective range of the separable potential

	Laguerre spinor basis	Oscillator spinor basis
$\kappa > 0$	$\psi_n(r) = a_n\, x^{\ell} e^{-x/2} \begin{pmatrix} (n+2\ell+1)L_n^{2\ell}(x) - (n+1)L_{n+1}^{2\ell}(x) \\ \frac{\lambda\omega}{4}\left[(n+2\ell+1)L_n^{2\ell}(x) + (n+1)L_{n+1}^{2\ell}(x)\right] \end{pmatrix}$ $a_n = \sqrt{\omega\,\Gamma(n+1)/\Gamma(2\ell+n+2)}$	$\psi_n(r) = b_n\, x^{\ell-1} e^{-x^2/2} \begin{pmatrix} \left(n+\ell+\frac{1}{2}\right) L_n^{\ell-1/2}(x^2) - (n+1)L_{n+1}^{\ell-1/2}(x^2) \\ \frac{\lambda\omega}{2} x \left[\left(n+\ell+\frac{1}{2}\right) L_n^{\ell-1/2}(x^2) + (n+1)L_{n+1}^{\ell-1/2}(x^2)\right] \end{pmatrix}$ $b_n = \sqrt{2\omega\,\Gamma(n+1)/\Gamma(n+\ell+3/2)}$
$\kappa < 0$	$\psi_n(r) = a_n\, x^{\ell+1} e^{-x/2} \begin{pmatrix} L_n^{2\ell+2}(x) - L_{n-1}^{2\ell+2}(x) \\ -\frac{\lambda\omega}{4}\left[L_n^{2\ell+2}(x) + L_{n-1}^{2\ell+2}(x)\right] \end{pmatrix}$	$\psi_n(r) = b_n\, x^{\ell+1} e^{-x^2/2} \begin{pmatrix} L_n^{\ell+3/2}(x^2) - L_{n-1}^{\ell+3/2}(x^2) \\ -\frac{\lambda\omega}{2} x \left[L_n^{\ell+3/2}(x^2) + L_{n-1}^{\ell+3/2}(x^2)\right] \end{pmatrix}$
$\kappa > 0$	$\bar{\psi}_n(r) = \frac{a_n}{4\ell}\, x^{\ell} e^{-x/2} \begin{pmatrix} (n+2\ell+1)L_n^{2\ell}(x) + (n+1)L_{n+1}^{2\ell}(x) \\ \frac{4}{\lambda\omega}\left[(n+2\ell+1)L_n^{2\ell}(x) - (n+1)L_{n+1}^{2\ell}(x)\right] \end{pmatrix}$	$\bar{\psi}_n(r) = \frac{b_n}{2\ell-1}\, x^{\ell} e^{-x^2/2} \begin{pmatrix} x\left[\left(n+\ell+\frac{1}{2}\right) L_n^{\ell-1/2}(x^2) + (n+1)L_{n+1}^{\ell-1/2}(x^2)\right] \\ \frac{2}{\lambda\omega}\left[\left(n+\ell+\frac{1}{2}\right) L_n^{\ell-1/2}(x^2) - (n+1)L_{n+1}^{\ell-1/2}(x^2)\right] \end{pmatrix}$
$\kappa < 0$	$\bar{\psi}_n(r) = \frac{a_n/4}{\ell+1}\, x^{\ell+1} e^{-x/2} \begin{pmatrix} L_n^{2\ell+2}(x) + L_{n-1}^{2\ell+2}(x) \\ -\frac{4}{\lambda\omega}\left[L_n^{2\ell+2}(x) - L_{n-1}^{2\ell+2}(x)\right] \end{pmatrix}$	$\bar{\psi}_n(r) = \frac{b_n}{2\ell+3} x^{\ell+2} e^{-x^2/2} \begin{pmatrix} x\left[L_n^{\ell+3/2}(x^2) + L_{n-1}^{\ell+3/2}(x^2)\right] \\ -\frac{2}{\lambda\omega}\left[L_n^{\ell+3/2}(x^2) - L_{n-1}^{\ell+3/2}(x^2)\right] \end{pmatrix}$

For example, in the Laguerre basis and for $M = 1$, we obtain the following separable potential

$$\tilde{V}(r,r') = \frac{\omega^{2\ell+1}\,V_{00}}{(\ell+1)^2}\frac{(rr')^{\ell}}{(2\ell+1)!}e^{-\omega(r+r')/2} \times \begin{pmatrix} \lambda^2\left(\ell+\frac{1}{2}-\frac{\omega r}{4}\right)\left(\ell+\frac{1}{2}-\frac{\omega r'}{4}\right) & \lambda r\left(\ell+\frac{1}{2}-\frac{\omega r'}{4}\right) \\ \lambda r'\left(\ell+\frac{1}{2}-\frac{\omega r}{4}\right) & rr' \end{pmatrix}, \qquad \kappa > 0 \tag{15a}$$

$$\tilde{V}(r,r') = \frac{\omega^{2\ell+1}\,V_{00}}{(\ell+1)^2}\frac{(rr')^{\ell+1}}{(2\ell+1)!}e^{-\omega(r+r')/2}\begin{pmatrix} (\lambda\omega/4)^2 & -\frac{1}{4}\lambda\omega \\ -\frac{1}{4}\lambda\omega & 1 \end{pmatrix}, \qquad \kappa < 0 \tag{15b}$$

where ω is a length scale parameter, which is a measure of the effective range of the potential. Including the separable potential (12) in the wave equation (2) gives

$$\sum_{m=0}^{\infty} d_m\,(H_0-\varepsilon)\,|\psi_m\rangle + \lambda^2\sum_{m,k=0}^{M-1} V_{km}d_m\,|\bar{\psi}_k\rangle = 0, \tag{16}$$

where H_0 is the reference Hamiltonian, that is the 2×2 matrix operator in (2) with $\varepsilon = 0$. Projecting on the left of (16) by $\langle\psi_n|$, we obtain the following equivalent matrix wave equation:

$$\sum_{m=0}^{\infty}\Im_{nm}(\varepsilon)d_m(\varepsilon) + \lambda^2\sum_{m=0}^{M-1}V_{nm}d_m(\varepsilon) = 0 \qquad ;n = 0,1,2,\ldots \tag{17}$$

where $\Im_{nm}(\varepsilon) \equiv (H_0)_{nm} - \varepsilon\Omega_{nm}$ and Ω is the matrix representation of the identity, i.e. the basis overlap matrix whose elements are $\Omega_{nm} = \langle\psi_n \mid \psi_m\rangle$. In the J-matrix method, the bases $\{\psi_n\}_{n=0}^{\infty}$ are chosen such that the matrix representations of H_0 and Ω (thus, so is $\Im$) are tridiagonal. Therefore, (17) could be written explicitly in the following matrix form

$$\left[\begin{pmatrix} \Im_{00} & \Im_{01} & & & \\ \Im_{10} & \Im_{11} & \Im_{12} & & 0 \\ & \Im_{21} & \Im_{22} & \Im_{23} & \\ & & .. & .. & .. \\ 0 & & .. & .. & .. \\ & & & .. & .. \end{pmatrix} + \lambda^2\begin{pmatrix} V_{00} & V_{01} & .. & V_{0,M-1} & 0 & .. \\ V_{10} & V_{11} & .. & V_{1,M-1} & 0 & .. \\ \vdots & \vdots & .. & \vdots & \vdots & .. \\ V_{M-1,0} & V_{M-1,1} & .. & V_{M-1,M-1} & 0 & .. \\ 0 & 0 & .. & 0 & 0 & .. \\ \vdots & \vdots & .. & \vdots & \vdots & .. \end{pmatrix}\right]\begin{pmatrix} d_0 \\ d_1 \\ \vdots \\ d_{M-1} \\ d_M \\ \vdots \end{pmatrix} = 0 \tag{18}$$

This relativistic J-matrix scattering problem has an *exact* numerical solution which is obtained by taking the $M\times M$ matrix representation of the separable potential as the exact short-range potential in the standard J-matrix scheme [2, 7]. To obtain this numerical solution, we proceed as follows. For an integer $N \geq M$ the $M\times M$ potential matrix, $\left\{\lambda^2 V_{nm}\right\}_{n,m=0}^{M-1}$, will be added to the $N\times N$ tridiagonal

Table 2 In the two spinor bases, $\Im_{nm}(\varepsilon)$ is the tridiagonal matrix representation of the wave operator, $\{s_n(\varepsilon)\}$ are the sine-like expansion coefficients of the asymptotic regular solution of the wave equation, and $\{c_n(\varepsilon)\}$ are the cosine-like expansion coefficients of the asymptotic irregular solution [6]. ${}_2F_1(a,b;c;z)$ is the hypergeometric series, ${}_1F_1(a;c;z)$ is the confluent hypergeometric series, and $C_n^{\nu}(z)$ is the ultra-spherical (Gegenbauer) polynomial

	Laguerre spinor basis	Oscillator spinor basis
$\Im_{nm}(\varepsilon)$	$\frac{\bar{\lambda}^2\omega^2}{\varepsilon+1}\left(\frac{2E}{\omega^2}+\frac{1}{4}\right)[-2\cos\theta\,(n+\ell+1)\,\delta_{nm}$ $+\sqrt{n\,(n+2\ell+1)}\delta_{n,m+1}+\sqrt{(n+1)\,(n+2\ell+2)}\delta_{n,m-1}]$	$\frac{\bar{\lambda}^2\omega^2}{\varepsilon+1}\left[-\left(\mu/\omega\right)^2\delta_{n,m}+\left(2n+\ell+\frac{3}{2}\right)\delta_{n,m}\right.$ $\left.+\sqrt{n\left(n+\ell+\frac{1}{2}\right)}\delta_{n,m+1}+\sqrt{(n+1)\left(n+\ell+\frac{3}{2}\right)}\delta_{n,m-1}\right]$
	$E(\varepsilon)=(\varepsilon^2-1)/2\bar{\lambda}^2$, $\cos\theta=\left(\frac{2E}{\omega^2}-\frac{1}{4}\right)/\left(\frac{2E}{\omega^2}+\frac{1}{4}\right)$	$\mu(\varepsilon)=\frac{1}{\bar{\lambda}}\left(\varepsilon^2-1\right)^{1/2}$
$s_n(\varepsilon)$	$s_n(\varepsilon)=\frac{a_n}{\omega\sqrt{2\pi}}2^{\ell+1}\Gamma(\ell+1)\,(\sin\theta)^{\ell+1}\,C_n^{\ell+1}(\cos\theta)$	$(-1)^n\frac{a_n}{\omega}\left(\frac{\mu}{\omega}\right)^{\ell+1}e^{-\mu^2/2\omega^2}L_n^{\ell+\frac{1}{2}}\left(\mu^2/\omega^2\right)$
$c_n(\varepsilon)$	$-\frac{a_n}{\pi\,\omega}2^{\ell+\frac{1}{2}}\Gamma\left(\ell+\frac{1}{2}\right)(\sin\theta)^{-\ell}$ $\times{}_2F_1\left(-n-2\ell-1,n+1;-\ell+\frac{1}{2};\sin^2\frac{\theta}{2}\right)$	$(-1)^n\frac{\sqrt{2}}{\pi\omega}a_n\,\Gamma\left(\ell+\frac{1}{2}\right)\left(\mu/\omega\right)^{-\ell}e^{-\mu^2/2\omega^2}$ ${}_1F_1\left(-n-\ell-\frac{1}{2};-\ell+\frac{1}{2};\mu^2/\omega^2\right)$

matrix representation of the reference Hamiltonian giving the $N \times N$ truncated total relativistic Hamiltonian where $(H_0)_{nm}$ is obtained from Table 2 as $\Im_{nm}(\varepsilon = 0)$ for the two spinor bases corresponding to the two types of separable potentials. The total Hamiltonian matrix is, then diagonalized and used in the calculation of the finite Green's function. This, together with the exact J-matrix asymptotic solution of the H_0-problem, leads to the relativistic phase shift in the usual way [2,4,7].

However, it turns out that without too much difficulty and with a reasonable effort we can obtain *analytic* solution for the relativistic phase shift for $M = 1, 2$, and 3. This is accomplished by considering a *new* relativistic J-matrix problem in which the reference Hamiltonian, $\hat{H}_0$, is the sum of the infinite tridiagonal kinetic energy term, H_0, and this finite separable potential matrix V. It is evident from the matrix wave equation (18) that the resulting recursion relation for the expansion coefficients of the new wavefunction is asymptotically ($n \geq M$) three-term. While, the first set of M equations for $\{d_n(\varepsilon)\}_{n=0}^{M-1}$ could be thought of as the initial relations for this new three-term recursion. That is, the separable potential acts like a source term for the recursion. The analytic solution of the recursion relation, with these newly emerging initial conditions due to the separable potential, gives the new sine-like, $d_n = \hat{s}_n$, and cosine-like, $d_n = \hat{c}_n$, expansion coefficients of the wavefunction. The asymptotic behavior of these coefficients gives the sought after phase shift as an energy dependent rotation angle of the original coefficients $\{s_n, c_n\}_{n \geq M-1}$. To obtain a more compact and transparent solution we write the problem in terms of the complex coefficients defined as

$$h_n^{\pm}(\varepsilon) = c_n(\varepsilon) \pm \mathrm{i}\, s_n(\varepsilon). \tag{19}$$

In this notation, the asymptotic form of (18) gives the following three-term recursion relation

$$\Im_{n,n-1}\hat{h}_{n-1}^{+} + \Im_{n,n}\hat{h}_n^{+} + \Im_{n,n+1}\hat{h}_{n+1}^{+} = 0, \qquad n \geq M \tag{20}$$

which is the same as that of the original J-matrix problem for $n \geq 1$ [1,2,7]. The initial relations, on the other hand, are replaced by the following:

$$\begin{pmatrix} \Im_{00} & \Im_{01} & & & & 0 \\ \Im_{10} & \Im_{11} & \Im_{12} & & & \\ & \Im_{21} & \Im_{22} & \Im_{23} & & \\ & & \ddots & \ddots & \ddots & \\ 0 & & & \Im_{M-1,M-2} & \Im_{M-1,M-1} & \Im_{M-1,M} \end{pmatrix} \begin{pmatrix} \hat{h}_0^{+} \\ \hat{h}_1^{+} \\ \hat{h}_2^{+} \\ \vdots \\ \hat{h}_{M-1}^{+} \\ \hat{h}_M^{+} \end{pmatrix} = -\bar{\lambda}^2 \begin{pmatrix} V_{00} & V_{01} & \cdots & V_{0,M-1} \\ V_{10} & V_{11} & \cdots & V_{1,M-1} \\ \vdots & \vdots & \cdots & \vdots \\ \vdots & \vdots & \cdots & \vdots \\ V_{M-1,0} & V_{M-1,1} & \cdots & V_{M-1,M-1} \end{pmatrix} \begin{pmatrix} \hat{h}_0^{+} \\ \hat{h}_1^{+} \\ \vdots \\ \vdots \\ \hat{h}_{M-1}^{+} \end{pmatrix} - \frac{\mathrm{i}\bar{\lambda}^2 w}{\hat{h}_0^{+} - \hat{h}_0^{-}} \begin{pmatrix} 1 \\ 0 \\ \vdots \\ \vdots \\ 0 \end{pmatrix} \tag{18'}$$

where $w(\varepsilon)$ is the Wronskian of the regular and irregular solutions of the free Dirac problem [2–4, 6]. That is, $w(\varepsilon) = -\frac{4}{\pi\lambda}\sqrt{\frac{\varepsilon-1}{\varepsilon+1}}$. To recover the original recursion, whose initial relation is

$$\Im_{00}h_0^+ + \Im_{01}h_1^+ = -\mathrm{i}\lambda^2 w/(h_0^+ - h_0^-), \tag{21}$$

we propose the following $2M - 1$ parameter transformation of the complex coefficients:

$$\hat{h}_m^{\pm} = \mu_m e^{\pm\mathrm{i}\sigma_m} h_m^{\pm} \quad ; m = 0, 1, .., M-2, \tag{22a}$$

$$\hat{h}_n^{\pm} = e^{\pm\mathrm{i}\tau} h_n^{\pm} \quad ; n \geq M-1 \tag{22b}$$

where $\{\mu_m\}$ and $\{\sigma_m\}$ are real constant parameters, and $\mu_m > 0$. Equation (22b) represents the asymptotic behavior of the expansion coefficients and gives the phase shift due to the separable potential as the energy dependent rotation angle, τ, of the original coefficients.

Substituting from (22a) and (22b) into (18′) and using the original recursion and its initial relation (as was done in Ref. [1]) we arrive at the analytic expressions for the scattering matrix $e^{2i\tau}$ giving the phase shift angle τ. In these expressions, which are given below, the J-matrix kinematical coefficients $T_n(\varepsilon)$ and $R_n^{\pm}(\varepsilon)$ are defined as

$$T_n \equiv \frac{c_n - is_n}{c_n + is_n} = \frac{h_n^-}{h_n^+} \quad ; \quad R_{n+1}^{\pm} \equiv \frac{c_{n+1} \pm is_{n+1}}{c_n \pm is_n} = \frac{h_{n+1}^{\pm}}{h_n^{\pm}} \quad ; \; n \geq 0 \tag{23}$$

The wavefunction expansion coefficients $\{s_n(\varepsilon)\}$ and $\{c_n(\varepsilon)\}$ for the H_0-problem in the Laguerre and oscillator spinor basis are given in Table 2 [6]. The tridiagonal matrix elements of $\Im(\varepsilon)$ for the two types of separable potentials are also listed in Table 2. The results of the calculation for the three lowest-term separable potentials are as follows:

M = 1:

$$e^{2\mathrm{i}\tau} = T_0 + (1 - T_0)\left[1 + \frac{\lambda^2 V_{00}}{\Im_{00} + \Im_{01}R_1^+}\right]^{-1} \tag{24}$$

M = 2:

$$e^{2\mathrm{i}\tau} = T_0 e^{-2\mathrm{i}\zeta} + \frac{1 - T_0}{\gamma}\left(\Im_{00} + \Im_{01}R_1^+\right)\left[\gamma\left(\Im_{00} + \lambda^2 V_{00}\right) + R_1^+\left(\Im_{01} + \lambda^2 V_{01}\right)\right]^{-1} \tag{25}$$

where $\gamma = \left(\Im_{01} - \lambda^2 V_{11} R_1^+\right)/\left(\Im_{01} + \lambda^2 V_{01}\right)$ and $\zeta = \arg(\gamma)$.

M = 3:

$$e^{2i\tau} = T_0 e^{-2i\xi} + \frac{1-T_0}{R_1^+\Lambda}\left(\Im_{00} + \Im_{01} R_1^+\right) \times \left\{R_1^+\Lambda\left(\Im_{00} + \bar{\lambda}^2 V_{00}\right)\right.$$

$$\left. + \bar{\lambda}^2 R_1^+ \left[\frac{\Im_{01} + \bar{\lambda}^2 V_{01}}{\Im_{12} + \bar{\lambda}^2 V_{12}}\left(\Im_{12}/\bar{\lambda}^2 - R_2^+ V_{22} - V_{02}\Lambda\right) + R_2^+ V_{02}\right]\right\}^{-1} \tag{26}$$

where $\xi = \arg\left(R_1^+\Lambda\right)$ and,

$$\Lambda(\varepsilon, V) = \frac{\left(\Im_{01}/R_1^+\right) + \Im_{11} - \bar{\lambda}^2 R_2^+ V_{12} + \frac{\Im_{11} + \bar{\lambda}^2 V_{11}}{\Im_{12} + \bar{\lambda}^2 V_{12}}\left(-\Im_{12} + \bar{\lambda}^2 R_2^+ V_{22}\right)}{\Im_{01} + \bar{\lambda}^2 V_{01} - \bar{\lambda}^2 V_{02}\frac{\Im_{11} + \bar{\lambda}^2 V_{11}}{\Im_{12} + \bar{\lambda}^2 V_{12}}}. \tag{27}$$

Table 2 shows that the angular momentum dependence of $\Im_{nm}(\varepsilon)$, $T_n(\varepsilon)$ and $R_n^{\pm}(\varepsilon)$ is on ℓ not κ. Therefore, the above expressions imply that the resulting phase shift for the separable potential corresponding to κ is the same as that which corresponds to $-\kappa - 1$. This is due to our particular construction of the separable potential in terms of the basis as given by (14). The nonrelativistic limit of these expressions is obtained by taking $\bar{\lambda} \to 0$.

One should note that $\{V_{nm}\}$ in the analytic expressions (24–27) above are the separable potential parameters given as input in (12) while in Ref. [1] they were calculated as shown in (4)–(10). Moreover, when evaluating $\Im_{nm}$, T_n and $R_n^{\pm}$ in the formulas above we should use their expressions for a general angular momentum as given in Table 2. Whereas, in Ref. [1] they were given only for $\ell = 0$. The graphical results in the Figs. 1–4 give the phase shift versus the relativistic energy variable $E(\varepsilon)$, which is defined by $E(\varepsilon) = (\varepsilon^2 - 1)/2\bar{\lambda}^2$. In the nonrelativsistic limit, $E(\varepsilon)$ becomes the systems's energy. The parameters $\{V_{nm}\}_{n,m=0}^{M-1}$ of the M-term separable potentials, for $M = 1, 2, 3$, and 4 are given by the $M \times M$ submatrix of following potential parameter matrix

$$V_{nm} = \begin{pmatrix} 5 & -1 & 0 & 2 \\ -1 & 3 & 2 & 1 \\ 0 & 2 & -1 & 0 \\ 2 & 1 & 0 & 1 \end{pmatrix}_{nm} \tag{28}$$

In all figures, the calculation was performed in the Laguerre basis. That is, the form factors for all these separable potentials are of the type defined by the choice in (1a). It is obvious that the difference between the relativistic and nonrelativistic results become prominent only at higher energies.

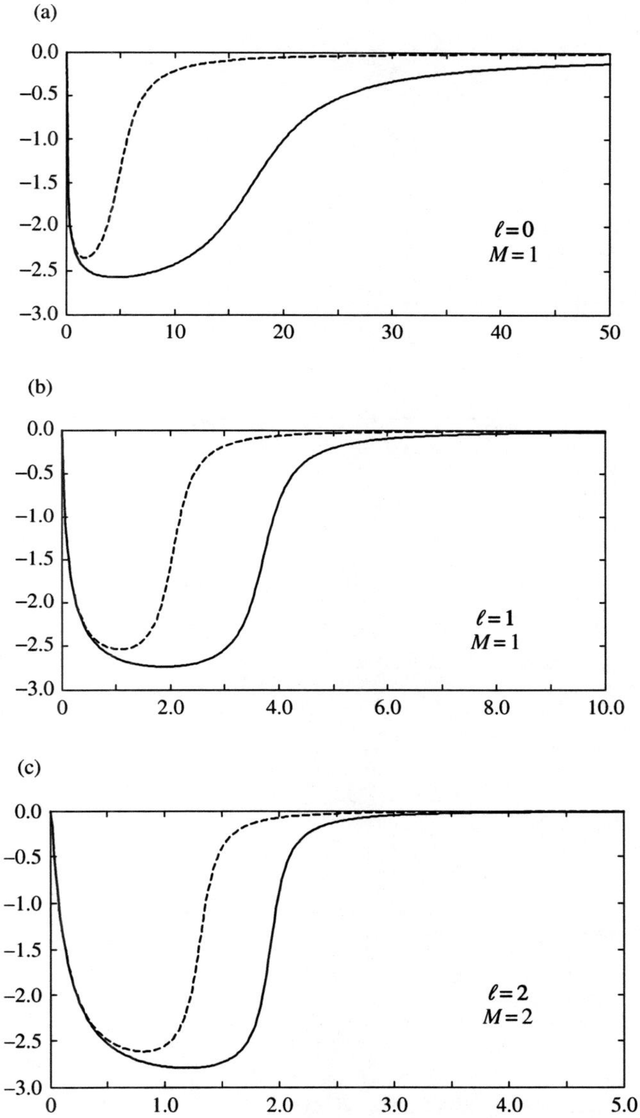

Fig. 1 The phase shift $\tau(\varepsilon)$ in radians versus the energy $E(\varepsilon)$, in units of mc^2. The solid (*dashed*) curve corresponds to the relativistic (nonrelativistic) results. The potential is one-term separable with form factors in the Laguerre basis as given by (15a) and (15b). The physical parameters are $\bar{\lambda} = 1$, $\omega = 1$, and $V_{00} = 5$ (in atomic units). The angular momentum quantum number is taken as: (*a*) $\ell = 0$ (i.e., $\kappa = -1$), (*b*) $\ell = 1$ (i.e., $\kappa = +1$ or $\kappa = -2$), and (*c*) $\ell = 2$ (i.e., $\kappa = +2$ or $\kappa = -3$)

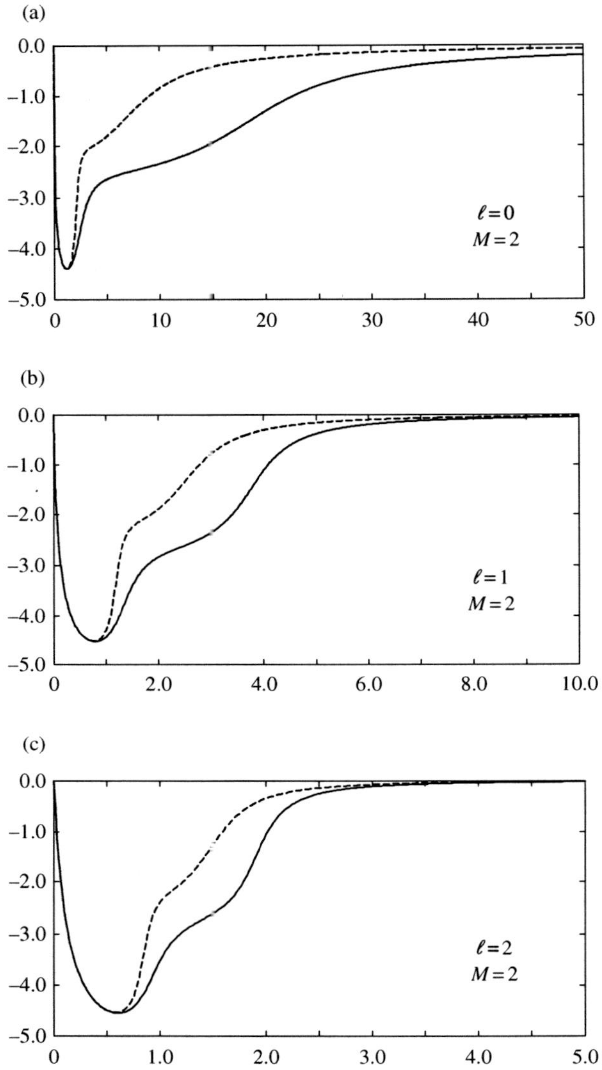

Fig. 2 Same as Fig. 1, except that the potential is now two-term separable with $V_{00} = 5$, $V_{11} = 3$, and $V_{01} = V_{10} = -1$

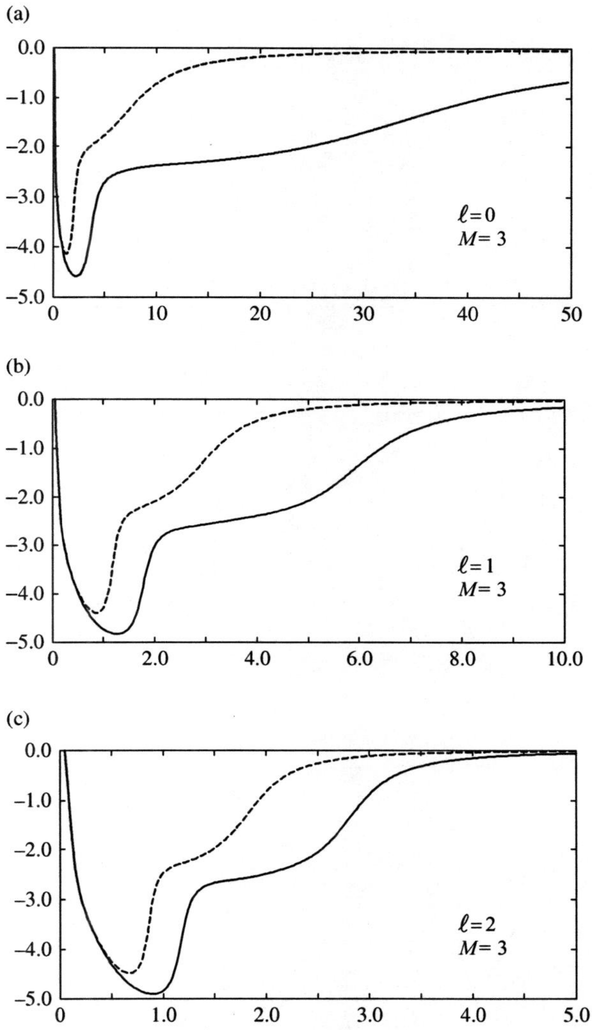

Fig. 3 Same as Fig. 1, except that the potential is now three-term separable with

$$V = \begin{pmatrix} 5 & -1 & 0 \\ -1 & 3 & 2 \\ 0 & 2 & -1 \end{pmatrix}$$

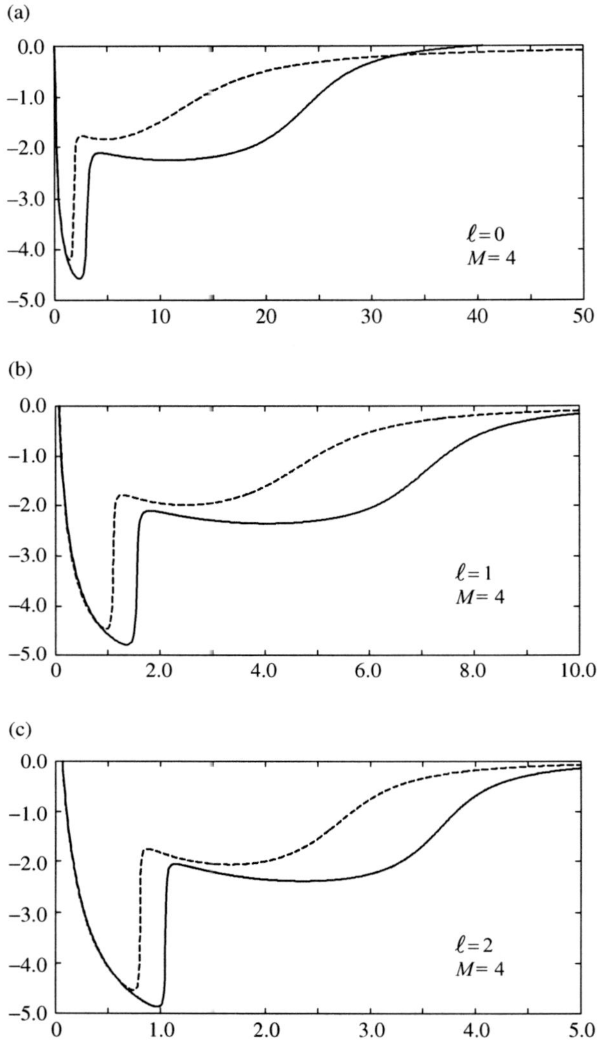

Fig. 4 Same as Fig. 1, except that the potential is now four-term separable with the potential parameters V_{nm} given by (28) and the phase shift is obtained numerically using the relativistic J-matrix method [2–4, 6]

Acknowledgments I am grateful to Prof. H. A. Yamani for fruitful discussions and suggestions for improvements on the original manuscript.

References

1. A. D. Alhaidari, J. Phys. A: Math. Gen. **34,** 11273 (2001); **37**, 8911 (2004)
2. A. D. Alhaidari, J. Math. Phys. **43**, 1129 (2002)
3. P. Horodecki, Phys. Rev. A **62**, 052716 (2000)
4. A. D. Alhaidari, H. A. Yamani, and M. S. Abdelmonem, Phys. Rev. A **63**, 062708 (2001)
5. W. Magnus, F. Oberhettinger, and R. P. Soni, *Formulas and Theorems for the Special Functions of Mathematical Physics*, 3rd edition (Springer-Verlag, New York, 1966) pp. 239–249
6. A. D. Alhaidari, H. Bahlouli, A. Al-Hasan, and M. S. Abdelmonem, Phys. Rev. A **75**, 062711 (2007)
7. H. A. Yamani and L. Fishman, J. Math. Phys. **16**, 410 (1975)

Accurate Evaluation of the S-Matrix for Multi-Channel Analytic and Non-Analytic Potentials in Complex L^2 Bases

H.A. Yamani and M.S. Abdelmonem

Abstract We describe an efficient and accurate scheme to compute the S-matrix elements for a given multi-channel analytic and non-analytic potentials in complex-scaled orthonormal Laguerre or oscillator bases using the J-matrix method. As examples of the utilization of the scheme, we evaluate the cross section of two-channel square wells in an oscillator basis and find the resonance position for the same potential using the Laguerre basis. We also find resonance positions of a two-channel analytic potential for several angular momenta ℓ using both bases. Additionally, we evaluate the effect of including the Coulomb term (z/r) when employing the Laguerre basis.

The work reported here is a continuation of the authors' effort [1–7] to utilize the advantages of the J-matrix method of scattering. In particular, we have provided a solution for the Lippman–Schwinger equation in a closed form for a model finite-rank potential. This yields a T-matrix and S-matrix in terms of certain coefficients that are related to the H_0-problem, with the matrix elements of Green's matrix being evaluated in the basis set chosen. In this chapter, we answer the question of how to accurately and easily evaluate the S-matrix elements for such a potential from analytic and non-analytic general potentials.

This chapter is organized as follows. In Sect. 2, we review the defining statements of the various functions that come into play in the calculation of the S-matrix elements. In Sect. 3, we explicitly outline the various methods available for the calculation of these functions. It is important to be certain of the accuracy of the potential element V_{nm} for high values of n and m. We specifically show that for the complex-scaled bases, the Gauss quadrature scheme of evaluating V_{nm} has several desired features. It is easy to implement, makes full use of the nature of the basis, and works for general potential. Furthermore, its accuracy is assured, even for high n and m indices, through the proper choice of a parameter K, which specifies the order of the quadrature scheme utilized. We plan to achieve the goal of accurately

H.A. Yamani
Ministry of Commerce & Industry, P.O. Box 5729, Riyadh 11127, Saudi Arabia
e-mail: haydara@sbm.net.sa

A.D. Alhaidari et al. (eds.), *The J-Matrix Method*, 83–102.

calculating the matrix elements V_{nm} in stages. In Sect. 3.1 we propose an approximate scheme to determine the matrix elements of a one-channel central analytic potential in the Laguerre basis using a real and complex scale parameter λ. As this scheme does not work for non-analytic potentials, in Sect. 3.2 we develop an approximate scheme that may be applied for both kinds of potentials. In Sect. 3.3 and 3.4 we generalize the scheme in the preceding two sub-sections to find the matrix elements $\left\langle \phi_n^{(\alpha)} \right| V^{\alpha\beta} \left| \phi_m^{(\beta)} \right\rangle$ of a multi-channel potential $V^{\alpha\beta}(r)$, where the scale parameters λ_α and λ_β associated with the channel basis are allowed to differ, and have the form

$$\lambda_\alpha = |\lambda_\alpha| e^{-i\theta_\alpha}, \text{ and } \quad \lambda_\beta = \left|\lambda_\beta\right| e^{-i\theta_\beta}.$$

The results of Sect. 3 that are associated with the use of the oscillator basis are summarized in Sect. 3.5.

Finally, in Sect. 4 we discuss the accuracy of the proposed scheme and illustrate its application by considering two examples. In the first we evaluate the cross section of two-channel square wells in the oscillator basis, and find the complex resonance energy of the same potential using the Laguerre basis. In the second example we find the complex resonance energy for a specific two-channel analytical potential (without a Coulomb term) for several angular momenta ℓ using both bases. We then consider the same problem with the Coulomb term included in the calculation using the Laguerre basis.

1 Review of the *J*-Matrix Method

Here we consider the scattering of a structureless, spinless particle by a target with M internal states labeled by the threshold energies $E_1, E_2, \ldots, E_M$. We assume that the given scattering potential, $V^{\alpha\beta}$, is of a short range. We also assume that the reference Hamiltonian of the particle, H_0, may include, in addition to the ℓth partial wave kinetic energy operator, the Coulomb term (z/r); that is,

$$H_0 = -\frac{1}{2}\frac{d^2}{dr^2} + \frac{\ell(\ell+1)}{2r^2} + \frac{z}{r}. \tag{1}$$

The multi-channel Schrödinger equation for the given scattering problem can be written as

$$\sum_{\beta=1}^{M} \left\{ [H_0 - (E - E_\alpha)] \delta_{\alpha\beta} + V^{\alpha\beta} \right\} \left| \psi_{\alpha\beta} \right\rangle = 0. \tag{2}$$

The basis chosen in each channel α is either the Laguerre basis

$$\phi_n^{(\alpha)}(r) = d_n^{(L)} e^{-\lambda_\alpha r/2} (\lambda_\alpha r)^{\nu/2} L_n^{\nu}(\lambda_\alpha r), \tag{3}$$

or the oscillator basis

$$\phi_n^{(\alpha)}(r) = d_n^{(o)} e^{-\lambda_\alpha^2 r^2/2} (\lambda_\alpha r)^{\nu+1/2} L_n^\nu\left(\lambda_\alpha^2 r^2\right). \tag{4}$$

In the Laguerre basis, $\nu = 2\ell + 2$ and the normalization factor is $d_n^{(L)} = \sqrt{\lambda_\alpha n!/\Gamma(n+\nu+1)}$. In the oscillator basis, $\nu = \ell + 1/2$ and the normalization factor is $d_n^{(o)} = \sqrt{2\lambda_\alpha n!/\Gamma(n+\nu+1)}$. Here, $L_n^\nu(\zeta)$ is the generalized Laguerre polynomial [8], λ is a free complex scale parameter, and ℓ is the partial wave angular momentum quantum number.

The J-matrix method of scattering finds the exact S-matrix to the model multi-channel scattering potential, $\tilde{V}$. The model potential is obtained from the given potential, V, by restricting its infinite matrix representation in a complete L^2 basis to a finite representation. That is, the model potential $\tilde{V}$ is defined as

$$\tilde{V}^{\alpha\beta} = \sum_{n=0}^{N_\alpha-1} \sum_{m=0}^{N_\beta-1} \left|\bar{\phi}_n^{(\alpha)}\right\rangle \left\langle \phi_n^{(\alpha)}\right| V^{\alpha\beta} \left|\phi_m^{(\beta)}\right\rangle \left\langle \bar{\phi}_m^{(\beta)}\right| \tag{5}$$

where $\left\{\left|\bar{\phi}_n^{(\alpha)}\right\rangle\right\}_{n=0}^{\infty}$ is the orthogonal complement to the basis $\left\{\left|\phi_n^{(\alpha)}\right\rangle\right\}_{n=0}^{\infty}$ in the αth channel in the sense that $\left\langle \bar{\phi}_n^{(\alpha)} \mid \phi_m^{(\alpha)}\right\rangle = \left\langle \phi_m^{(\alpha)} \mid \bar{\phi}_n^{(\alpha)}\right\rangle = \delta_{nm}$.

The exact multi-channel S-matrix can be written as [9]

$$S_{\alpha\beta}(E) = T_{N_\alpha}\delta_{\alpha\beta} - \sqrt{\left(\frac{J_{N_\alpha-1,N_\alpha}^{(\alpha)}}{J_{N_\beta-1,N_\beta}^{(\beta)}}\right)\left(\frac{R_{N_\alpha}^{+}}{R_{N_\beta}^{+}}\right)}\sqrt{T_{N_\alpha-1} - T_{N_\alpha}} \times \left(\frac{D_{\alpha\beta}^{(+)}(E)}{\Delta^{(+)}(E)}\right)\sqrt{T_{N_\beta-1} - T_{N_\beta}} \tag{6}$$

Here

$$\Delta^{(+)}(E) = \begin{vmatrix} (1 - Y_{1,1}) & -Y_{1,2} & \dots & -Y_{1,M} \\ -Y_{2,1} & (1 - Y_{2,2}) & \dots & -Y_{2,M} \\ \dots & \dots & \dots & \dots \\ -Y_{M,1} & -Y_{M,2} & \dots & (1 - Y_{M,M}) \end{vmatrix}, \tag{7}$$

and

$$D_{\alpha\beta}^{(+)}(E) = \begin{vmatrix} 0 & \delta_{\alpha,1} & \delta_{\alpha,2} & \dots\ \delta_{\alpha,M} \\ \delta_{\beta,1} & & & \\ \delta_{\beta,2} & & \Delta^{(+)}(E) & \\ \dots & & & \\ \delta_{\beta,M} & & & \end{vmatrix}, \tag{8}$$

where

$$Y_{\alpha,\beta} = -g^{(\alpha,\beta)}_{N_\alpha-1,N_\beta-1} \; J^{(\beta)}_{N_\beta-1,N_\beta} \; R^{+}_{N_\beta} \tag{9}$$

and in (8) we show the first row and column, while the remainder is represented by $\Delta^{(+)}(E)$ of equation (7). Additionally, $R^{\pm}_{N}(\alpha)$ and $T_{N-1}(\alpha)$ are the J-matrix ratios associated with H_0, which are given by

$$R^{\pm}_{n}(\alpha) = \frac{c^{(\alpha)}_{n} \pm i s^{(\alpha)}_{n}}{c^{(\alpha)}_{n-1} \pm i s^{(\alpha)}_{n-1}}, \text{ and } T_n(\alpha) = \frac{c^{(\alpha)}_{n} - i s^{(\alpha)}_{n}}{c^{(\alpha)}_{n} + i s^{(\alpha)}_{n}}. \tag{10}$$

$\{s^{(\alpha)}_n\}$ is the set of coefficients of the expansion of the sine-like eigenfunction of H_0 in the basis $\{|\phi^{(\alpha)}_n\rangle\}$. Similarly, $\{c^{(\alpha)}_n\}$ is the set of coefficients of the expansion of the asymptotically cosine-like eigenfunction of H_0 in the same basis [1]. Furthermore, $J^{(\beta)}_{N_\beta-1,N_\beta} = \langle\phi_{N_\beta-1}|\left(H_0 - E + E_\beta\right)|\phi_{N_\beta}\rangle$ and $g^{(\alpha,\beta)}_{N_\alpha-1,N_\beta-1}$ represent the $(N_\alpha - 1, N_\beta - 1)$ element of the (α,β) sub-matrix of g, the inverse of the total $N_c \times N_c$ Hamiltonian matrix. The matrix element $g^{(\alpha,\beta)}_{N_\alpha-1,N_\beta-1}$ takes the following explicit forms:

(i) In the Laguerre basis

$$g^{(\alpha,\beta)}_{N_\alpha-1,N_\beta-1}(E) = \left(\frac{\tilde{a}_{N_\alpha-1}\tilde{a}_{N_\beta-1}}{a_{N_\alpha-1}a_{N_\beta-1}}\right) \sum_{\mu=1}^{N_c} \frac{\Lambda^{(\alpha)}_{N_\alpha-1,\mu}\Lambda^{(\beta)}_{N_\beta-1,\mu}}{E_\mu - E}, \tag{11}$$

where

$$\tilde{a}_n = \sqrt{\frac{\lambda n!}{\Gamma(n+\nu+1)}} \text{ and } a_n = \sqrt{\frac{\lambda n!}{\Gamma(n+\nu)}}. \tag{12}$$

(ii) In the oscillator basis

$$g^{(\alpha,\beta)}_{N_\alpha-1,N_\beta-1}(E) = \sum_{\mu=1}^{N_c} \frac{\Lambda^{(\alpha)}_{N_\alpha-1,\mu}\Lambda^{(\beta)}_{N_\beta-1,\mu}}{E_\mu - E}. \tag{13}$$

In both bases, E_μ is the eigenvalue of the full Hamiltonian, and

$$\left(\Lambda^{(1)}_{0,\mu}, \Lambda^{(1)}_{1,\mu}, \ldots, \Lambda^{(1)}_{N_1-1,\mu}, \Lambda^{(2)}_{0,\mu}, \ldots, \Lambda^{(2)}_{N_2-1,\mu}, \ldots, \Lambda^{(M)}_{0,\mu}, \ldots, \Lambda^{(M)}_{N_M-1,\mu}\right) \tag{14}$$

is the associated eigenvector. Here, $N_c = \sum_{\alpha=1}^{M} N_\alpha$ is the dimension of the Hamiltonian matrix. This is the same as the total number of L^2 functions used to diagonalize the scattering Hamiltonian.

Thus, the calculation of the exact $S(E)$ for the model potential requires:

(1) The matrix element $J_{N-1,N}(E)$, the J-matrix ratios $R_N^{\pm}(E)$ and $T_{N-1}(E)$, and
(2) The matrix element $g_{N_\alpha-1,N_\beta-1}^{(\alpha,\beta)}(E)$.

2 $J_{N-1,N}(E)$, $R_N^{\pm}(E)$, $T_{N-1}(E)$ and the Matrix Elements of H_0 in Both Bases

We first find expressions for the matrix elements of the J matrix, which are symmetric and tridiagonal in both bases.

(1) In the Laguerre basis:

$$J_{n,n} = \lambda z - \left(E - E_\alpha - \lambda^2/8\right)(2n + 2\ell + 2) \tag{15}$$

$$J_{n,n-1} = \left(E - E_\alpha + \lambda^2/8\right)\sqrt{n\,(n + 2\ell + 1)} \tag{16}$$

(2) In the oscillator basis:

$$J_{n,n} = \left(2n + \ell + 3/2\right)\left(\lambda^2/2\right) - (E - E_\alpha) \tag{17}$$

$$J_{n,n-1} = \left(\lambda^2/2\right)\sqrt{n\,(n + \ell + 1/2)} \tag{18}$$

We then find expressions for the J-matrix ratios $R_N^{\pm}(E)$ and $T_{N-1}(E)$ in the Laguerre or oscillator bases. In order to do that, we first define the following functions

$$u_n = -J_{n,n}/J_{n,n-1}, \quad \text{and} \quad v_n = -J_{n,n-1}/J_{n,n+1} \tag{19}$$

It is subsequently determined that the recursion relation for $R_N^{\pm}(E)$ is

$$R_{n+1}^{\pm} = u_n + \frac{v_n}{R_n^{\pm}} \qquad \text{for} \quad n \geq 1. \tag{20}$$

The J-matrix ratio, T_n, can be written as:

$$T_n = T_0\left(\prod_{j=1}^{n}\frac{R_j^-}{R_j^+}\right). \tag{21}$$

Thus, for the evaluation of the ratios $R_N^{\pm}(E)$ and $T_{N-1}(E)$, we only need to calculate $R_1^{\pm}(E)$ and $T_0(E)$, where

$$R_1^{\pm} = (c_1 \pm i s_1)/(c_0 \pm i s_0) \quad \text{and} \quad T_0 = (c_0 - i s_0)/(c_0 + i s_0). \tag{22}$$

The explicit forms of $R_1^{\pm}(E)$ and $T_0(E)$ in the Laguerre and oscillator bases are presented as continued fraction expressions elsewhere [9].

We may then determine the matrix elements of the reference Hamiltonian in both bases. In addition to the ℓth partial wave kinetic energy operator, the reference Hamiltonian H_0 may include the Coulomb term (z/r),

$$H_0(r,z) = -\frac{1}{2}\frac{d^2}{dr^2} + \frac{\ell(\ell+1)}{2r^2} + \frac{z}{r}. \tag{23}$$

The bases (3) and (4) remain complete in $L^2(0,\infty)$ for complex λ, as long as $\mathrm{Re}(\lambda) > 0$ (Laguerre) and $\mathrm{Re}(\lambda^2) > 0$ (oscillator). If we write $\lambda = |\lambda| e^{-i\theta}$, the restriction will be $-\pi/2 < \theta < \pi/2$ for the Laguerre basis and $-\pi/4 < \theta < \pi/4$ for the oscillator basis. In fact, by considering θ as a complex rotation angle, the above bases with $\lambda = |\lambda| e^{-i\theta}$ become orthonormal complex-scaled bases. We note that the matrix elements of the reference Hamiltonian $\langle\phi_n^\theta| H_0 |\phi_m^\theta\rangle$ are calculated once, and used with every model potential that is considered. The matrix elements of H_0 have the following representation:

(i) For the Laguerre basis:

$$(H_0)_{nm} = \frac{\lambda^2}{4}\left\{\frac{c_{n>}}{c_{n<}}\left(1+\frac{2n_<}{\nu+1}+\frac{4(z/\lambda)}{\nu}\right) - \frac{1}{2}\delta_{nm}\right\}, \tag{24}$$

where $c_n = \sqrt{\lambda n!/\Gamma(n+\nu+1)}$, $\nu = 2\ell+2$, and $n_>$ $(n_<)$ is the larger (smaller) of the two indices n and m, respectively.

(ii) For the oscillator basis ($z = 0$ only):

$$(H_0)_{n,m} = (\lambda^2/2)\left[(2n+\nu+1)\delta_{nm} + \sqrt{n(n+\nu)}\left(\delta_{n,m+1}+\delta_{n,m-1}\right)\right], \tag{25}$$

where $\nu = \ell + 1/2$.

3 Evaluation of Potential Matrix Elements Using the Gauss Quadrature Scheme

3.1 The Case of a Single-Channel Central Analytic Potential

The potential matrix element $V_{nm} = \langle\phi_n| V |\phi_m\rangle$ in the Laguerre basis can be written explicitly as

$$V_{nm} = \int_0^\infty \phi_n(r)V(r)\phi_m(r)\,dr. \tag{26}$$

Depending on the form of the potential, some skill may be required to find reliable numerical values for the above matrix elements, especially for large values of

n and m. If the scale parameter of the basis is real, the above matrix element V_{nm} may be cast in the form

$$V_{nm} = \frac{c_n c_m}{\lambda} \int_0^{\infty} e^{-\zeta} \zeta^{\nu} L_n^{\nu}(\zeta) L_m^{\nu}(\zeta) V(\zeta/\lambda) d\zeta. \tag{27}$$

Instead of working with the polynomials $L_n^{\nu}(\zeta)$, we prefer to work with the polynomial $p_n(\zeta)$ satisfying the symmetric three-term recursion relation [10]

$$\zeta p_n(\zeta) = b_n p_{n+1}(\zeta) + a_n p_n(\zeta) + b_{n-1} p_{n-1}(\zeta); \quad n \geq 0, \tag{28}$$

with the definition $b_{-1} = 0$ and $p_0(\zeta) = 1$. In this case, we have explicitly

$$\begin{aligned} a_n &= 2n + \nu + 1, \\ b_n &= -\sqrt{(n+1)(n+\nu+1)}, \text{ and} \\ L_n^{\nu}(\zeta) &= \sqrt{\frac{\Gamma(n+\nu+1)}{n!\Gamma(\nu+1)}} p_n(\zeta). \end{aligned} \tag{29}$$

The polynomials $p_n^{\nu}(z)$ may be calculated as

$$p_n(z) = \prod_{k=0}^{n-1} \left(\frac{u_k(z)}{b_k} \right); \quad n \geq 1;$$

where $u_0(z) = (z - a_0)$ and

$$u_k(z) = (z - a_k) - b_{k-1}^2 / u_{k-1}(z); \;\; k \geq 1. \tag{30}$$

This process is numerically very stable. The matrix element of equation (27) now takes on the simple form

$$V_{nm} = \int_0^{\infty} \frac{\rho(\zeta)}{\Gamma(\nu+1)} p_n(\zeta) p_m(\zeta) V(\zeta/\lambda) d\zeta, \tag{31}$$

where $\rho(\zeta) = e^{-\zeta} \zeta^{\nu}$ is the density associated with the orthonormal polynomials $\{p_n(\zeta)\}_{n=0}^{\infty}$, as given by

$$\int_0^{\infty} \rho(\zeta)\, p_n(\zeta)\, p_m(\zeta)\, d\zeta = \delta_{nm}. \tag{32}$$

It is known [10, 11] from the theory of orthogonal polynomials that associated with each density $\rho(\zeta)$ and order K, a set of abscissas and weights $\left\{\zeta_k^{(K)}, \omega_k^{(K)}\right\}_{k=0}^{K-1}$ satisfy the following properties

(i) $$p_K\left(\zeta_k^{(K)}\right) = 0, \text{ and}$$

(ii) $$\int_0^\infty \rho(\zeta) f(\zeta) d\zeta \cong \sum_{k=0}^{K-1} \omega_k^{(K)} f\left(\zeta_k^{(K)}\right). \tag{33}$$

The above approximation is exact if $f(\zeta)$ is a polynomial of the order $\leq 2K-1$. Equation (33) is known as the Gauss quadrature approximation. Here, $\zeta_k^{(K)}$, is the kth eigenvalue of the finite Jacobi matrix

$$\mathrm{J}^{(K)} = \begin{pmatrix} a_0 & b_0 & 0 & 0 & \ldots & 0 & 0 & 0 \\ b_0 & a_1 & b_1 & 0 & \ldots & 0 & 0 & 0 \\ 0 & b_1 & a_2 & b_2 & \ldots & 0 & 0 & 0 \\ \ldots & \ldots & \ldots & \ldots & \ldots & \ldots & \ldots & \ldots \\ 0 & 0 & 0 & 0 & \ldots & b_{K-3} & a_{K-2} & b_{K-2} \\ 0 & 0 & 0 & 0 & \ldots & 0 & b_{K-2} & a_{K-1} \end{pmatrix}. \tag{34}$$

The associated weight $\omega_k^{(K)}$ is given by

$$\omega_k^{(K)} = \left[\left\langle \hat{e}_0 \middle| \vec{v}_k^{(K)} \right\rangle\right]^2 \Gamma(\nu+1), \tag{35}$$

where $\left|\vec{v}_k^{(K)}\right\rangle$ is the kth eigenvector associated with eigenvalue $\zeta_k^{(K)}$. Here $|\hat{e}_n\rangle$ is a unit column vector of dimension $K \times 1$, composed entirely of zeros except for a single element of unity in the nth row. Thus, the potential matrix integral (31) can be approximated as

$$V_{nm} \cong \sum_{k=0}^{K-1} \left[\left\langle \hat{e}_0 \middle| \vec{v}_k^{(K)} \right\rangle\right]^2 p_n\left(\zeta_k^{(K)}\right) p_m\left(\zeta_k^{(K)}\right) V\left(\zeta_k^{(K)}/\lambda\right). \tag{36}$$

Furthermore, it can be shown that for an order $n < K$, the quantity $p_n\left(\zeta_k^{(K)}\right)$ has the form

$$p_n\left(\zeta_k^{(K)}\right) = \left\langle \hat{e}_n \middle| \vec{v}_k^{(K)} \right\rangle \Big/ \left\langle \hat{e}_0 \middle| \vec{v}_k^{(K)} \right\rangle; \quad n < K. \tag{37}$$

Therefore, if we take $K > n, m$, the result (36) may be simplified to

$$V_{nm} \cong \sum_{k=0}^{K-1} \left\langle \hat{e}_n \middle| \vec{v}_k^{(K)} \right\rangle \left\langle \hat{e}_m \middle| \vec{v}_k^{(K)} \right\rangle V\left(\zeta_k^{(K)}/\lambda\right); \quad K > n, m. \tag{38}$$

Computationally, this is an extremely simple formula. Performing the diagonalization of the finite matrix $J^{(K)}$ of equation (34) once, we have the eigenvalues $\left\{\zeta_k^{(K)}\right\}_{k=0}^{K-1}$ and eigenvectors $\left\{\left|\vec{v}_k^{(K)}\right\rangle\right\}_{k=0}^{K-1}$. The summand in equation (38) is the value of the potential at the scaled positions $\left(\zeta_k^{(K)}/\lambda\right)$, weighted by the product of the nth and mth coefficient of the eigenvector $\left|\vec{v}_k^{(K)}\right\rangle$. Since the argument of the potential in the sum (38) is real, this approximation works just as well for both analytic and non-analytic central potentials.

The corresponding result of equation (38) for a complex scale parameter having the form $\lambda = |\lambda|\, e^{-i\theta}$ can be easily obtained. In this case, a change of variable in the definition (26) for the matrix elements yields an integral similar to equation (27), except that the contour of integration is along the ray $re^{-i\theta}$. Due to the restriction $0 \le \theta < \pi/2$ on the values of θ, the application of the Cauchy integral theorem shows that the integral along the stated ray is equivalent to the integral (27) along the real axis. Hence, even with a complex scale parameter λ, we also have the approximation

$$V_{nm} \cong \sum_{k=0}^{K-1} \left\langle \hat{e}_n \middle| \vec{v}_k^{(K)} \right\rangle \left\langle \hat{e}_m \middle| \vec{v}_k^{(K)} \right\rangle V\left(\zeta_k^{(K)} \middle/ |\lambda|\, e^{-i\theta}\right); \; K > n, m. \tag{39}$$

This approximation is numerically very stable. In contrast to equation (38), this result works only for analytic potentials.

The evaluation of the matrix elements of one-channel potentials in the oscillator basis with real λ can proceed in a fashion similar to the above Laguerre case. In fact, the equation for the matrix element V_{nm}, corresponding to equation (31), is represented by

$$V_{nm} = \int_0^\infty \frac{\rho(\zeta)}{\Gamma(\nu+1)}\, p_n(\zeta)\, p_m(\zeta)\; V\left(\sqrt{\zeta}\middle/\lambda\right) d\zeta, \tag{40}$$

where $\rho(\zeta) = e^{-\zeta}\zeta^{\nu}$ and the polynomial p_n satisfy the same three term recursion relation of equation (28), with the exception that $\nu = \ell + 1/2$ in this case. Thus, the result corresponding to equation (38) may be shown by the equation

$$V_{nm} \cong \sum_{k=0}^{K-1} \left\langle \hat{e}_n \middle| \vec{v}_k^{(K)} \right\rangle \left\langle \hat{e}_m \middle| \vec{v}_k^{(K)} \right\rangle V\left(\sqrt{\zeta_k^{(K)}}\middle/\lambda\right). \tag{41}$$

Again, since the argument of the potential in the sum (41) is real, this approximation works just as well for both analytic and non-analytic central potentials.

In a similar way, if the scale parameter is now complex and has the form $\lambda = |\lambda|\, e^{-i\theta}$, where $0 \le \theta < \pi/4$, an expression analogous to equation (39) may take the form

$$V_{nm} \cong \sum_{k=0}^{K-1} \left\langle \hat{e}_n \left| \vec{v}_k^{(K)} \right.\right\rangle \left\langle \hat{e}_m \left| \vec{v}_k^{(K)} \right.\right\rangle V\left(\sqrt{\zeta_k^{(K)}} \Big/ |\lambda|\, e^{-i\theta} \right). \tag{42}$$

3.2 The Case of a Single-Channel Central Non-Analytic Potential

We generalize the results in the previous subsection to the case of non-analytic potential. We start with the Laguerre basis in which the scale parameter is complex and has the form $\lambda = |\lambda|\, e^{-i\theta}$, where $0 \leq \theta < \pi/2$. After a simple manipulation, the integral (31) can be cast in the form

$$V_{nm} = \int_0^\infty \frac{\rho(\zeta)}{\Gamma(\nu+1)} \left[F(\zeta, \theta) \right] p_n(\zeta \gamma_\theta)\, p_m(\zeta \gamma_\theta)\, V(\zeta / |\lambda| \cos\theta)\, d\zeta, \tag{43}$$

where

$$F(\zeta, \theta) = \left[(\gamma_\theta)^{\nu+1}\, e^{i\zeta \tan\theta} \right],$$

and

$$\gamma_\theta = \left(\frac{e^{-i\theta}}{\cos\theta} \right). \tag{44}$$

Using the real set of abscissas and weights $\left\{ \zeta_k^{(K)}, \omega_k^{(K)} \right\}_{k=0}^{K-1}$, we are able to obtain an approximation to the above integral, as

$$V_{nm} \cong \sum_{k=0}^{K-1} \left[\left\langle \hat{e}_0 \left| \vec{v}_k^{(K)} \right.\right\rangle \right]^2 \left[F\left(\zeta_k^{(K)}, \theta \right) \right] p_n\left(\zeta_k^{(K)} \gamma_\theta \right) p_m\left(\zeta_k^{(K)} \gamma_\theta \right)$$
$$\times V\left(\zeta_k^{(K)} / |\lambda| \cos\theta \right) \tag{45}$$

If on the other hand we assume $n, m < K$, we can make use of equation (37) to write

$$V_{nm} \cong \sum_{k=0}^{K-1} \left\langle \hat{e}_n \left| \vec{v}_k^{(K)} \right.\right\rangle \left\langle \hat{e}_m \left| \vec{v}_k^{(K)} \right.\right\rangle F_{nm}\left(\zeta_k^{(K)}, \theta \right) V\left(\zeta_k^{(K)} / |\lambda| \cos\theta \right), \quad n, m < K \tag{46}$$

where

$$F_{nm}(\zeta, \theta) = \left\{ \frac{p_n(\zeta \gamma_\theta)}{p_n(\zeta)} \frac{p_m(\zeta \gamma_\theta)}{p_m(\zeta)} \right\} F(\zeta, \theta). \tag{47}$$

The following points may be noted concerning the above expression. First, it makes use of the eigenvalues and eigenvectors of the finite tridiagonal matrix of equation (34). Second, the ratio $p_n\left(\zeta_k^{(K)}\gamma_\theta\right)/p_n\left(\zeta_k^{(K)}\right)$ is easy to obtain using the recursion relation of equation (30). Furthermore, this process is numerically very stable. Third, the factor $F_{nm}\left(\zeta_k^{(K)},\theta\right)$ reduces to unity when the rotation angle θ vanishes. Finally, the potential is sampled at the real points $\left\{\zeta_k^{(K)}/\left(|\lambda|\cos\theta\right)\right\}_{k=0}^{K-1}$, making the result (46) applicable to both analytic and non-analytic potentials.

The evaluation of the matrix elements of one-channel potentials in the oscillator basis with complex-scaled parameter $\lambda = |\lambda|\, e^{-i\theta}$, where $0 \le \theta < \pi/4$, can proceed in a fashion similar to that of the above Laguerre case. In fact, the equation for the matrix element V_{nm} corresponding to equation (43) is

$$V_{nm} = \int_0^\infty \frac{\rho(\zeta)}{\Gamma(\nu+1)}\left[F(\zeta,\theta)\right] p_n(\zeta\gamma_\theta)\, p_m(\zeta\gamma_\theta)\, V\left(\sqrt{\zeta/|\lambda|^2\cos 2\theta}\right) d\zeta \qquad (48)$$

where $\rho(\zeta) = e^{-\zeta}\zeta^\nu$. The polynomial p_n satisfies the same three term recursion relation of equation (28), with the exception that here $\nu = \ell + 1/2$ in this case. Furthermore,

$$F(\zeta,\theta) = \left[(\gamma_\theta)^{\nu+1}\, e^{i\zeta\tan 2\theta}\right], \text{ and } \quad \gamma_\theta = \left(\frac{e^{-2i\theta}}{\cos 2\theta}\right). \qquad (49)$$

Thus, the result corresponding to equation (45) may be shown by the expression

$$V_{nm} \cong \sum_{k=0}^{K-1}\left[\left\langle \hat{e}_0 \middle| \vec{v}_k^{(K)}\right\rangle\right]^2 \left[F\left(\zeta_k^{(K)},\theta\right)\right] p_n\left(\zeta_k^{(K)}\gamma_\theta\right) p_m\left(\zeta_k^{(K)}\gamma_\theta\right) \times V\left(\sqrt{\zeta_k^{(K)}/|\lambda|^2\cos 2\theta}\right) \qquad (50)$$

On the other hand, the result that corresponds to equation (46) is now given by

$$V_{nm} \cong \sum_{k=0}^{K-1}\left\langle \hat{e}_n \middle| \vec{v}_k^{(K)}\right\rangle \left\langle \hat{e}_m \middle| \vec{v}_k^{(K)}\right\rangle F_{nm}\left(\zeta_k^{(K)},\theta\right) V\left(\sqrt{\zeta_k^{(K)}/|\lambda|^2\cos 2\theta}\right);\quad n,m < K \qquad (51)$$

where

$$F_{nm}(\zeta,\theta) = \left\{\frac{p_n(\zeta\gamma_\theta)}{p_n(\zeta)}\frac{p_m(\zeta\gamma_\theta)}{p_m(\zeta)}\right\} F(\zeta,\theta). \qquad (52)$$

3.3 The Case of a Multi-Channel Central Analytic Potential

We generalize the results obtained in the last section to the case of a multi-channel potential $V^{\alpha\beta}(r)$, which is dependent on the channel indices α and β. We take all channel bases to be of the Laguerre form (3) and we allow the channel scale parameter λ_α to differ from λ_β. The matrix element $V_{nm}^{\alpha\beta}$ is now given by

$$V_{nm}^{\alpha\beta} = c_n^{(\alpha)} c_m^{(\beta)} \int_0^\infty r^\nu \left(\lambda_\alpha \lambda_\beta\right)^{\nu/2} e^{-(\lambda_\alpha+\lambda_\beta)r/2} L_n^\nu(\lambda_\alpha r)\, L_m^\nu\left(\lambda_\beta r\right) V^{\alpha\beta}(r)\, dr \qquad (53)$$

With $\zeta = \left(\lambda_\alpha + \lambda_\beta\right) r/2$, and the usual definition of $p_n(\zeta)$ in terms of $L_n^\nu(\zeta)$ given by equation (29), we may write equation (53) as

$$V_{nm}^{\alpha\beta} = \int_0^\infty \frac{\rho(\zeta)}{\Gamma(\nu+1)} p_n(\zeta)\, p_m(\zeta)\, F_{nm}^{\alpha\beta}(\zeta)\, V^{\alpha\beta}\left(\zeta/\lambda_{\alpha\beta}\right) d\zeta, \qquad (54)$$

where

$$\rho(\zeta) = e^{-\zeta}\zeta^\nu, \text{ and}$$
$$F_{nm}^{\alpha\beta}(\zeta) = \left\{ \frac{p_n\left(\lambda_\alpha\zeta/\lambda_{\alpha\beta}\right)}{p_n(\zeta)} \frac{p_m\left(\lambda_\beta\zeta/\lambda_{\alpha\beta}\right)}{p_m(\zeta)} \right\}. \qquad (55)$$

A consideration of $\lambda_{\alpha\beta} = \left(\lambda_\alpha + \lambda_\beta\right)/2$ with $\lambda_\alpha = |\lambda_\alpha| e^{-i\theta_\alpha}$ and $\lambda_\beta = \left|\lambda_\beta\right| e^{-i\theta_\beta}$ results in the restriction $0 \le \theta_\alpha, \theta_\beta < \pi/2$. We note that the integral in equation (54) is similar to that of equation (31) except that the potential function $V(\zeta/\lambda)$ is now replaced by the product $F_{nm}^{\alpha\beta}(\zeta)\, V^{\alpha\beta}\left(\zeta/\lambda_{\alpha\beta}\right)$. Therefore, the technique we used to approximate the integral of equation (33) may be used to approximate the integral of equation (54). Thus, the corresponding multi-channel result in the Laguerre basis (3) is given by

$$V_{nm}^{\alpha\beta} \cong \sum_{k=0}^{K-1} \left\langle \hat{e}_n \left| \vec{v}_k^{(K)} \right.\right\rangle \left\langle \hat{e}_m \left| \vec{v}_k^{(K)} \right.\right\rangle F_{nm}^{\alpha\beta}\left(\zeta_k^{(K)}\right) V^{\alpha\beta}\left(\zeta_k^{(K)}/\lambda_{\alpha\beta}\right); \quad K > n,m \qquad (56)$$

It is to be noted that when $\lambda_\alpha = \lambda_\beta = \lambda$, the parameter $\lambda_{\alpha\beta}$ reduces to λ. At the same time, the function $F_{nm}^{\alpha\beta}$ reduces to F_{nm} of equation (47). In fact the above result reduces to that of the one-channel case (46).

As has been noticed in the one-channel case, the similarities and differences between the evaluation of the one-channel potential matrix elements in the Laguerre and oscillator bases are preserved in the multi-channel case. Thus, the matrix elements of multi-channel potentials, $V_{nm}^{\alpha\beta}$, in the oscillator basis are now given by

$$V_{nm}^{\alpha\beta} = \int_0^\infty \frac{\rho(\zeta)}{\Gamma(\nu+1)} p_n(\zeta)\, p_m(\zeta)\, \tilde{F}_{nm}^{\alpha\beta}(\zeta)\, V^{\alpha\beta}\left(\sqrt{\zeta}/\lambda_{\alpha\beta}\right) d\zeta, \tag{57}$$

where

$$\rho(\zeta) = e^{-\zeta}\zeta^{\nu},$$

$$\tilde{F}_{nm}^{\alpha\beta}(\zeta) = \left\{ \frac{p_n\left(\lambda_\alpha^2\zeta/\lambda_{\alpha\beta}^2\right)}{p_n(\zeta)} \frac{p_m\left(\lambda_\beta^2\zeta/\lambda_{\alpha\beta}^2\right)}{p_m(\zeta)} \right\}, \tag{58}$$

and $\left(\lambda_{\alpha\beta}\right)^2 = \left(\lambda_\alpha^2 + \lambda_\beta^2\right)/2$ with $\lambda_\alpha = |\lambda_\alpha|\, e^{-i\theta_\alpha}$, and $\lambda_\beta = \left|\lambda_\beta\right| e^{-i\theta_\beta}$. The restriction on θ_α and θ_β in this case is $0 \leq \theta_\alpha, \theta_\beta < \pi/4$. Also, the corresponding multi-channel results in the oscillator basis (4) are given by

$$V_{nm}^{\alpha\beta} \cong \sum_{k=0}^{K-1} \left\langle \hat{e}_n \left| \vec{v}_k^{(K)} \right.\right\rangle \left\langle \hat{e}_m \left| \vec{v}_k^{(K)} \right.\right\rangle \tilde{F}_{nm}^{\alpha\beta}\left(\zeta_k^{(K)}\right) V^{\alpha\beta}\left(\sqrt{\zeta_k^{(K)}}/\lambda_{\alpha\beta}\right); \quad K > n, m \tag{59}$$

3.4 The Case of a Multi-Channel Central Non-Analytic Potential

If the multi-channel potential $V^{\alpha\beta}(r)$ is non-analytic, we use the substitutions

$$\lambda_{\alpha\beta}^{(c)} = \left\{ |\lambda_\alpha| \cos\theta_\alpha + \left|\lambda_\beta\right| \cos\theta_\beta \right\}/2,$$

$$\lambda_{\alpha\beta}^{(s)} = \left\{ |\lambda_\alpha| \sin\theta_\alpha + \left|\lambda_\beta\right| \sin\theta_\beta \right\}/2, \tag{60}$$

$$\gamma_\theta^\alpha = \frac{|\lambda_\alpha|\, e^{-i\theta_\alpha}}{\lambda_{\alpha\beta}^{(c)}}, \text{ and } \quad \gamma_\theta^\beta = \frac{\left|\lambda_\beta\right| e^{-i\theta_\beta}}{\lambda_{\alpha\beta}^{(c)}}. \tag{61}$$

to cast equation (54) in the form

$$V_{nm}^{\alpha\beta} = \int_0^\infty \frac{\rho(\zeta)}{\Gamma(\nu+1)} \left[F^{\alpha\beta}(\zeta,\theta)\right] p_n\left(\zeta\gamma_\theta^\alpha\right) p_m\left(\zeta\gamma_\theta^\beta\right) V\left(\zeta/\lambda_{\alpha\beta}^{(c)}\right) d\zeta, \tag{62}$$

where

$$F^{\alpha\beta}(\zeta,\theta) = \left\{\gamma_\theta^\alpha \gamma_\theta^\beta\right\}^{(\nu+1)/2} \exp\left(i\lambda_{\alpha\beta}^{(s)}\zeta/\lambda_{\alpha\beta}^{(c)}\right) \tag{63}$$

with $\zeta = \lambda_{\alpha\beta}^{(c)} r$ and the usual definition of $p_n(\zeta)$ in terms of $L_n^{\nu}(\zeta)$ given by equation (29). In analogy with equation (45) we may explicitly write the approximation

$$V_{nm}^{\alpha\beta} \cong \sum_{k=0}^{K-1} \left[\left\langle \hat{e}_0 \left| \vec{v}_k^{(K)} \right.\right\rangle\right]^2 \left[F^{\alpha\beta}\left(\zeta_k^{(K)}, \theta\right)\right] p_n\left(\zeta_k^{(K)} \gamma_\theta^\alpha\right) p_m\left(\zeta_k^{(K)} \gamma_\theta^\beta\right) V\left(\zeta_k^{(K)} / \lambda_{\alpha\beta}^{(c)}\right). \tag{64}$$

On the other hand, in analogy to the argument used to reach equation (46), we also have

$$V_{nm}^{\alpha\beta} \cong \sum_{k=0}^{K-1} \left\langle \hat{e}_n \left| \vec{v}_k^{(K)} \right.\right\rangle \left\langle \hat{e}_m \left| \vec{v}_k^{(K)} \right.\right\rangle F_{nm}^{\alpha\beta}\left(\zeta_k^{(K)}, \theta\right) V\left(\zeta_k^{(K)} / \lambda_{\alpha\beta}^{(c)}\right), \tag{65}$$

where

$$F_{nm}^{\alpha\beta}(\zeta, \theta) = \left\{ \frac{p_n\left(\zeta \gamma_\theta^\alpha\right)}{p_n(\zeta)} \frac{p_m\left(\zeta \gamma_\theta^\beta\right)}{p_m(\zeta)} \right\} F^{\alpha\beta}(\zeta, \theta). \tag{66}$$

The above results reduce to the one-channel result of the previous section when $\lambda_\alpha = \lambda_\beta$. Again the multi-channel potential $V^{\alpha\beta}(r)$ is sampled at real points $\left\{\zeta_k^{(K)} / \lambda_{\alpha\beta}^{(c)}\right\}_{k=0}^{K-1}$. We note that the formula (65) is valid for analytic as well as non-analytic potentials.

The matrix elements of the multi-channel non-analytic radial potential $V^{\alpha\beta}(r)$ in the oscillator basis (4), is given by

$$V_{nm}^{\alpha\beta} = d_n^{(\alpha)} d_m^{(\beta)} \int_0^\infty \left(\lambda_\alpha \lambda_\beta r^2\right)^{\nu+1/2} e^{-\lambda_\alpha^2 r^2/2} e^{-\lambda_\beta^2 r^2/2} L_n^{\nu}\left(\lambda_\alpha^2 r^2\right) L_m^{\nu}\left(\lambda_\beta^2 r^2\right) V^{\alpha\beta}(r)\, dr \tag{67}$$

If we define

$$\begin{aligned}
\left(\lambda_{\alpha\beta}^{(c)}\right)^2 &= \left\{|\lambda_\alpha|^2 \cos 2\theta_\alpha + \left|\lambda_\beta\right|^2 \cos 2\theta_\beta\right\}/2, \\
\left(\lambda_{\alpha\beta}^{(s)}\right)^2 &= \left\{|\lambda_\alpha|^2 \sin 2\theta_\alpha + \left|\lambda_\beta\right|^2 \sin 2\theta_\beta\right\}/2, \\
\gamma_\theta^\alpha &= \left(\frac{|\lambda_\alpha| e^{-i\theta_\alpha}}{\lambda_{\alpha\beta}^{(c)}}\right)^2, \text{ and } \gamma_\theta^\beta = \left(\frac{\left|\lambda_\beta\right| e^{-i\theta_\beta}}{\lambda_{\alpha\beta}^{(c)}}\right)^2
\end{aligned} \tag{68}$$

and set $\zeta = \left(\lambda_{\alpha\beta}^{(c)}\right)^2 r^2$, then we obtain

$$V_{nm}^{\alpha\beta} = \int_0^\infty \frac{\rho(\zeta)}{\Gamma(\nu+1)} \left[F^{\alpha\beta}(\zeta, \theta)\right] p_n\left(\zeta \gamma_\theta^\alpha\right) p_m\left(\zeta \gamma_\theta^\beta\right) V\left(\sqrt{\zeta} / \left(\lambda_{\alpha\beta}^{(c)}\right)\right) d\zeta \tag{69}$$

where

$$F^{\alpha\beta}(\zeta,\theta)=\left\{\gamma_\theta^\alpha\gamma_\theta^\beta\right\}^{(\nu+1)/2}\exp\left(i\lambda_{\alpha\beta}^{(s)}\zeta/\lambda_{\alpha\beta}^{(c)}\right). \tag{70}$$

This matrix element can be approximated using arguments similar to those employed in the Laguerre case. A result that is analogous to equation (50) may take the form

$$V_{nm}^{\alpha\beta}\cong\sum_{k=0}^{K-1}\left[\left\langle\hat{e}_0\,\middle|\,\vec{v}_k^{(K)}\right\rangle\right]^2\left[F^{\alpha\beta}\left(\zeta_k^{(K)},\theta\right)\right]p_n\left(\zeta_k^{(K)}\gamma_\theta^\alpha\right)p_m\left(\zeta_k^{(K)}\gamma_\theta^\beta\right)$$
$$\times V\left(\sqrt{\zeta_k^{(K)}}/\lambda_{\alpha\beta}^{(c)}\right) \tag{71}$$

and an equation that is compatible to (51) may be represented by

$$V_{nm}^{\alpha\beta}\cong\sum_{k=0}^{K-1}\left\langle\hat{e}_n\,\middle|\,\vec{v}_k^{(K)}\right\rangle\left\langle\hat{e}_m\,\middle|\,\vec{v}_k^{(K)}\right\rangle F_{nm}^{\alpha\beta}\left(\zeta_k^{(K)},\theta\right)V\left(\sqrt{\zeta_k^{(K)}}/\lambda_{\alpha\beta}^{(c)}\right), \tag{72}$$

where

$$F_{nm}^{\alpha\beta}(\zeta,\theta)=\left\{\frac{p_n\left(\zeta\gamma_\theta^\alpha\right)}{p_n(\zeta)}\frac{p_m\left(\zeta\gamma_\theta^\beta\right)}{p_m(\zeta)}\right\}F^{\alpha\beta}(\zeta,\theta). \tag{73}$$

4 Discussion and Numerical Examples

The scheme employed to find an approximate value of the potential matrix element V_{nm} given by equation (31) is based on property (33). Specifically, it can be written as

$$V_{nm}=\int_0^\infty\rho(\zeta)f(\zeta)d\zeta=\sum_{k=0}^{K-1}\omega_k^{(K)}f\left(\zeta_k^{(K)}\right)+R_N, \tag{74}$$

where $f(\zeta)=p_n(\zeta)\,p_m(\zeta)\,V(\zeta/\lambda)/\Gamma(\nu+1)$ and R_N is the remainder. The magnitude of the remainder is dependent on the behavior of the function $f(\zeta)$. Since we consider a general potential, it is not possible to obtain a tighter bound on the remainder than the bound already available from the classical treatment of the Gauss quadrature scheme associated with the theory of orthogonal polynomials [12]. Instead, we prefer to illustrate the scheme by evaluating specific examples.

Example 1. Scattering and Resonance Information for a Two-Channel Square Well Potential

The method of complex scaling has proven to be very powerful in locating resonances [1] in the complex energy plane as well as in carrying out calculations of multi-channel scattering [2]. It involves the evaluation of a given system Hamiltonian, H, in a finite rotated basis, $\left\{\left|\phi_n^\theta\right\rangle\right\}_{n=0}^N$, or alternatively [3, 4] the evaluation of a complex-scaled Hamiltonian, H^θ, in the non-rotated basis, $\{|\phi_n\rangle\}_{n=0}^N$. Both schemes are equivalent when the potential involved is analytic. However, only the first scheme works for non-analytic potentials, such as the square well [2]. In the later scheme the rotated basis is obtained from the non-rotated one by the transformation

$$\left|\phi_n^\theta\right\rangle = U^{-1}(\theta)\left|\phi_n\right\rangle \tag{75}$$

where $U(\theta)$ acts on any function $f(r)$ as

$$U(r) f(r) = e^{i\theta/2} f\left(re^{i\theta}\right) \tag{76}$$

We propose to test the method developed in the previous section by applying it in the calculation of s-wave scattering for the two-channel square well problem as previously considered by Rescigno and Reinhardt [2]

$$\begin{aligned} V^{\alpha\beta}(r) &= V_0^{\alpha\beta} \quad r \leq 1 \\ &= 0 \quad r > 1. \end{aligned}$$

The target energies have only two internal states $\tilde{E}_1 = 0.0$ and $\tilde{E}_2 = 2.0$ a.u. The potential strengths are taken to be

$$V_0^{\alpha\beta} = \begin{pmatrix} -2.0 & -1.0 \\ -1.0 & -2.0 \end{pmatrix}. \tag{77}$$

The L^2 basis set consisted of 30 oscillator functions for each channel. The free parameter $\lambda_\alpha = 1.0$ for both channels and a rotation angle, θ, of 0.0 rad. were chosen. The matrix elements of the reference Hamiltonian are given by equation (25). We also chose the order of approximation K to equal 60. Figure 1 plots the results for the un-normalized elastic cross sections $\pi\left|1 - S_{11}\right|^2$ and $\pi\left|1 - S_{22}\right|^2$, and the un-normalized inelastic cross section, $\pi\left|S_{12}\right|^2$. The results compare well with the un-normalized exact cross sections that can be obtained analytically for this problem [13]. Furthermore, we find the position of the resonance of the same potential using the complex rotation method. We chose the Laguerre basis with $N_\alpha = N_\beta = N$. The scaling parameters $|\lambda_\alpha|$ and $\left|\lambda_\beta\right|$ are considered to be the same and equal 6.0. The rotation angles are taken to be $\theta_\alpha = \theta_\beta = 0.1$ rad. We also chose the order of approximation K to equal 60. We then evaluate the matrix elements of the potential,

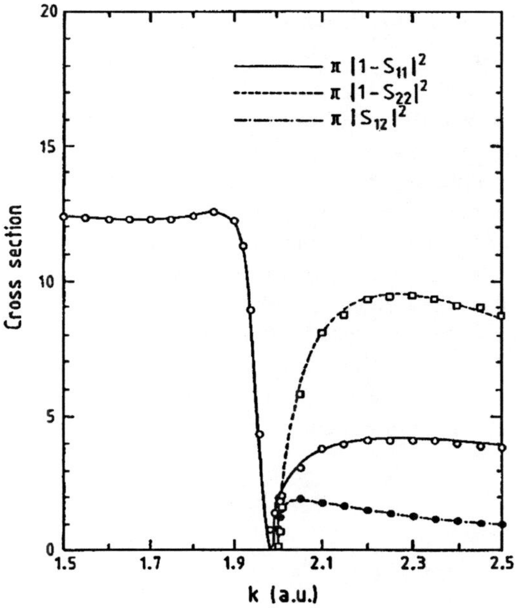

Fig. 1 Un-normalized elastic and inelastic cross sections for s-wave scattering from two coupled square wells. Potential parameters are those of equation (77) and the potential matrix elements are calculated using equation (72). Exact results for the un-normalized elastic cross section for channels 1 and 2 are indicated as solid and dashed lines respectively; for the un-normalized inelastic cross section are shown as dashed-dotted lines. The order of approximation K is taken as 60. The basis set consisted of 30 oscillator functions for each channel. The free parameter $\lambda_1 = \lambda_2 = 1.0$ was used for both channels, as was a rotation angle of $\theta_1 = \theta_2 = 0.0$ radians. The calculated un-normalized elastic cross sections for channels 1 and 2 are indicated by open circles and squares respectively, while the un-normalized inelastic cross sections are indicated by solid circles

$V_{nm}^{\alpha\beta}$. The $2N \times 2N$ matrix of the full Hamiltonian is then diagonalized for different values of N up to $N = 40$ for each channel. A stable resonance state for $\ell = 0$ is found whose energy converges to the exact value of $\varepsilon = 1.9247 - i0.09010$ with increasing N, as shown in Table 1.

Table 1 The position of resonance for s-wave scattering from two coupled square wells of equation (77) as a function of basis-set size N in comparison with the exact result. The order of approximation K is taken as 60. We use the Laguerre basis with a complex scale parameter $\lambda = |\lambda|\, e^{-i\theta}$. Here, $|\lambda_1| = |\lambda_2| = 6.0$, rotational angles $\theta_1 = \theta_2 = 0.1 rad$

N	ε
10	1.936 – i0.0640
20	1.933 – i0.0976
30	1.922 – i0.0902
40	1.924 – i0.0901
Exact	1.937 – i0.0853

Example 2. Resonance Information for a Two-Channel Analytic Potential

In this example, we find the complex resonance energy for a specific two-channel analytical potential (with no Coulomb term) for several angular momenta ℓ using both bases. We then consider the same potential when the Coulomb term is included using the Laguerre basis. The matrix elements of the reference Hamiltonian are given by equation (24). We apply the proposed method to the characterization of the s-wave narrow resonance for a two-channel problem for a case in which $z = 0$ while using both the Laguerre and oscillator bases, a model two-channel problem previously considered by Noro and Taylor [14] and Mandelshtam et al. [15]. The problem consists of the scattering of a structureless particle with a charge of $z = 0$ by a target that has only two internal states with threshold energies 0.0 and 0.1 a.u. The matrix elements of the interaction potential are taken to be

$$V^{\alpha\beta} = V_0^{\alpha\beta} r^2 e^{-r},$$

where the potential strengths are

$$V_0^{\alpha\beta} = \begin{pmatrix} -1.0 & -7.5 \\ -7.5 & 7.5 \end{pmatrix}. \tag{78}$$

We chose bases with $N_\alpha = N_\beta = N$. The scaling parameters $|\lambda_\alpha|$ and $|\lambda_\beta|$ are considered the same and equal 5.0 for the Laguerre basis and 1.5 for the oscillator basis. The rotation angles are taken to be $\theta_\alpha = \theta_\beta = 0.5$ rad. We also chose the order of approximation K equal 60. We then evaluate the matrix elements of the potential, $V_{nm}^{\alpha\beta}$. The $2N \times 2N$ matrix of the full Hamiltonian is then diagonalized for different values of N, up to $N = 50$ for each channel. A stable resonance state for $\ell = 0$ is found, whose energy converges to the value $\varepsilon = 4.7682 - i\,0.000710$ with increasing N, as shown in Table 2. This result compares well with that of Mandelshtam et al. [15]. We notice that the accuracy of the results using the oscillator basis is a little better than that observed when using the Laguerre basis. The calculations with higher angular momenta $\ell = 1$ and 2 were also included in Table 2. It is noted that the calculations converge quickly with increasing N for both the oscillator and Laguerre bases. Furthermore, the accuracy of calculations using the oscillator basis is similar to the accuracy of calculations using the Laguerre basis.

As a further example, we consider the same potential (78) for an $\ell = 0$ case in which $z = +1.0$ and -1.0 within a Laguerre basis, which has the advantage of being able to account fully for H_0 and the Coulomb term (z/r). The results are shown in Table 3.

The variety of examples presented in this chapter show that the Gauss quadrature scheme is a natural and accurate way of evaluating potential matrix elements in complex scaled Laguerre and oscillator bases.

Table 2 Results for the two-channel potential of equation (78) as a function of basis-set size N for different angular momenta ℓ and $z = 0$. The order of approximation K is taken as 60. We use bases with a complex scale parameter $\lambda = |\lambda| e^{i\theta}$. Here, $\theta_1 = \theta_2 = 0.1 rad.$, $|\lambda_1| = |\lambda_2| = 1.5$ for the oscillator basis, and $|\lambda_1| = |\lambda_2| = 5.0$ for the Laguerre basis. The potential matrix elements are calculated using equation (56) for the Laguerre case and equation (59) for the oscillator case. The result of the calculation using $\ell = 0$ is compared with that of Mandelshtam et al. [15]

ℓ	N	ε (Oscillator)	ε (Laguerre)
0	10	4.768194 – i0.0007085	4.768303 – i0.0020759
0	20	4.768194 – i0.0007099	4.768199 – i0.0007107
0	30	4.768194 – i0.0007098	4.768197 – i0.0007101
0	40	4.768194 – i0.0007098	4.768197 – i0.0007101
0	50	4.768194 – i0.0007098	4.768197 – i0.0007101
Ref [15]		4.7682 – i0.000710	
1	10	6.70372 – i 0.125652	6.70288 – i 0.125735
1	20	6.70372 – i0.125653	6.70372 – i0.125653
1	30	6.70372 – i0.125653	6.70372 – i0.125653
1	40	6.70372 – i0.125653	6.70372 – i0.125653
1	50	6.70372 – i0.125653	6.70372 – i0.125653
2	10	7.99099 – i0.739165	7.99116 – i0.739241
2	20	7.99097 – i0.739155	7.99097 – i0.739155
2	30	7.99097 – i0.739155	7.99097 – i0.739155
2	40	7.99097 – i0.739155	7.99097 – i0.739155
2	50	7.99097 – i0.739155	7.99097 – i0.739155

Table 3 Results for s-wave scattering of the two-channel potential of equation (78) as a function of basis-set size N with $z = \pm 1$. The order of approximation K is taken as 60. We use the Laguerre basis with a complex scale parameter $\lambda = |\lambda| e^{-i\theta}$. Here, $|\lambda_1| = |\lambda_2| = 5.0$, rotational angles $\theta_1 = \theta_2 = 0.5 rad$. The potential matrix elements are calculated using equation (56)

N	ε	
	$z = -1$	$z = +1$
10	2.81003 – i2.498e – 4	6.27899 – i1.8507e – 2
20	2.81150 – i1.832e – 4	6.27804 – i1.8432e – 2
30	2.81150 – i1.814e – 4	6.27804 – i1.8433e – 2
40	2.81150 – i1.814e – 4	6.27804 – i1.8433e – 2

Acknowledgments The support of this work by King Fahd University of Petroleum and Minerals is gratefully acknowledged.

References

1. For a review of this method see, Reinhardt W P 1982 *Ann. Rev. Phys. Chem.* **33**, 223
2. Rescigno T N and Reinhardt W P 1973 *Phys. Rev.* **A8**, 2828
3. Rescigno T N and McCurdy C W 1986 *Phys. Rev.* **A34** 1882
4. Yamani H A and Abdelmonem M S 1996 *J. Phys. A: Math. Gen.* **29**, 6991
5. Arickx F, Broeckhove J, Van Leuven P, Vasilevsky V, and Filippov 1994 *Am. J. Phys.* **62**, 362
6. Alhaidari A D, Bahlouli H, Abdelmonem M S, Al-Ameen F, and Al-Abdulaal T (2007) Phys. Lett. A **364**, 372

7. Igashov S Yu, "*Oscillator basis in the J-matrix method: convergence of expansions, asymptotics of expansion coefficients and boundary conditions*", Part II, Chapter 1, this volume
8. Magnus W, Oberhettinger F, and Soni R P 1966, *Formulas and Theorems for the Special Functions of Mathematical Physics* (New York: Springer-Verlag)
9. Yamani H A and Abdelmonem M S 1997 *J. Phys. B: At. Mol. Opt. Phys.* **30**, 1633
10. Akhiezer N I 1965, *The Classical Moment Problem* (Einburgh: Oliver and Boyd)
11. Szego G 1939, *Orthogonal Polynomials* (New York: American Mathematical Society)
12. Krylov V I 1962, *Approximate Calculation of Integrals* (New York: The Macmillan Company)
13. Newton R G 1966, *Scattering Theory of Waves and Particles* (New York: McGraw-Hill), p. 543
14. Noro T and Taylor H S 1980 *J. phys.* **B13**, L377
15. Mandelshtam V.A. Ravuri T. R. and Taylor H. S. 1993 *Phys. Rev. Lett.* **70**, 1932

J-Matrix and Isolated States

A.M. Shirokov and S.A. Zaytsev

Abstract We show that a quantum system with nonlocal interaction can have bound states of unusual type—Isolated States (IS). IS is a bound state that is not in correspondence with the S-matrix pole. IS can have a positive as well as a negative energy and can be treated as a generalization of the bound states embedded in continuum on the case of discrete spectrum states. The formation of IS in the spectrum of a quantum system is studied using a simple rank–2 separable potential with harmonic oscillator form factors. Some physical applications are discussed, in particular, we propose a separable NN potential supporting IS that describes the deuteron binding energy and the s-wave triplet and singlet scattering phase shifts. We use this potential to examine the so-called problem of the three-body bound state collapse discussed in literature. We show that the variation of the two-body IS energy causes drastic changes of the binding energy and of the spectrum of excited states of the three-nucleon system.

1 Introduction

Quantum mechanical bound states are known to have the wave functions decreasing rapidly at large distances r. Usually the bound states are possible at negative energies ($E < 0$) only while at positive energies ($E > 0$) the system has only continuum spectrum states with wave functions oscillating at large distances r. Nevertheless von Neumann and Wigner showed long ago [1] that a quantum system can have a bound state at positive energy $E > 0$. Such states are conventionally refered to as '*continuum bound states*' or as '*bound states embedded in continuum*' (BSEC). Von Neumann and Wigner used a local potential in their study of BSEC. BSECs are also natural when the interaction is nonlocal (see [2] and references therein) or in the case of multichannel scattering (see, e.g., [3–7]).

A.M. Shirokov
Skobeltsyn Institute of Nuclear Physics, Moscow State University, Moscow 119992, Russia
e-mail: shirokov@nucl-th.sinp.msu.ru

A.D. Alhaidari et al. (eds.), *The J-Matrix Method*, 103–115.

A phenomenological nonlocal nucleon-nucleon (NN) potential supporting BSEC was suggested by Tabakin [8]. The potential predicts the NN data fairly well. However it was found that Tabakin potential generates an extremely large binding energy for the three-nucleon system [9, 10]. Such a '*bound state collapse*' was investigated by several groups of workers [11–16] with Tabakin and similar nonlocal potentials. All these groups associated the bound state collapse with the two-body BSEC and suggested various interpretations of the origin of such a puzzling phenomenon.

In some of these studies [13, 14, 16], BSEC was interpreted as an S-matrix pole on the real positive energy axis. This is obviously a mistake since it is well known from textbooks [17] that the unitarity condition for scattering requires $|S(E)| = 1$ for real $E > 0$ in the case of elastic scattering ($|S(E)| \leq 1$ for real $E > 0$ in the case of inelastic scattering), hence the S-matrix $S(E)$ cannot have poles on the real positive energy axis. BSEC appears to be a very interesting state: contrary to the conventional discrete spectrum states it is a bound state which is not associated with any of the S-matrix poles [18]. We introduce *Isolated States* (IS) that are, by definition, bound states that do not correspond to the S-matrix poles. Isolated states are a generalization of BSECs: any BSEC is IS, however the energy E_I of IS can be also negative, in particular, the ground state of the system can be an Isolated State. Generally, if the S-matrix $S(E)$ is known than we obtain the energies of the discrete spectrum states by associating the S-matrix poles at negative energies with the discrete spectrum states. The information about the IS energy cannot be extracted from the S-matrix, such bound states are *isolated* from the continuum spectrum states.

In the next Section we present a nonlocal interaction supporting IS. Within the J-matrix formalism, it is easy to formulate a realization of this interaction which makes it possible to find a simple analytical expression for the S-matrix. This interaction is used to study the IS formation when the Hamiltonian parameters are varied, the IS contribution to the Levinson theorem, etc.

We address the bound state collapse problem in Section 3. We fit the parameters of our exactly solvable nonlocal interaction model supporting IS to the NN scattering data. The IS energy E_I is arbitrary because it is not related to the S-matrix. We show that varying the IS energy E_I (note that the S-matrix is unaffected by this variation) we produce great changes of the three-body binding energy: some E_I values bring us to the ^{3}H system with three bound states with extremely large (few GeV) binding energies; with larger E_I values we obtain the ^{3}H nucleus with two bound states (the binding energy of the ground state is of the order of few hundreds MeV); the further increase of E_I results in the further decrease with E_I of the binding energy of the ^{3}H nucleus which has a single bound state; if the IS energy E_I is large enough the ^{3}H nucleus become unbound. The experimental value of the ^{3}H binding energy can be exactly reproduced if some particular positive energy E_I is taken when IS appears to be BSEC. Therefore the problem of the three-body bound state collapse does not exists as a general problem: this problem arose only due to the use of very restricted models of interaction supporting BSEC for the construction of NN potentials, i.e. this problem is inherent for such interaction models only.

We present in Section 4 a short discussion and compare our results with the results of other authors who studied the three-body bound state collapse problem.

2 Simple Potential Model Supporting Isolated State

The radial wave function $\Psi^l_E(r)$ satisfies the Schrödinger equation

$$(T^l + V^l - E)\Psi^l_E(r) = 0, \tag{1}$$

where E is the energy, l is the angular momentum, T^l is the kinetic energy operator and V^l is the potential energy.

Let $\{|i\rangle\}$, $i = 0, 1, 2, \ldots$ be a complete L^2 basis. The Hamiltonian matrix $H_{ij} \equiv \langle i|H|j\rangle$ is generally infinite and the wave function $\Psi^l_E(r)$ at any energy E can be expressed generally as an infinite expansion in basis functions,

$$\Psi^l_E(r) = \sum_{i=0}^{\infty} \alpha_i(E)\,|i\rangle. \tag{2}$$

However, at some particular energy E_I the infinite Hamiltonian matrix H_{ij} can have a finite eigenvector, and the wave function $\Psi_I(r)$ at this energy is expressed as a finite expansion in basis functions,

$$\Psi_I(r) = \sum_{i=0}^{M} \alpha_i(E_I)\,|i\rangle. \tag{3}$$

The wave function $\Psi_I(r)$ rapidly decreases with distance r since it is a superposition of a finite number of L^2 functions. Therefore at energy E_I we have a bound state of a particular type hereafter refered to as an *Isolated State* (IS).

Clearly, we have the IS solutions of the type (3) in the case when the Hamiltonian matrix H_{ij} is block-diagonal,

$$H_{ij} = H^{(1)}_{ij} \oplus H^{(2)}_{ij}, \tag{4}$$

where the $(M+1) \times (M+1)$ submatrix $H^{(1)}_{ij}$ is defined in a finite-dimensional subspace spanned by the basis functions $|i\rangle$ with $i = 0, 1, \ldots, M$. The infinite-dimensional submatrix $H^{(2)}_{ij}$ is defined in the orthogonal supplement to this subspace. Any eigenvector of the submatrix H_{ij} gives rise to the wave function of the type (3), i.e. each of the submatrix H_{ij} eigenvectors is associated with IS.

The scattering state wave functions with oscillating asymptotics at energies $E > 0$, can be expressed only as a superposition of an infinite number of L^2 functions,

$$\Psi^l_E(r) = \sum_{i=M+1}^{\infty} \alpha^{(2)}_i(E)\,|i\rangle, \tag{5}$$

where $\left\{\alpha_i^{(2)}(E)\right\}$ are eigenvectors of the infinite submatrix $H_{ij}^{(2)}$. The S-matrix and scattering phase shifts are defined through the asymptotics of the functions (5), hence they are governed by the structure of the submatrix $H_{ij}^{(2)}$ only. The energy E_I of IS and its other features are dictated by the structure of the other submatrix $H_{ij}^{(1)}$ that is generally independent from $H_{ij}^{(2)}$. Therefore we cannot expect that the S-matrix has a pole at the IS energy E_I. We can define the Isolated State as the bound state that is not associated with any of the S-matrix poles. The IS energy E_I can be positive, and in this case it appears to be the so-called *bound state embedded in continuum* (BSEC). Any BSEC is IS. The IS energy E_I can be also negative, in particular, the ground state of the system can be isolated. Thus IS can be treated as a generalization of BSEC on the case of arbitrary (negative or positive) energy.

The J-matrix formalism [19] makes it possible to study IS properties and to formulate a simple exactly-solvable model of a system possessing IS. In this contribution we use the oscillator basis; the exactly-solvable model of IS can be also easily formulated by means of the J-matrix formalism with the Laguerre basis.

The idea of the block-diagonal structure of the Hamiltonian matrix (4) can be easily realized if the interaction between the particles is described by a separable nonlocal potential of the rank $N+1$,

$$V^l = \sum_{n,n'=0}^{N} V_{nn'}^l \, |\varphi_{nl}(r)\rangle\langle\varphi_{n'l}(r')|, \tag{6}$$

with the harmonic oscillator form factors

$$\varphi_{nl}(r) = (-1)^n \left[\frac{2n!}{r_0\Gamma(n+l+\frac{3}{2})}\right]^{\frac{1}{2}} \left(\frac{r}{r_0}\right)^{l+1} \exp\left(-\frac{r^2}{2r_0^2}\right) L_n^{l+\frac{1}{2}}\left(\frac{r^2}{r_0^2}\right). \tag{7}$$

Here $r_0 = \sqrt{\hbar/m\omega}$ is the oscillator radius and $L_n^{\alpha}(x)$ is the Laguerre polynomial.

In the J-matrix method, the wave function has a form of series in terms of L^2 functions (7),

$$\Psi_E^l(r) = \sum_{n=0}^{\infty} X_n(E)\,\varphi_{nl}(r). \tag{8}$$

The coefficients $X_n(E)$ for $n \geq N$ are given by the formula

$$X_n(E) = S_{nl}(p)\cos\delta_l + C_{nl}(p)\sin\delta_l\ , \tag{9}$$

where $p = \sqrt{2E/\hbar\omega}$ is the momentum, $S_{nl}(p)$ and $C_{nl}(p)$ are the eigenvectors of the infinite tridiagonal matrix of the kinetic energy $T_{nn'}^l$. The following analytical expressions [19] can be used to calculate $S_{nl}(p)$ and $C_{nl}(p)$:

$$S_{nl}(p) = \left[\frac{2\Gamma(n+l+\frac{3}{2})}{\Gamma(n+1)} \right]^{\frac{1}{2}} \frac{p^{l+1}}{\Gamma(l+\frac{3}{2})} \exp(-\frac{p^2}{2})\, {}_1F_1(-n,\, l+\frac{3}{2};\, p^2)\,, \tag{10}$$

$$C_{nl}(p) = \left[\frac{2\Gamma(n+1)}{\Gamma(n+l+\frac{3}{2})} \right]^{\frac{1}{2}} \frac{(-1)^l}{p^l\Gamma(-l+\frac{1}{2})} \exp(-\frac{p^2}{2})\, {}_1F_1(-n-l-\frac{1}{2},\, -l+\frac{1}{2};\, p^2)\,. \tag{11}$$

The phase shift δ_l in the partial wave with the angular momentum l can be calculated as

$$\tan\delta_l = -\frac{S_{Nl}(p) - \wp_{NN}(E)\, S_{N+1,l}(p)}{C_{Nl}(p) - \wp_{NN}(E)\, C_{N+1,l}(p)}, \tag{12}$$

where

$$\wp_{nn'}(E) = -\sum_{\mu} \frac{U_n^{\mu}\, U_{n'}^{\mu}}{\varepsilon_{\mu} - E} T^l_{n',n'+1}, \tag{13}$$

U_n^{μ} $(n = 0, 1, \ldots, N)$ are the eigenvectors and ε_{μ} are the corresponding eigenvalues of the truncated Hamiltonian matrix $\tilde{H}^N_{nn'} = T^l_{nn'} + V^l_{nn'}$ $(n, n' = 0, 1, \ldots, N)$. The coefficients $X_n(E)$ for $n \leq N$ can be found by the formula

$$X_n(E) = \wp_{nN}(E)\, X_{N+1}(E). \tag{14}$$

We are considering the case when the Hamiltonian matrix is block-diagonal. We note that the kinetic energy matrix in the oscillator basis is tridiagonal. Hence with the interaction (6) we can obtain the Hamiltonian matrix in the oscillator basis of the type (4) that has the structure shown in Fig. 1. Solid lines schematically show the infinite tridiagonal kinetic energy tail of the Hamiltonian matrix, the rest non-zero matrix elements are displayed by two boxes representing submatrices $H^{(1)}$ and $H^{(2)}$ (we include the infinite tridiagonal kinetic energy tail in the submatrix $H^{(2)}$).

Obviously, the eigenvectors of the submatrix $H^{(1)}$ are the eigenvectors of the entire Hamiltonian matrix H, too. The corresponding wave functions are decreasing with r similarly to the bound state wave functions since they are superpositions (3)

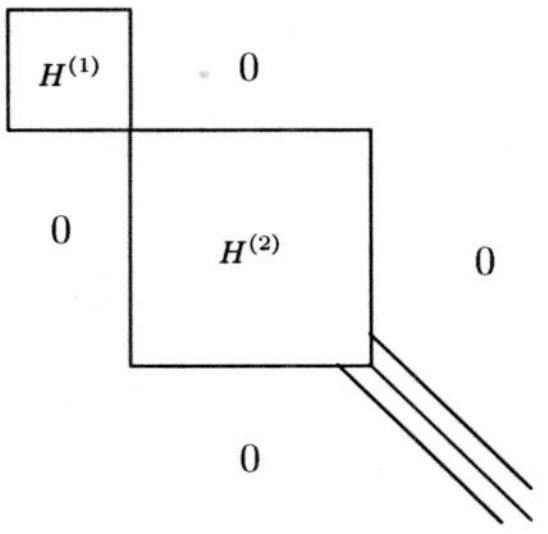

Fig. 1 The structure of the Hamiltonian matrix

of a finite number of the oscillator functions $|i\rangle = \varphi_{nl}(r)$ given by (7). However these wave functions have an unusual asymptotics $\sim \exp[-r^2/(2r_0^2)]$ instead of the standard one $\sim \exp(-\alpha r)$. All eigenstates of the submatrix $H^{(1)}$ are ISes. It is obvious that generally the submatrix $H^{(1)}$ may have positive or/and negative eigenvalues $\varepsilon_\mu^{(1)}$. If $\varepsilon_\mu^{(1)} > 0$, the corresponding state appears to be BSEC. If $\varepsilon_\mu^{(1)} < 0$, the corresponding state appears to be a bound state of a specific type.

The continuum spectrum states with the oscillating asymptotically wave functions as well as the conventional bound states with the $e^{-\alpha r}$–type wave function asymptotics, can be expressed only as infinite series (5) of the oscillator functions $|i\rangle = \varphi_{nl}(r)$ given by Eq. (7). They are generated by the submatrix $H^{(2)}$. The scattering and conventional bound state wave functions (5) are obviously orthogonal to the IS wave functions (3). Thus the scattering and usual bound state wave functions have node(s) at a small distance ($\sim r_0$) in the presence of IS(es) similarly to the wave functions generated by the potentials with forbidden states [20].

The asymptotic behavior of the scattering state wave functions is characterized completely by the S-matrix [3]. Hence the structure of the S-matrix is governed by the infinite-dimensional submatrix $H^{(2)}$ that is independent from the submatrix $H^{(1)}$. The IS energies $\varepsilon_\mu^{(1)}$, on the other hand, are controled by the submatrix $H^{(1)}$ and are independent from the submatrix $H^{(2)}$. Varying matrix elements $H_{nn'}^{(1)}$ of the submatrix $H^{(1)}$ one causes variation of the IS energies $\varepsilon_\mu^{(1)}$ without affecting the S-matrix. Thus, the energy of IS $\varepsilon_\mu^{(1)}$ is not in correspondence with the location of the S-matrix poles. Using symmetry properties of the S-matrix as a function of the complex momentum p [3], it is easy to show [18] that the energy of BSEC is not in correspondence with any of the S-matrix poles. An interesting new point, so far as we know never discussed in literature, is the appearance of the discrete spectrum states, i.e., of ISes with negative energy $\varepsilon_\mu^{(1)} < 0$, that are divorced from the S-matrix poles. So, IS being the state with the asymptotically decreasing wave function and with the energy at which the S-matrix does not have a pole, can been treated as a generalization of BSEC on the case of the discrete spectrum states.

We examine in more detail the formation of IS in the spectrum of a quantum system with nonlocal interaction using as an example a simple analytically solvable model. The simplest realization of the situation depicted in Fig. 1, corresponds to the case when the submatrix $H^{(1)}$ is a 1×1 matrix and the separable potential (6) is of the rank 2, i.e. $N = 1$. In this case, IS arises due to the cancellation of the potential energy matrix elements $V_{01} = V_{10}$ and the kinetic energy matrix elements $T_{01} = T_{10} = -V_{01}$ that results in $H_{01} = H_{10} = 0$. The IS energy ε_0 is equal to the diagonal matrix element H_{00}, $\varepsilon_0 = H_{00}$. It should be stressed that H_{00} can take an arbitrary value in our model. Using Eqs. (10–13) we obtain the following expression for the phase shift δ_l:

$$\tan\delta_l = -\frac{S_{0l}(p)\left\{[V_{11}(\varepsilon_0 - E) - \beta](T_{00}^l - E) + \left(T_{01}^l\right)^2(\varepsilon_0 - E)\right\}}{C_{0l}(p)\left\{[V_{11}(\varepsilon_0 - E) - \beta](T_{00}^l - E) + \left(T_{01}^l\right)^2(\varepsilon_0 - E)\right\} - \frac{p[V_{11}(\varepsilon_0 - E) - \beta]}{\pi\, S_{0l}(p)}}, \tag{15}$$

where $\beta \equiv H_{01}^2$.

Suppose $\varepsilon_0 > 0$ and $V_{11} > 0$. The evolution of the *s* wave phase shift $\delta_0(p)$ as $\beta = H_{01}^2 = H_{10}^2$ tends to zero is shown in Fig. 2. If $\beta \neq 0$ there is a resonance at the energy $E \approx \varepsilon_0$ of the width Γ that decreases as β is reduced. When $\beta = 0$, BSEC arises as the resonance of the zero width producing the jump of the height π of the phase shift δ_0 at the energy $E = \varepsilon_0$. This spurious jump should be eliminated that results in the $\delta_0(0)$ increase by an extra π if we suppose, as usual, that $\delta_0(\infty) = 0$. Thus when applying the Levinson theorem [3] to the system pertaining IS, one should count IS as a usual discrete spectrum bound state. Such behavior of the phase shift is typical for systems pertaining BSEC that have been studied in various models [3, 6, 7, 18, 21]. Thus our model represents an alternative simple analytical approach in the study of BSEC.

As for the *S*-matrix, it is given by the following expression:

$$S_l = -\frac{C_{0l}^{(-)}(p)\left\{\left[V_{11}^l(\varepsilon_0 - E) - \beta\right](T_{00}^l - E) + \left(T_{01}^l\right)^2(\varepsilon_0 - E)\right\} - \frac{p[V_{11}^l(\varepsilon_0 - E) - \beta]}{\pi S_{0l}(p)}}{C_{0l}^{(+)}(p)\left\{\left[V_{11}^l(\varepsilon_0 - E) - \beta\right](T_{00}^l - E) + \left(T_{01}^l\right)^2(\varepsilon_0 - E)\right\} - \frac{p[V_{11}^l(\varepsilon_0 - E) - \beta]}{\pi S_{0l}(p)}}, \tag{16}$$

where $C_{nl}^{(\pm)}(p) = C_{nl}(p) \pm i S_{nl}(p)$. The single *S*-matrix pole on the unphysical sheet tends to the real energy $E = \varepsilon_0$ as β tends to zero. However in the limit $\beta = 0$, the factor $(\varepsilon_0 - E)$ in the numerator of Eq. (16) cancels the same factor in the denominator and the singularity at the energy $E = \varepsilon_0$ disappears:

$$S_l = -\frac{C_{0l}^{(-)}(p)\left[V_{11}^l(T_{00}^l - E) + \left(T_{01}^l\right)^2\right] - \frac{pV_{11}^l}{\pi S_{0l}(p)}}{C_{0l}^{(+)}(p)\left[V_{11}^l(T_{00}^l - E) + \left(T_{01}^l\right)^2\right] - \frac{pV_{11}^l}{\pi S_{0l}(p)}}, \tag{17}$$

This illustrates the mechanism of the *S*-matrix pole loss in the limit $\beta \to 0$ when the resonance transforms into BSEC. The nontrivial result is that if IS is a bound state ($\varepsilon_0 < 0$), it does not generate the *S*-matrix pole, too.

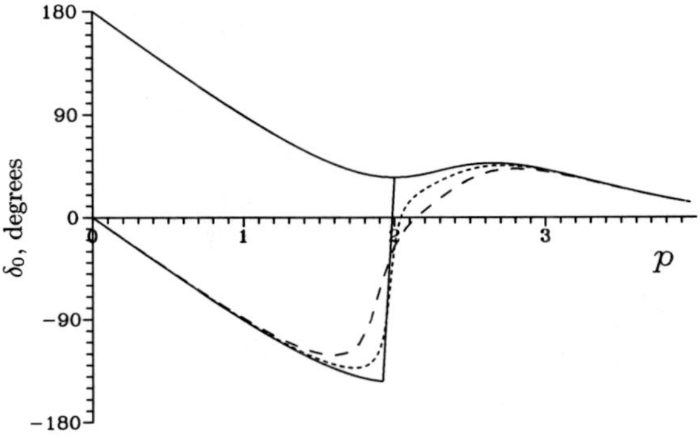

Fig. 2 The evolution of the phase shift $\delta_0(p)$ as $\beta = H_{01}^2 = H_{10}^2$ tends to zero. Dashed, dotted, and solid curves are the phase shifts obtained with $\beta = \beta_1$, β_2 and β_3, respectively; $\beta_1 > \beta_2 > \beta_3 = 0$

The above results can be easily generalized as the following statement.

Let all eigenvalues of the truncated Hamiltonian matrix H^N be non-degenerate. Then Isolated States occur in the spectrum of the system with the rank–$(N+1)$ separable interaction (6) if and only if the truncated matrix $\tilde{H}^N$ and its principal minor $\tilde{H}^{N-1}$ of the rank $(N-1)$ have common eigenvalues. The number ν of the common eigenvalues is equal to the number of the Isolated States. These eigenvalues and the corresponding eigenfunctions are just the energies and the wave functions of the Isolated States.

The equivalent formulation is

The system with the nonlocal interaction (6) has the Isolated State at the energy ε_μ if and only if ε_μ is the eigenvalue of the truncated Hamiltonian matrix $\tilde{H}^N$ and the corresponding eigenvector U^μ has the last component $U_N^\mu = 0$.

3 Phenomenological NN Potential with Isolated State and the Three-Nucleon System

The simplest rank-2 separable potential (6) supporting IS discussed in the previous Section, was used to fit the NN singlet $^1s_0$ and triplet $^3s_1$ scattering phase shifts. The oscillator function parameter $\hbar\omega = 500$ MeV. To ensure the existence of IS, we should set the off-diagonal potential energy matrix elements $V_{01} = V_{10} = -T_{10}$. The phase shifts are independent of the matrix element V_{00} that governs the IS energy. The only parameter responsible for the phase shifts is the matrix element V_{11}.

The singlet phase shifts obtained with $V_{11}^s = -0.7315\,\hbar\omega$ and the triplet phase shifts obtained with $V_{11}^t = -0.81512\,\hbar\omega$ are shown in Figs. 3 and 4 respectively. These values of V_{11}^s and V_{11}^t are seen to reproduce with a reasonable accuracy the scattering data.

The deuteron ground state energy should be obtained by the calculation of the S-matrix pole as is discussed in Refs. [22, 23]. Since in our case the S-matrix is given by the expression (17), the S-matrix poles can be calculated by solving the equation

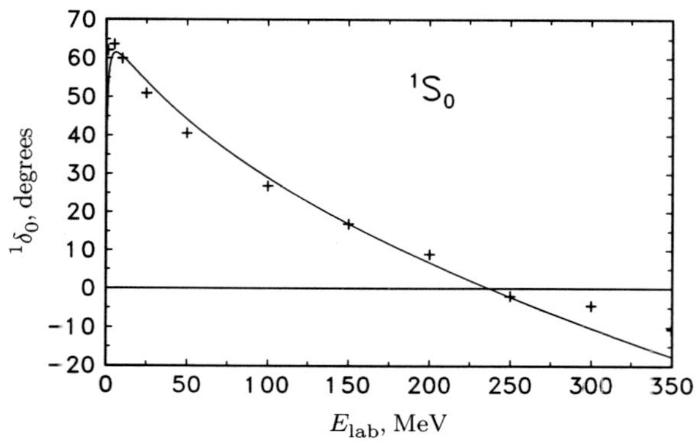

Fig. 3 Singlet s wave phase shifts. Solid line — phenomenological potential with IS; + — experimental data

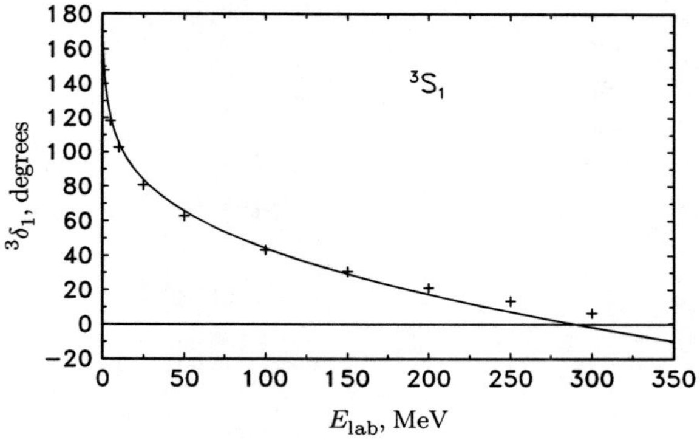

Fig. 4 Triplet s wave phase shifts. Solid line — phenomenological potential with IS; + — experimental data

$$C_{0l}^{(+)}(p)\left[V_{11}^{l}(T_{00}^{l}-E)+\left(T_{01}^{l}\right)^{2}\right]-\frac{pV_{11}^{l}}{\pi\,S_{0l}(p)}=0. \tag{18}$$

The bound state (deuteron) should be searched for in the triplet $^3s_1$ wave, i.e. $l=0$ and V_{11}^t should be used as V_{11}^l in Eq. (18). We are searching for negative E_d value and $p_d=i\sqrt{2|E_d|/\hbar\omega}$ fitting Eq. (18). The deuteron wave function can be calculated using the J-matrix formalism (see also the discussion in Refs. [22, 23] and references therein). Our very simple potential with the only fitting parameter V_{11}^t provides a good description of the deuteron energy $E_d=-2.22496\,\text{MeV}$ and rms radius $\sqrt{\langle r^2\rangle}=1.87$ fm (the respective experimental values are $E_d^{\text{exp}}=-2.224575\,\text{MeV}$ and $\sqrt{\langle r_{\text{exp}}^2\rangle}=1.9676$ fm).

The obtained singlet and triplet s wave NN potentials are used in the calculation of the triton bound states. As in the other studies of the three-body bound state collapse [9–16], we do not allow for the interaction in other partial waves. We perform a conventional variational calculation of the $S=T=J=\frac{1}{2}$ three-nucleon states with the three-body oscillator basis allowing for all components with the total number of oscillator quanta $N=2n+l\leq 32$. The energies of the ground state and of the lowest excited states are shown in Fig. 5 for different E_I values (we are varying both the triplet V_{00}^t and the singlet V_{00}^s potential energy matrix elements so that $V_{00}^s=V_{00}^t$; in other words the IS energy E_I^t in the triplet state is equal to the IS energy E_I^s in the singlet state: $E_I=E_I^s=E_I^t$). It is seen that the variation of the IS energy E_I (that does not affect the phase shifts and the deuteron properties) results in the drastic changes of the triton binding energy and of the spectrum of excited $S=T=J=\frac{1}{2}$ states. When the IS energy E_I is small enough, the three-nucleon system collapses (the binding energy becomes extremely large); two excited states are bound in addition to the ground state. The triton ground state energy increases with the IS energy E_I; the same is true for the energies of the excited states. At some E_I value the second excited state becomes unbound; at some larger E_I value the first excited state becomes unbound, too; however the triton binding energy is still too large. Nevertheless the further increase of E_I results in the increase of the

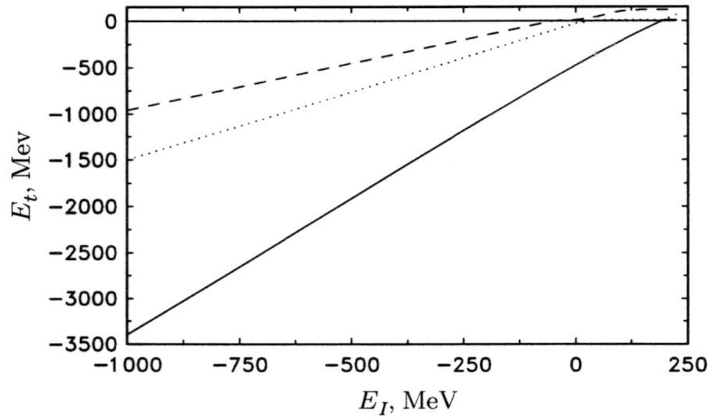

Fig. 5 The triton ground and excited state energies vs the IS energy E_I

triton ground state energy and at some E_I value the three-nucleon system becomes unbound.

We see that we really have the three-body bound state collapse for some values of the IS energy E_I. However the collapse disappears for larger E_I values. Therefore our general conclusion is that the three-body bound state collapse problem does not exist as a general problem. In the previous studies of the three-body bound state collapse, very restricted potential models were used that did not allow to change the BSEC energy $E_{\rm BSEC}$. The $E_{\rm BSEC}$ value used in these studies caused the extreme overbinding of the trinucleon. This is clearly the problem of the particular potentials used and not the general problem of the three-body bound state collapse inherent for all two-body potentials supporting BSEC. In our model any trinucleon binding energy can be obtained by phase equivalent variation of the IS energy E_I. In particular, we can fit the E_I value to the triton binding energy. Setting $E_I = E_I^s = E_I^t = 189.525$ MeV, we obtain the triton binding energy of $E_t = 8.4731$ MeV in our $32\hbar\omega$ variational calculations. The accuracy of the variational approach is improved if the three-body S-matrix pole is calculated as is discussed in Ref. [22]: in this case the triton binding energy of $E_t = 8.4748$ MeV is obtained.

4 Discussion

We suppose that the most important feature of BSEC is that this bound state is not in correspondence with the S-matrix pole. We introduce IS as a generalization of BSEC on the case of arbitrary (positive or negative) energy. We propose a simple model of non-local interaction elucidating the formation of IS. To investigate in detail the formation and the features of IS, it is natural to employ the J-matrix formalism that makes it possible to obtain analytical expressions for the S-matrix and other observables, i.e. to formulate the exactly solvable model of IS.

The general results given above can be straightforwardly extended on the coupled channel case, on the case of the separable finite-rank nonlocal potentials with the Laguerre form factors, etc.

It is interesting that IS naturally appeares in the standard nuclear shell model. It is well-known (the Dubovoy–Flores theorem) [24, 25] that the lowest eigenvalue $\varepsilon_0^N(\omega)$ obtained in the shell model calculation in the $N\hbar\omega$ model space, coincides with the lowest eigenvalue $\varepsilon_0^{N-2}(\omega)$ obtained in the shell model calculation in the $(N-2)\hbar\omega$ model space if the parameter $\hbar\omega$ of the oscillator basis minimizes the eigenvalue $\varepsilon_0^{N-2}(\omega)$. According to the statement formulated in Sect. 3, the ground state obtained in the conventional variational nuclear shell-model calculations appears to be IS.

Using our exactly solvable model of IS, we obtain a simple nonlocal NN potential describing the singlet and triplet s wave phase shifts and deuteron properties. With this NN interaction we study the triton properties and examine the three-body bound state collapse problem. We can vary the IS energy E_I in our model without violating the description of the phase shift and deuteron properties, i.e. without altering the on-shell properties of the interaction. However the off-shell properties of the interaction are modified when E_I is varied and this is clearly seen in the strong E_I-dependence of the triton ground state energy E_t. The three-body system collapses at some E_I values and does not at other E_I values. We can even fit the position of IS to the triton binding energy. Therefore we conclude that the discussion of the three-body bound state collapse in literature [11–16] arose only due to the use of unsuccessful NN potential supporting BSEC; other two-body interactions supporting BSEC provide a correct description of the three-nucleon binding.

Various interpretations of the nature of the three-body bound state collapse have been suggested in literature [11–16]. We cannot accept most of them.

For example, Nakaichi-Maeda [15] and Pantis et al [16] supposed that the three-body bound state collapse arises due to the nodel behavior of the deuteron wave function in the case of potentials supporting BSEC. Delfino et al [14] supposed that collapse originates from the structure of the BSEC wave function which is identical with that of the Pauli forbidden state. In our model, we shifted the IS energy changing neither the deuteron wave function nor the IS wave function. We show that the three-body bound state collapse can be eliminated by the increase of the IS energy only, therefore we cannot agree that the collapse originates from the structure of the deuteron or IS wave function.

Delfino et al [13] supposed that the three-body bound state collapse is a manifestation of the Thomas effect (the increase of the triton binding energy when the range of the NN potential tends to zero). The range of the general nonlocal interaction cannot be established unambiguosly. Delfino et al suggested to use the average kinetic energy $\langle T \rangle$ of the two-body bound state as an indicator of the potential range and demonstrated that BSEC is accompanied by an increase of $\langle T \rangle$. We cannot agree with this conclusion. In our model the bound state wave functions are unaltered when the IS energy is varied, hence the variation of IS energy does not cause changes of $\langle T \rangle$ of any of the bound state. Therefore with the same values of $\langle T \rangle$ of two-body bound states, we obtain either the collapsed triton at some E_I values or normally-bound (or just unbound) triton at other E_I values.

Rupp et al [12] note that the triton binding energy increases when the BSEC energy grows. We obtain an opposite result in our investigation: the triton binding energy decreases with the BSEC energy. We note here that in our studies we shift BSEC phase equivalently while Rupp et al base their conlusion on the results obtained with a number of potentials providing different phase shifts. In other words, they use potentials with different on-shell properties that can mask any effect of the BSEC energy variation.

It is interesting that it is easy to change the BSEC (or any other bound state) energy phase equivalently. The BSEC wave function Ψ_{BSEC} fits the Schrödinger equation

$$H\Psi_{\mathrm{BSEC}} = E_{\mathrm{BSEC}}\Psi_{\mathrm{BSEC}}. \tag{19}$$

Let us define a new Hamiltonian

$$H' = H + \lambda|\Psi_{\mathrm{BSEC}}\rangle\langle\Psi_{\mathrm{BSEC}}|. \tag{20}$$

The wave functions $\Psi'^{l}_{E}(r)$ of all the rest (bound or continuum spectrum) eigenstates of the Hamiltonian H, fit the Schrödinger equation

$$H\Psi'^{l}_{E}(r) = E\Psi'^{l}_{E}(r). \tag{21}$$

Clearly $\Psi'^{l}_{E}(r)$ will be also the eigenfunctions of the Hamiltonian H' since Ψ_{BSEC} is orthogonal to any $\Psi'^{l}_{E}(r)$. At the same time, the BSEC energy E_{BSEC} will be increased by λ. We are sure that if Rupp et al employed this method of varying the BSEC energy, they would obtain the natural result that the triton binding energy decreases with E_{BSEC}.

The Hamiltonian (20) can be defined with the projector $|\Psi'^{l}_{E_b}\rangle\langle\Psi'^{l}_{E_b}|$ on any other bound state $\Psi'^{l}_{E_b}(r)$ at the energy E_b. In this case we obtain another simple model of IS. Really, none of the continuum spectrum wave functions $\Psi'^{l}_{E}(r)$ is altered by the projector $|\Psi'^{l}_{E_b}\rangle\langle\Psi'^{l}_{E_b}|$. Hence the Hamiltonians H and H' provide the same S-matrix. Thus the projector $\lambda|\Psi'^{l}_{E_b}\rangle\langle\Psi'^{l}_{E_b}|$ increases the energy E_b of the bound state by λ but does not shift the S-matrix pole at the negative energy E_b. As a result the conventional bound state at the energy E_b is transformed into IS. We note here that the Hamiltonians with the projection operators $\lambda|\Psi'^{l}_{E_b}\rangle\langle\Psi'^{l}_{E_b}|$ are widely used, e.g., to project out Pauli forbidden states (see, e.g., [26]).

We do not suppose that it is needed to criticize other interpretations of the origin of the three-body bound state collapse suggested in various papers. Our general conclusion is that the collapse is inherent for very restricted models of potentials supporting BSEC only and does not appear if other two-body interactions with BSEC are used. Therefore for the proper interpretation of the three-body bound state collapse origin, one should carefully study the respective potential model restrictions and reveal what is wrong with them. Up to the best of our knowledge, it was never done.

Acknowledgments This work was supported in part by the State Program "Russian Universities", by the Russian Foundation of Basic Research grants No 02-02-17316 and No 05-02-17429 and by the US DOE grants DE-FG02-87ER40371 and DE-FC02-07ER41457.

References

1. J. V. von Neumann and E. Wigner, Phys. Z. **30**, 465 (1929).
2. B. Mulligan, L. G. Arnold, B. Bagchi, and T. O. Krause, Phys. Rev. **C 13**, 2131 (1976); B. Bagchi, T. O. Krause, and B. Mulligan, Phys. Rev. **C 15**, 1623 (1977).
3. R. G. Newton, *Scattering Theory of Waves and Particles*, 2nd. ed. (Springer-Verlag, New York, 1982).
4. L. Fonda and R. G. Newton, Ann. Phys. (NY) **10**, 490 (1960).
5. H. Friedrich and D. Wintgen, Phys. Rev. **A 31**, 3964 (1985); Phys. Rev. **A 33**, 3231 (1985).
6. L. S. Cederbaum, R. S. Friedman, V. M. Ryaboy, and N. Moiseyev, Phys. Rev. Lett. **90**, 013001 (2003).
7. M. Grupp, G. Nandi, R. Walser, and W. P. Schleich, Phys. Rev. **A 73**, 050701 (2006).
8. F. Tabakin, Phys. Rev. **174**, 1208 (1968).
9. J. E. Beam, Phys. Lett. **B 30**, 67 (1969).
10. V. A. Alessandrini and C. A. Garcia Canal, Nucl. Phys **A 133**, 950 (1969).
11. S. Sofianos, N. J. McGurk, and H. Fiedeldey, Z. Phys. **A 286**, 87 (1978); G. Pantis, H. Fiedeldey, and D. W. L. Sprung, Z. Phys. **A 291**, 367 (1979); **294**, 101 (1980).
12. G. Rupp, L. Streit, and J. A. Tjon, Phys. Rev. **C 31**, 2285 (1985).
13. A. Delfino, S. K. Adhikari, L. Tomio, and T. Federico, Phys. Rev. **C 46**, 471 (1992).
14. A. Delfino, S. K. Adhikari, L. Tomio, and T. Federico, Phys. Rev. **C 46**, 1612 (1992).
15. S. Nakaichi-Maeda, Phys. Rev. **C 51**, 1633 (1995).
16. G. Pantis, I. E. Lagaris, and S. A. Sofianos, Phys. Rev. **C 63**, 044009 (2001).
17. L. D. Landau and E. M. Lifshitz, *Quantum Mechanics: Non-relativistic Theory* (Pergamon Press, Oxford, New York, 1977).
18. L. A. Roriz and A. Delfino, Canad. J. Phys. **67**, 37 (1989).
19. H. A. Yamani and L. Fishman, J. Math. Phys. **16**, 410 (1975).
20. Yu. L. Dorodnykh, V. G. Neudatchin, and N. P. Yudin, Yad. Fiz. [Sov. J. Nucl. Phys.] **48**, 1796 (1988).
21. B. N. Zakhariev and A. A. Suzko A.A., *Direct and Inverse Problems* (Springer, Heidelberg, 1990).
22. Yu. A. Lurie and A. M. Shirokov, Ann. Phys. (NY) **312**, 284 (2004); see also *contribution to this volume*.
23. A. M. Shirokov, A. I. Mazur, S. A. Zaytsev, J. P. Vary, and T. A. Weber, Phys. Rev. **C 70**, 044005 (2004); see also *contribution to this volume*.
24. M. Dubovoy and J. Flores, Revista Mex. Fiz. **17**, 289 (1968).
25. L. Majling, J. Rizek, and Z. Pluhar, Bull. Acad. Sci. USSR, Phys. ser. **38**, 2141 (1974).
26. V. I. Kukulin, V. N. Pomerantsev, Kh. D. Razikov, V. T. Voronchev, and G. G. Ryzhikh, Nucl. Phys. **A 586**, 151 (1995).

On the Regularization in J-Matrix Methods

J. Broeckhove, V.S. Vasilevsky, F. Arickx and A.M. Sytcheva

Abstract We investigate the effects of the regularization procedure used in the J-Matrix method. We show that it influences the convergence, and propose an alternative regularization approach.We explicitly perform some model calculations to demonstrate the improvement.

1 Introduction

The J-Matrix (JM) method has become a popular tool in solving quantum mechanical scattering problems in atomic, nuclear and molecular systems. It is based on a square-integrable form of the wave function. The basis used for the representation of the wave function has to reduce the reference Hamiltonian, which is responsible for the asymptotic behavior of the solutions, to a tridiagonal or Jacobi matrix form. The scattering boundary conditions can then be expressed in terms of the expansion coefficients of the wave function. The Schrödinger equation has to be solved by matching the interior to the asymptotic region, which can now be done on the expansion coefficients. This leads to a set of linear equations, the solutions of which are the expansion coefficients of the interior representation, and the scattering parameters (phase shift or S-matrix).

The JM approach proved to be successful for 2-, 3- and many-particle systems in atomic and molecular [1, 2, 3, 4, 5, 6, 7], and nuclear physics [8, 9, 10, 11, 12, 13, 14, 15, 16, 17, 18, 19, 20, 21, 22, 23].

Despite the successes, a slow convergence as a function of the number of basis functions in the internal region leads to the need of large matrices. The calculation of matrix elements of the potential energy operator usually represents the main computational cost in a realistic many-body application. An improvement on the method leading to faster convergence with a smaller size of the Hamiltonian matrix is therefore necessary. There have already been attempts, based on asymptotic

J. Broeckhove
University of Antwerp, Group Computational Modeling and Programming, Antwerp, Belgium
e-mail: Jan.Broeckhove@ua.ac.be

A.D. Alhaidari et al. (eds.), *The* J*-Matrix Method*, 117–134.

properties of the potential [24, 25, 26], that led to sometimes dramatic reductions of the size of the model space. Other possibilities lie in computational methods for faster evaluation of potential matrix elements. Still a number of problems remain to solve the convergence issue.

In this paper we analyze the behavior of the irregular (or Neumann-like) solution of the free-motion Schrödinger equation, and the corresponding expansion coefficients. Because of the square-integrable representation, a regularizing boundary condition has to be introduced to solve this equation. The standard regularization considered in the JM approach [27, 28, 29] will be shown to put a limit on the minimal size of the mode space, and thus to have an important impact on the convergence. An analysis of the regularized asymptotic solution leads to a new regularization procedure that improves the size limitation. The enhancement in convergence of this procedure will be explicitly demonstrated on a number of potential problems.

We consider a JM formulation based on the oscillator basis, and a free-motion Hamiltonian. The analysis and the new regularization procedure can be easily extended to other bases, with analogous conclusions.

2 *J*-Matrix Methods

A Schrödinger equation for a continuum wave function with a spherically symmetric, non-coulombic potential,

$$\left(-\frac{\hbar^2}{2m}\left[\frac{1}{r^2}\frac{\partial}{\partial r}\left(r^2\frac{\partial}{\partial r}\right)-\frac{l(l+1)}{r^2}\right]+V(r)-E\right)\Psi_l(r)=0 \tag{1}$$

must have a solution that is matched asymptotically with the free-space Bessel and Neumann functions

$$\Psi_l(r\to\infty)\to\sqrt{\frac{2}{\pi}}\,j_l(kr)-\tan\delta_l(k)\,\sqrt{\frac{2}{\pi}}\,n_l(kr), \tag{2}$$

where $k^2=\frac{2mE}{\hbar^2}$. This match of a solution in the interaction region, where the effect of the potential is felt, with the asymptotic reference states determines the phase shift at momentum k corresponding to energy E. We use the traditional spherical Bessel and spherical Neumann function definitions [30, 31] with the delta-function normalization.

The JM method reformulates this problem into an algebraic setting by the use of a basis functions. In our case these are the harmonic oscillator functions

$$\Phi_{nl}(r\,|\,b)=(-1)^n\frac{1}{b^{3/2}}N_{nl}\,(\frac{r}{b})^l L_n^{l+1/2}\left((\frac{r}{b})^2\right)\exp\left(-\frac{1}{2}(\frac{r}{b})^2\right) \tag{3}$$

$$N_{nl}=\sqrt{\frac{2\Gamma(n+1)}{\Gamma(n+l+3/2)}},$$

where the $L_n^\alpha(r)$ are Laguerre polynomials, N_{nl} is a norm-factor and b is the oscillator length. We take a functional notation in which the variable of the function is separated from the parameter by a "|"; it will be used throughout the paper. While it is straightforward to turn (1), using the superposition ansatz

$$\Psi_l(r|k) = \sum_{n=0}^{\infty} C_{nl}(k,b)\Phi_{nl}(r \mid b) \tag{4}$$

into a matrix equation, more care must be taken when mapping the boundary condition (2). The Bessel and Neumann functions are the solutions of the free-space reference Hamiltonian.

The regular Bessel solution

$$(T_l - E)\,\Psi_l^B(r|k) = 0 \tag{5}$$

$$\Psi_l^B(r \mid k) = \sqrt{\frac{2}{\pi}}\, j_l(kr) \tag{6}$$

presents no problems and can be reconstructed as the solution of

$$\sum_{m=0}^{\infty} \langle \Phi_{nl}(b)|T_l - E|\Phi_{ml}(b)\rangle\, C_{ml}^B(k,b) = 0, \tag{7}$$

where the parameter of the basis states in brackets is indicated as a variable notation; this notation is used throughout the chapter. This is a three-term recurrence relation that can be solved explicitly [27, 28, 29]:

$$\begin{aligned} C_{nl}^B(k,b) &= N_{nl}\, b^{3/2}\, (kb)^l \exp\left(-\frac{1}{2}(kb)^2\right) L_n^{l+1/2}\left((kb)^2\right) \\ &= \frac{2}{N_{nl}} \frac{b^{3/2}}{\Gamma(l+\frac{3}{2})} (kb)^l \exp(-\frac{(kb)^2}{2})\,{}_1F_1\left(-n,\, l+\frac{3}{2}; (kb)^2\right), \end{aligned} \tag{8}$$

where the ${}_1F_1$ stands for Kummer's function [32]. For large n, the asymptotic behavior of the coefficients is given by

$$C_{nl}^B(k,b) \to b\sqrt{2R_{nl}}\Psi_l^B(R_{nl} \mid k), \tag{9}$$

where the $R_{nl} = b\sqrt{4n+2l+3}$ are the oscillator turning points.

The Neumann solution is irregular

$$(T_l - E)\,\Psi_l^N(r|k) = 0 \tag{10}$$

$$\Psi_l^N(r \mid k) = \sqrt{\frac{2}{\pi}}\, n_l(kr) \tag{11}$$

and cannot be mapped directly because the oscillator states cannot represent the singular behavior at the origin. This is not a fundamental problem because we only require these reference solutions to formulate the asymptotic conditions i.e. for $r \rightarrow \infty$. An equivalent solution, i.e. one that behaves asymptotically as the Neumann function but is regular at the origin, will do just as well. In some applications one uses a simple cutoff procedure. In the JM method it is of course important that the procedure translates easily to the algebraic representation of the problem. One takes the solution of

$$(T_l - E)\overline{\Psi}_l^N (r|k, b) = \beta\, \Phi_{0l}(r \mid b) \tag{12}$$

$$\overline{\Psi}_l^N (r \mid k, b) \rightarrow \sqrt{\frac{2}{\pi}}\, n_l(kr) \text{ for } r \rightarrow \infty. \tag{13}$$

The regularized Neumann function can be determined from this equation using the Green's function technique [27],[29] (see also next section). The coefficient β is fixed by the boundary condition. This regularized Neumann function can be reconstructed as a solution of the equation

$$\sum_{m=0}^{\infty} \langle \Phi_{nl}(b)|T_l - E|\Phi_{ml}(b)\rangle\, C_{ml}^N(k, b) = \beta\, \delta_{n,0}. \tag{14}$$

It is a three-term recurrence relation with a simple inhomogeneous contribution and can again be solved explicitly

$$C_{nl}^N (k, b) = \frac{(-)^{l+1} N_{nl} b^{\frac{3}{2}}}{\Gamma(-l + \frac{1}{2})} (kb)^{-l-1} \exp\left(-\frac{(kb)^2}{2}\right) \\ {}_1F_1\left(-n - l - \frac{1}{2}, -l + \frac{1}{2}; (kb)^2\right). \tag{15}$$

For large n, the asymptotic behavior is

$$C_{nl}^N(k, b) \rightarrow b\sqrt{2R_{nl}}\, \overline{\Psi}_l^N (r \mid k, b) \rightarrow b\sqrt{2R_{nl}}\Psi_l^N(R_{nl} \mid k). \tag{16}$$

We use these asymptotic values (9) and (16) to compute the $C_{nl}^B(k, b)$ and $C_{nl}^N(k, b)$ coefficients by seeding the recurrence relations at some large index value, and recurring back towards small n. This turns out to be faster and numerically more efficient than directly evaluating the analytical expressions.

Thus we are led to a formulation of the scattering problem where we represent the solution as follows. The wave function has two essential components. In the internal region the effects of the potential are felt. We assume it has a finite range and that this component can be approximated by a finite combination of the L^2 basis states. In the asymptotic region the solution must match the boundary condition which we now express with the Bessel and regularized Neumann functions.

$$\Psi_l(r|k) = \Psi_l^I(r|k) + \Psi_l^B(r|k) - \tan\delta_l \overline{\Psi}_l^N(r|k, b) \tag{17}$$

$$\Psi_l^I(r|k) \to 0 \quad \text{for } r \to \infty. \tag{18}$$

In the algebraic representation this becomes

$$C_{nl}(k, b) = C_{nl}^I(k, b) + C_{nl}^B(k, b) - \tan\delta_l\, C_{nl}^N(k, b) \tag{19}$$

$$C_{nl}^I(k, b) \to 0 \quad \text{for } n \to \infty. \tag{20}$$

The equation for the internal wave function and the phase shift is obtained by inserting (17) into the Schrödinger equation, projected onto the basis set:

$$\left\langle \Phi_{nl}(b) | T_l + V - E | \Psi_l^I(k) + \Psi_l^B(k) - \tan\delta_l\, \overline{\Psi}_l^N(k, b) \right\rangle = 0 \text{ for all } n \tag{21}$$

Taking advantage of the properties of the asymptotic reference states (5) and (12) this simplifies to:

$$\begin{aligned} &\left\langle \Phi_{nl}(b) | T_l + V - E | \Psi_l^I(k) \right\rangle - \tan\delta_l \left(\beta\delta_{n0} + \left\langle \Phi_{nl}(b) | V | \overline{\Psi}_l^N(k, b) \right\rangle \right) \\ &= - \left\langle \Phi_{nl}(b) | V | \Psi_l^B(k) \right\rangle \end{aligned} \tag{22}$$

In the approach of [27, 29] the algebraic Schrödinger equation is approximated by assuming that the interaction matrix for potential V can be truncated at some large $n = N$. This N defines a sharp boundary between the internal and asymptotic regions in the coefficient space. The resulting problem can be solved with $C_{nl}^I(k, b) = 0$ for $n \geq N$. One has an $N + 1$ by $N + 1$ matrix equation in the unknowns $\{C_{0l}^I, C_{1,l}^I, \ldots, C_{N-1l}^I, \tan\delta_l\}$:

$$\begin{aligned} &\sum_{m=0}^{N-1} \langle \Phi_{nl}(b) | T_l + V - E | \Phi_{ml}(b) \rangle\, C_{ml}^I(k, b) \\ &\qquad - \tan\delta_l(k) \left(\beta\delta_{n0} + \sum_{m=0}^{N-1} \langle \Phi_{nl}(b) | V | \Phi_{ml}(b) \rangle\, C_{ml}^N(k, b) \right) \\ &= - \sum_{m=0}^{N-1} \langle \Phi_{nl}(b) | V | \Phi_{ml}(b) \rangle\, C_{ml}^B(k, b). \end{aligned} \tag{23}$$

Convergence of the results is achieved by extending the interaction region i.e. increasing N.

3 The Regularized Neumann Function

The regularized Neumann function is a solution of the inhomogeneous differential equation (see details for the definition of $\overline{\Psi}_l^N(r|k,b)$ in [27, 28, 29]):

$$(T_l - E)\,\overline{\Psi}_l^N(r \mid k,b) = \beta\,\Phi_{0l}(r \mid b)\,, \tag{24}$$

where

$$\beta = -\frac{\hbar^2}{m\pi k}\frac{1}{\Phi_{0l}(k|1/b)}. \tag{25}$$

The integral representation of $\overline{\Psi}_l^N(r|k,b)$ in coordinate space can be written as:

$$\overline{\Psi}_l^N(r \mid k,b) = \beta \int_0^\infty G(r,r')\,\Phi_{0l}(r'|\,b)dr', \tag{26}$$

where $G(r,r')$ is the Green's function, explicitized in this case as:

$$G(r,r') = \frac{-\beta mk}{\hbar^2}\,j_l(kr_<)\,n_l(kr_>)r^2. \tag{27}$$

In what follows we will restrict ourselves to the case of zero angular momentum. The regularized Neumann function can then be calculated analytically starting from the expression above, by straightforward integration of a combination of sines, cosines and gaussians. One finds

$$\begin{aligned}\overline{\Psi}_0^N(r \mid k,b) = &-\frac{1}{\sqrt{2\pi}kr}[\exp(-ikr)\,\mathrm{erf}\left(\frac{kr}{\sqrt{2}kb}-\frac{ikb}{\sqrt{2}}\right)\\ &+\exp(ikr)\,\mathrm{erf}\left(\frac{kr}{\sqrt{2}kb}+\frac{ikb}{\sqrt{2}}\right)]\end{aligned} \tag{28}$$

From this expression, we can see that the oscillator length has a strong impact on the behavior of $\overline{\Psi}_0^N(r \mid k,b)$. Indeed, when kb tends to zero, and if we take only the leading term in the series approximation of the complex error function for small complex argument, then

$$\begin{aligned}\overline{\Psi}_0^N(r \mid k,b) &\simeq -\frac{2}{\sqrt{2\pi}kr}\cos(kr)\cdot\mathrm{erf}\left(\frac{kr}{\sqrt{2}kb}\right)\\ &= \Psi_0^N(r \mid k)\cdot\mathrm{erf}\left(\frac{r}{\sqrt{2}b}\right).\end{aligned} \tag{29}$$

Thus the larger is $kr/kb = r/b$, the closer the regularized function $\overline{\Psi}_0^N(r|k,b)$ is to the original Neumann function $\Psi_0^N(r|k)$. For instance, when $r \geq 5b$, the functions $\overline{\Psi}_0^N(r|k,b)$ and $\Psi_0^N(r|k)$ are equal within single precision computation. When kb becomes large then

$$\begin{aligned}\overline{\Psi}_0^N(r|k,b) &\approx -\sqrt{\frac{2}{\pi}}\frac{1}{kr}\cos(kr) + \frac{2}{\pi}\frac{kb}{(kr)^2+(kb)^4}\exp\left\{-\frac{1}{2}\frac{(kr)^2}{(kb)^2} - \frac{1}{2}(kb)^2\right\}\\ &= \Psi_0^{(N)}(r|k) + \frac{2}{\pi}\frac{kb}{(kr)^2+(kb)^4}\exp\left\{-\frac{1}{2}\frac{(kr)^2}{(kb)^2} - \frac{1}{2}(kb)^2\right\}\end{aligned} \tag{30}$$

In this case kr should be large compared to kb to suppress the second term in this expression, so that the function $\overline{\Psi}_0^N(r|k,b)$ coincides with $\Psi_0^N(r|k)$. The larger is kb, the larger the coordinate r has to be to reduce the difference between $\overline{\Psi}_0^N(r|k,b)$ and $\Psi_0^N(r|k)$. For values of $r \leq b$, the regularized Neumann function can be represented as

$$\overline{\Psi}_0^N(r \mid k,b) = -\frac{1}{\pi kb}\left[-2\exp\left(\frac{(kb)^2}{2}\right) - i\sqrt{2\pi}kb\,\mathrm{erf}\left(\frac{ikb}{\sqrt{2}}\right)\right] + \ldots \tag{31}$$

by taking the leading term of the Taylor expansion of (28). Unlike the appearance, this expression is real because the error function is imaginary at an imaginary argument. One remarks (confirmed later by Fig. 1) that in the limit for $kb \to 0$ one has

$$\overline{\Psi}_0^N(0 \mid k,b) \simeq -\frac{1}{kb} \longrightarrow -\infty \quad \text{for } kb \to 0 \tag{32}$$

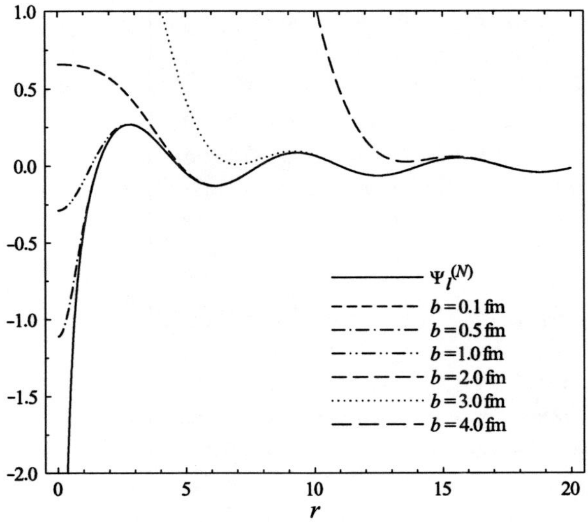

Fig. 1 Neumann Ψ_l^N and regularized $\overline{\Psi}_l^N$ functions, for $l = 0$, $k = 1$ and various oscillator radii b. ($\overline{\Psi}_l^N$ for $b = 0.1$ fm coincides with Ψ_l^N on this scale)

whereas when $kb \to \infty$, the function behaves as:

$$\overline{\Psi}_0^N(0|k,b) \simeq \frac{\exp((kb)^2/2)}{kb} \longrightarrow +\infty \quad \text{for } kb \to \infty \tag{33}$$

The second situation in particular may be the source of numerical difficulties in actual calculations. Perhaps more to the point for the JM method is the fact that this behavior is reflected in the corresponding expansion coefficients $C_{n0}^N(k,b)$ for $kb \ll 1$ and for $kb \gg 1$. Those coefficients then become very large at small values of n. From the analytical expressions (15) for $C_{00}^N(k,b)$, and using the expressions in [32] for the limiting behavior of Kummer's function, one finds for instance:

$$C_{00}^N(k,b) = O\left(\frac{1}{kb}\right) \longrightarrow +\infty \quad \text{for } kb \to 0 \tag{34}$$

and

$$C_{00}^N(k,b) = O\left(\frac{\exp\left(\frac{1}{2}(bk)^2\right)}{kb}\right) \longrightarrow +\infty \quad \text{for } kb \to \infty \tag{35}$$

Care is needed in these situations to avoid numerical difficulties associated with the large values of these coefficients..

In Fig. 1 we display the behavior of the $\overline{\Psi}_0^N(r|k,b)$ function for $k = 1$ and for values of the oscillator radii $b = 0.1$ up to $b = 4.0$ fm. One clearly notices that for small values of b the functions $\overline{\Psi}_0^N(r|k,b)$ and $\Psi_0^N(r|k)$ are very close to each other in the whole range of r, with the exception of small r. The larger is b (or better, in general, kb), the larger is the value of the coordinate r where the functions $\overline{\Psi}_0^N(r|k,b)$ and $\Psi_0^N(r|k)$ are equal within required numerical precision.

The present analysis has shown that the combined parameters for the momentum k, and the oscillator length b (more specifically the product kb) have a great impact on the behavior of the function $\overline{\Psi}_l^N(r|k,b)$. For increasing b, given a fixed k, there is an increasing region of the coordinate r where the regularized and original Neumann functions differ considerably. Small values of b thus seem preferable for obtaining a faster convergence in the JM method.

4 The *J*-Matrix Method Revisited

In this section we will define new expansion coefficients for the regularized Neumann functions in order to increase the convergence rate of the JM method. For this aim we rewrite the (24) in the following way

$$(T_l - E)\overline{\Psi}_l^N(r|k,b_0) = \beta_0\,\Phi_{0l}(r|b_0)\,. \tag{36}$$

We have chosen an oscillator state $\Phi_{0l}(r|b_0)$ for the regularization procedure that has a different oscillator radius b_0 than the oscillator radius b of the basis for the expansion (4), and a corresponding β_0. The expansion coefficients of the (new) regularized Neumann function (36)

$$C_{nl}^{N,b}(k,b_0) = \left\langle \Phi_{nl}(b)\,|\,\overline{\Psi}_l^N(k,b_0)\right\rangle \tag{37}$$

now satisfy the following set of algebraic inhomogeneous equations

$$\sum_{m=0}^{\infty} \langle \Phi_{nl}(b)\,|T_l - E\,|\,\Phi_{ml}(b)\rangle\, C_{ml}^{N,b}(k,b_0) = \beta_0 \langle \Phi_{nl}(b)\,|\Phi_{0l}(b_0)\rangle \tag{38}$$

i.e. a three-term inhomogeneous recurrence relation, with a source containing the overlap $\langle \Phi_{nl}(b)\,|\Phi_{0l}(b_0)\rangle$. The latter can be easily calculated (see Appendix A):

$$\langle \Phi_{nl}(b)\,|\,\Phi_{0l}(b_0)\rangle = \left(\frac{2bb_0}{b^2-b_0^2}\right)^j \sqrt{\frac{\Gamma(n+j)}{n!\Gamma(j)}} \left(\frac{b^2-b_0^2}{b^2+b_0^2}\right)^n, \tag{39}$$

where $j = l + 3/2$. This overlap decreases, for $b > b_0$, for large values of n as

$$\langle \Phi_{nl}(b)\,|\,\Phi_{0l}(b_0)\rangle \approx \left(\frac{2b_0}{b}\right)^j \sqrt{\frac{n^{j-1}}{\Gamma(j)}} \exp\left(-(2n+j)\frac{b_0^2}{b^2}\right). \tag{40}$$

The larger is the ratio b_0/b, the faster the overlap tends to zero. When the ratio is small, the overlap slowly approaches to zero. In the region of n, where the overlap is negligibly small, the expansion coefficients $C_{nl}^{N,b}(k,b_0)$ coincide with $C_{nl}^N(k,b)$, defined by the formula (15).

The system of linear equations (38) can now be solved in the following way. We start the three-term recurrence relation in a region of n where the overlap is negligibly small, and take the original asymptotic $C_{nl}^N(k,b)$ values as seeding values, then solve (38) for $C_{nl}^{N,b}(k,b_0)$ by stepping down the recurrence towards $n = 0$.

Thus we are led to a formulation of the scattering problem where we represent the solution as follows. The wave function has two essential components. In the internal region the effects of the potential are felt. We assume it has a finite range and that this component can be approximated by a finite combination of the L^2 basis states. In the asymptotic region the solution must match the boundary condition which we now express with the Bessel and new regularized Neumann functions.

$$\Psi_l(r|k) = \Psi_l^I(r|k) + \Psi_l^B(r|k) - \tan\delta_l(k)\overline{\Psi}_l^N(r\,|\,k,b_0) \tag{41}$$

$$\Psi_l^I(r|k) \to 0 \quad \text{for } r \to \infty \tag{42}$$

In the algebraic representation this becomes

$$C_{nl}(k,b) = C^{I}_{nl}(k,b) + C^{B}_{nl}(k,b) - \tan\delta_l(k)\; C^{N,b}_{nl}(k,b_0) \tag{43}$$

$$C^{I}_{nl}(k,b) \rightarrow 0 \quad \text{for } n \rightarrow \infty \tag{44}$$

The equation for the internal wave function and the phase shift is obtained by inserting (17) into the Schrödinger equation, projected onto the basis set:

$$\left\langle \Phi_{nl}(b) \right| T_l + V - E \left| \Psi^{I}_{l}(k) + \Psi^{B}_{l}(k) - \tan\delta_l\, \overline{\Psi}^{N}_{l}(k,b_0) \right\rangle = 0 \tag{45}$$

Taking advantage of the properties of the asymptotic reference states (5) and (12) this simplifies to:

$$\begin{aligned}\langle \Phi_{nl}(b)|T_l + V - E|\Psi^{I}_{l}(k)\rangle \\ - \tan\delta_l \left(\beta_0 \left\langle \Phi_{nl}(b) \middle| \overline{\Psi}^{N}_{l}(k,b_0)\right\rangle + \left\langle \Phi_{nl}(b) \middle| V \middle| \overline{\Psi}^{N}_{l}(k,b_0)\right\rangle\right) \\ = -\left\langle \Phi_{nl}(b) \middle| V \middle| \Psi^{B}_{l}(k)\right\rangle\end{aligned} \tag{46}$$

In the approach of [27, 29] the algebraic Schrödinger equation is approximated by assuming that the interaction matrix for potential V can be truncated at some large $n = N$. This N defines a sharp boundary between the internal and asymptotic regions in the coefficient space. The resulting problem can be solved with $C^{I}_{nl}(k,b) = 0$ for $n \geq N$. One has an $N+1$ by $N+1$ matrix equation in the unknowns$\{C^{I}_{0l}, C^{I}_{1,l}, \ldots, C^{I}_{N-1l}, \tan\delta_l\}$:

$$\begin{aligned}\sum_{m=0}^{N-1} \langle \Phi_{nl}(b)|\, T_l + V - E \,|\Phi_{ml}(b)\rangle\; C^{I}_{ml}(k,b) \\ - \tan\delta_l \left(\beta_0 C^{N,b}_{nl}(k,b_0) + \sum_{m=0}^{N-1} \langle \Phi_{nl}(b)|V\,|\Phi_{ml}(b)\rangle\; C^{N,b}_{ml}(k,b_0)\right) \\ = -\sum_{m=0}^{N-1} \langle \Phi_{nl}(b)|V\,|\Phi_{ml}(b)\rangle\; C^{B}_{ml}(k,b)\end{aligned} \tag{47}$$

Convergence of the results is achieved by extending the interaction region i.e. increasing N.

5 Some Examples

In this section we present some detailed results for radial one-dimensional model potentials. We compare the application of the traditional JM regularization procedure to the one introduced in Section 4 for Gauss, exponential, Yukawa and

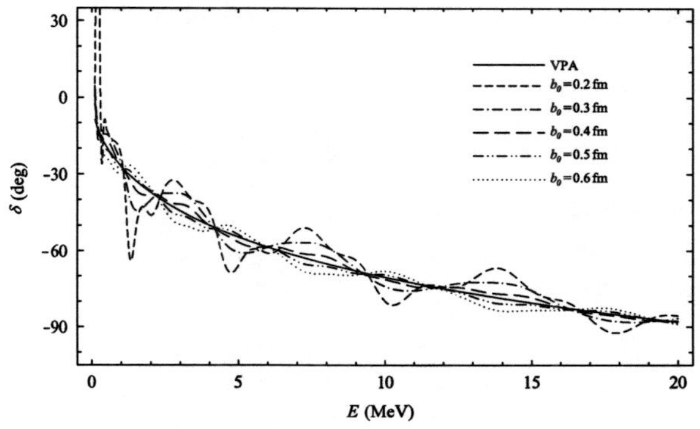

Fig. 2 s-wave phase shift for the square-well potential, obtained with $b = 3$ fm and varying renormalization widths b_0

square-well potentials. We choose comparable parameters, in particular a depth of $V_0 = -80$ MeV and width $a = 1.0$ fm, for all of these potentials.

All calculations are made in the standard JM approach, i.e. truncating the potential matrix beyond the boundary condition matching point N [1, 27, 33]. For the traditional regularization we consider the standard $C_{nl}^{B}(k, b)$ and $C_{nl}^{N}(k, b)$ asymptotic expansion forms, whereas for the new regularization (J-Matrix Regularized (JMR) method) the $C_{nl}^{B}(k, b)$ and $C_{nl}^{N,b}(k, b_0)$ asymptotic forms are used. We have set the matching point to $N = 25$ in all cases. To obtain a fair comparison of the convergence properties for the two regularization properties, we compare the s-wave phase shifts to each other and to the "exact" result, obtained with the Variable Phase approach [34, 35].

In Figs. 2 and 3 we show how the solutions obtained with the new regularization (JMR) depend on its defining value b_0. We have done this for both the Yukawa an square-well potentials, as it is on these potentials that the effects are more noticeable. It is also seen that the optimal value of b_0 strongly depends on the functional form of the potential, more particularly on the behavior at the origin.

In Figs. 4, 5, 6 and 7 the comparison of the JM, JMR and VPA s-wave phase shifts for all four model potentials is displayed, for three different values of the oscillator length $b = 1.0$, 2.0 and 3.0 fm. For uniformity the value for the regularization

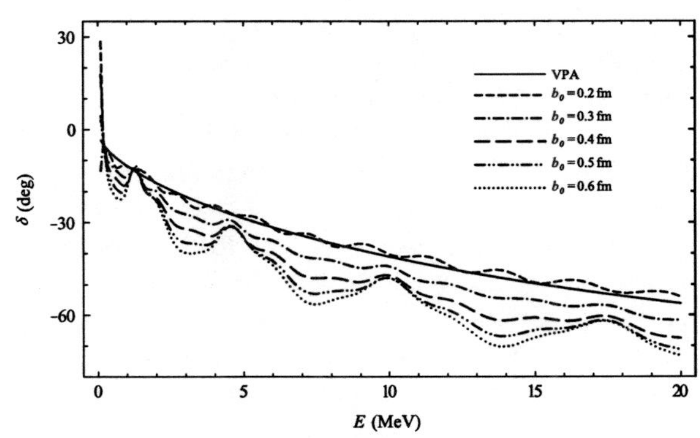

Fig. 3 s-wave phase shift for the Yukawa potential, obtained with $b = 3$ fm and varying renormalization widths b_0

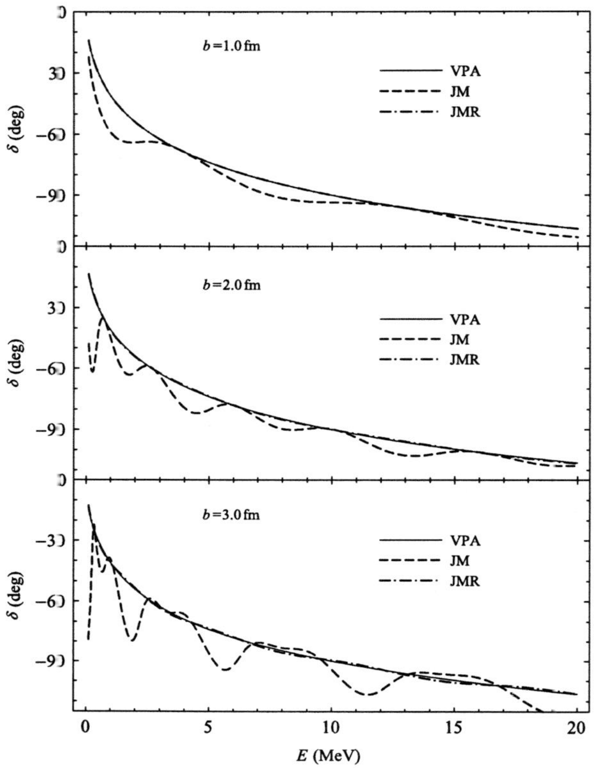

Fig. 4 Comparison of s-wave pahse shifts for JM and JMR for a Gauss potential ($V_0 = 80\,\text{MeV}$, $a = 1.0\,\text{fm}$). The matching point is $N = 25$ in both calculations. The Regularization parameter is $b_0 = 0.6\,\text{fm}$

parameter was chosen to be $b_0 = 0.6$ fm in all cases, as it is a near optimal value to limit the differences between the renormalized $\overline{\Psi}_L^N(r|k, b_0)$ and true asymptotic Neumann $\Psi_L^N(r|k)$ functions in coordinate representation (see Fig. 1).

We notice that, for all potentials, the results deviate more from the exact (VPA) results with increasing b. This is an indication that a larger value for the matching point N should be chosen for truly convergent results. The phase shifts obtained by JMR however remain much closer to the VPA results. The potential effects are more pronounced for the square-well (Fig. 6) and Yukawa potential (Fig. 7). The regularization effect is in all cases seen to be more important for large values of b where the convergence problem becomes an important issue.

It is also clear that convergence, both for JM and JMR, can be more easily achieved for the Gauss and exponential potentials than for the square-well and the Yukawa potentials. This undoubtedly has to do with the more regular behavior at the origin of the former potentials. The JMR results can be further improved if a more judicious choice is made for the regularization parameter b_0. Indeed, from Figs. 2 and 3 it is seen that a value of b_0 around 0.4 fm (square-well) or 0.2 fm (Yukawa) would improve the results even more drastically.

It should also be clear that combining the new regularization method introduced here with other methods for improving convergence, e.g. using semiclassical

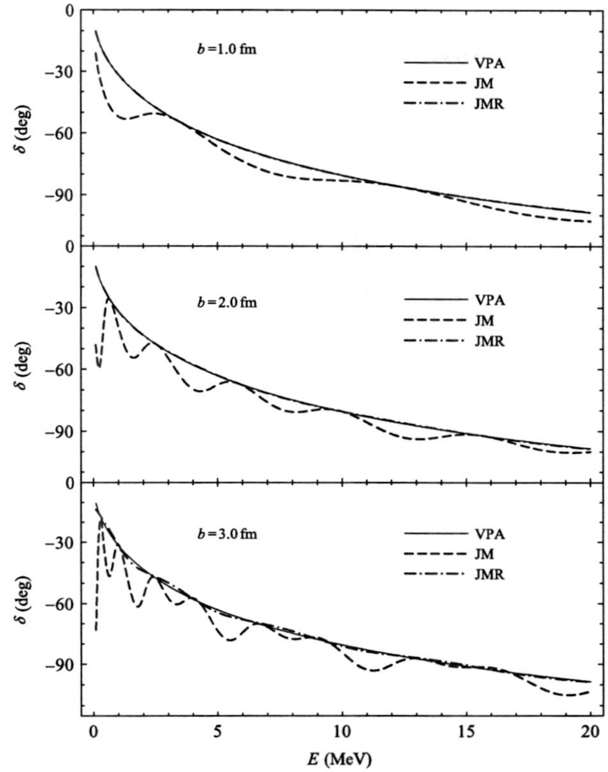

Fig. 5 Comparison of s-wave pahse shifts for JM and JMR for an exponential potential ($V_0 =$ 80 MeV, $a = 1.0$ fm). The matching point is $N = 25$ in both calculations. The Regularization parameter is $b_0 = 0.6$ fm

potential considerations such as [24], [25], will carry the merits of both approaches, and can dramatically reduce the size of potential matrices.

6 Conclusions

We have investigated the regularization procedure of the JM method. We have shown that the differences between the regularized Neumann and Neumann functions can have a significant impact on the convergence of the JM method. We have proposed a new approach to the regularization. It is computationally only slightly more involved but it yields significant improvement in the convergence of the phase shifts obtained by the method. We have explicitly demonstrated the latter point through calculations on several model potentials.

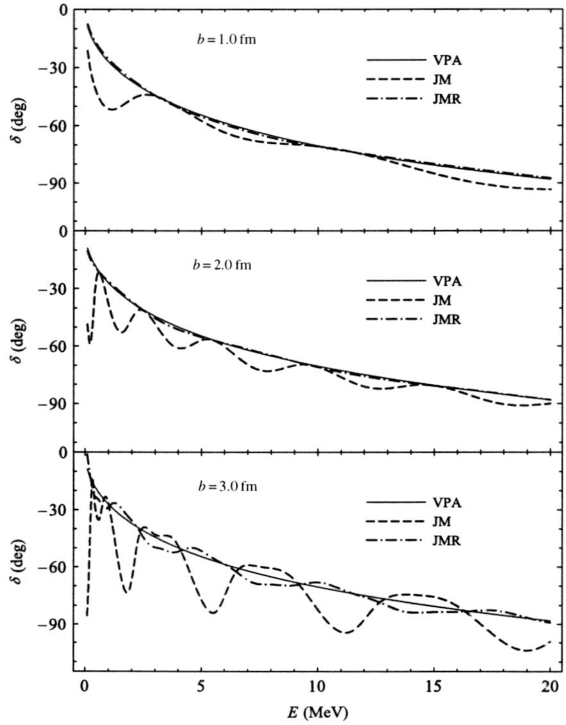

Fig. 6 Comparison of s-wave pahse shifts for JM and JMR for a square-well potential ($V_0 = 80\,\text{MeV}$, $a = 1.0\,\text{fm}$). The matching point is $N = 25$ in both calculations. The Regularization parameter is $b_0 = 0.6\,\text{fm}$

Appendix A: Overlap

To calculate the overlap between oscillator functions of different widths we use the technique of generating functions. It is known that the function

$$\Phi_l(r \mid \varepsilon, b) = \sqrt{\frac{2}{\Gamma(l+3/2)} \frac{1}{b^{3/2}}} (1+\varepsilon)^{-(l+3/2)} \left(\frac{r}{b}\right)^{l+3/2} \exp\left\{-\frac{1}{2}\frac{1-\varepsilon}{1+\varepsilon}\left(\frac{r}{b}\right)^2\right\} \tag{A1}$$

generates a complete set of the oscillator functions of width b

$$\Phi_l(r \mid \varepsilon, b) = \sum_{n=0}^{\infty} \frac{\varepsilon^n}{\mathcal{N}_{nl}} \Phi_{nl}(r, b), \tag{A2}$$

where

$$\mathcal{N}_{nl} = \sqrt{\frac{n!\Gamma(j)}{\Gamma(n+j)}}, \quad j = l + 3/2 \tag{A3}$$

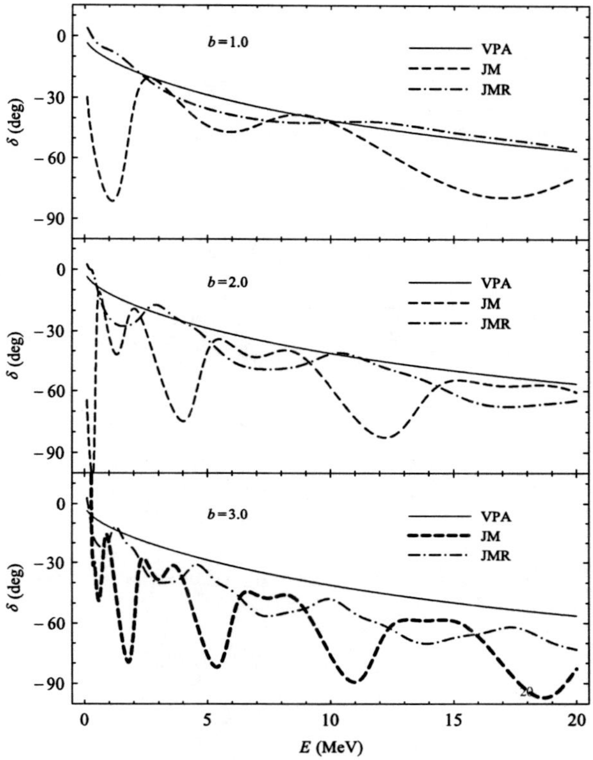

Fig. 7 Comparison of s-wave pahse shifts for JM and JMR for a Yukawa potential ($V_0 = 80\,\mathrm{MeV}$, $a = 1.0\,\mathrm{fm}$). The matching point is $N = 25$ in both calculations. The Regularization parameter is $b_0 = 0.6\,\mathrm{fm}$

Using the relation

$$\frac{\mathcal{N}_{nl}}{n!}\left[\left(\frac{d}{d\varepsilon}\right)^n \Phi_l(r\,|\,\varepsilon, b)\right]|_{\varepsilon=0} = \Phi_{nl}\,(r\,|\,b) \tag{A4}$$

on matrix elements of the generator state, one can extract the oscillator function matrix elements for definite quantum number n.

It is straightforward to calculate the overlap of two generating functions. One finds

$$\langle\Phi_l(\varepsilon, b)\,|\,\Phi_l(\widetilde{\varepsilon}, \widetilde{b})\rangle = \left[\sqrt{\gamma}\frac{z}{\Delta}\right]^{\,j}, \tag{A5}$$

where

$$\gamma = \left(\frac{\widetilde{b}}{b}\right)^2, \qquad z = \frac{2}{(1+\gamma)}, \quad \alpha = \frac{(1-\gamma)}{(1+\gamma)}, \quad \Delta = 1 - \alpha\varepsilon + \alpha\widetilde{\varepsilon} - \varepsilon\widetilde{\varepsilon} \tag{A6}$$

Closer inspection of the expression will reveal that it is actually symmetric in the width parameters as it should be. By applying relation (A4) to this overlap we obtain the required overlap of the oscillator functions:

$$\frac{N_{nl}}{n!}\frac{\mathcal{N}_{\widetilde{n}l}}{\widetilde{n}!}\left[\left(\frac{d}{d\varepsilon}\right)^{n}\left(\frac{d}{d\widetilde{\varepsilon}}\right)^{\widetilde{n}}\left\langle\Phi_{l}(\varepsilon,b)\mid\Phi_{l}(\widetilde{\varepsilon},\widetilde{b})\right\rangle\right]_{\varepsilon=\widetilde{\varepsilon}=0}=\left\langle\Phi_{nl}(b)\mid\Phi_{\widetilde{n}l}(\widetilde{b})\right\rangle \quad (A7)$$

It can be represented in the following form:

$$\begin{aligned}
&\left\langle\Phi_{nl}(|\,b)\mid\Phi_{\widetilde{n}l}(|\,\widetilde{b})\right\rangle\\
&=N_{nl}\mathcal{N}_{\widetilde{n}l}\left(z\sqrt{\gamma}\right)^{j}\alpha^{n+\widetilde{n}}\frac{\Gamma(n+j)\Gamma(\widetilde{n}+j)}{n!\widetilde{n}!\Gamma(j)\Gamma(j)}\\
&\sum_{k=0}^{\min(n,\widetilde{n})}\frac{n!\widetilde{n}!\Gamma(j)}{k!(n-k)!(\widetilde{n}-k)!\Gamma(k+j)}\left[-\frac{\left(1-\alpha^{2}\right)}{\alpha^{2}}\right]^{k}\\
&=N_{nl}\mathcal{N}_{\widetilde{n}l}\left(z\sqrt{\gamma}\right)^{j}\alpha^{n+\widetilde{n}}\frac{\Gamma(n+j)\Gamma(\widetilde{n}+j)}{n!\widetilde{n}!\Gamma(j)\Gamma(j)}\\
&{}_{2}F_{1}\left(-n,-n\,;\,j\,;-\frac{\left(1-\alpha^{2}\right)}{\alpha^{2}}\right)
\end{aligned} \quad (A8)$$

For the particular case where one of the quantum numbers is zero one finds:

$$\begin{aligned}
\left\langle\Phi_{nl}(|\,b)\mid\Phi_{\widetilde{n}l}(|\,\widetilde{b})\right\rangle&=\mathcal{N}_{nl}\left(z\sqrt{\gamma}\right)^{j}\alpha^{n+\widetilde{n}}\frac{\Gamma(n+j)}{n!\Gamma(j)}\\
&=\left(\frac{2b\widetilde{b}}{b^{2}+\widetilde{b}^{2}}\right)^{j}\sqrt{\frac{\Gamma(n+j)}{n!\Gamma(j)}}\left(\frac{b^{2}-\widetilde{b}^{2}}{b^{2}+\widetilde{b}^{2}}\right)^{n}.
\end{aligned} \quad (A9)$$

References

1. E. J. Heller and H. A. Yamani, "J-matrix method: Application to s-wave electron-hydrogen scattering," *Phys. Rev.*, vol. **A9**, pp. 1209–1214, 1974.
2. W. P. Reinhardt, D. W. Oxtoby, and T. N. Rescigno, "Computation of elastic scattering phase shifts via analytic continuation of fredholm determinants constructed using an L^2 basis," *Phys. Rev. Lett.*, vol. **28**, p. 401403, 1972.
3. D. A. Konovalov and I. E. McCarthy, "Convergent J-matrix calculation of electron-helium resonances," *J. Phys. B Atom. Mol Phys.*, vol. 28, pp. L139–L145, Mar. 1995.
4. J. T. Broad and W. P. Reinhardt, 'One- and two-electron photoejection from H^-: A multi-channel J-matrix calculation," *Phys. Rev. A*, vol. 14, pp. 2159–2173, Dec. 1976.
5. V. A. Knyr and L. Y. Stotland, "The three-body problem and J-matrix method," *Phys. Atom. Nucl.*, vol. 56, pp. 886–889, July 1993.
6. V. A. Knyr, V. V. Nasyrov, and Y. V. Popov, "Application of the J-matrix method for describing the $(e, 3e)$ reaction in the Helium Atom," *J. Exp. Theor. Phys.*, vol. 92, pp. 789–794, May 2001.
7. I. Cacelli, R. Moccia, and A. Rizzo, "Gaussian-type-orbital basis sets for the calculation of continuum properties in molecules The differential photoionization cross section of molecular nitrogen," *Phys. Rev. A*, vol. 57, pp. 1895–1905, Mar. 1998.
8. V. S. Vasilevsky and I. Y. Rybkin, "Astrophysical S-factor of the $t(t, 2n)^4He\ {}^3H({}^3H, 2n)^4He$ reactions," *Sov. J. Nucl. Phys.*, vol. **50**, p. 411, 1989.

9. V. S. Vasilevsky, I. Y. Rybkin, and G. F. Filippov, "Theoretical analysis of the mirror reactions $d(d,n)^3He$ and $d(d,p)^3H$ and resonance states of the ^4He nucleus," *Sov. J. Nucl. Phys.*, vol. **51**, p. 71, 1990.
10. V. Vasilevsky, G. Filippov, F. Arickx, J. Broeckhove, and P. V. Leuven, "Coupling of collective states in the continuum: An application to ^{4}He," *J. Phys. G: Nucl. Phys.*, vol. **G18**, pp. 1227–1242, 1992.
11. V. S. Vasilevsky, A. V. Nesterov, F. Arickx, and P. V. Leuven, "Three-cluster model of six-nucleon system," *Phys. Atom. Nucl.*, vol. **60**, pp. 343–349, 1997.
12. V. S. Vasilevsky, A. V. Nesterov, F. Arickx, and J. Broeckhove, "Algebraic model for scattering in three-s-cluster systems. I. Theoretical background," *Phys. Rev.*, vol. **C63**, p. 034606, 2001.
13. V. S. Vasilevsky, A. V. Nesterov, F. Arickx, and J. Broeckhove, "Algebraic model for scattering in three-s-cluster systems. II. Resonances in three-cluster continuum of ^6He and ^6Be," *Phys. Rev.*, vol. **C63**, p. 034607, 2001.
14. V. S. Vasilevsky, A. V. Nesterov, F. Arickx, and J. Broeckhove, "S factor of the $^3H(^3H,2n)^4He$ and $^3He(^3He,2p)^4He$ reactions using a three-cluster exit channel," *Phys. Rev.*, vol. **C63**, p. 064604, 2001.
15. A. Sytcheva, J. Broeckhove, F. Arickx, and V. S. Vasilevsky, "Influence of monopole and quadrupole channels on the cluster continuum of the lightest p-shell nuclei," *J. Phys. G Nucl. Phys.*, vol. 32, pp. 2137–2155, Nov. 2006.
16. V. Vasilevsky, F. Arickx, J. Broeckhove, and T. Kovalenko, "A microscopic model for cluster polarization, applied to the resonances of ^7Be and the reaction $^6Li\left(p,^3He\right)^4He$," in *Proc. of the 24 International Workshop on Nuclear Theory, Rila Mountains, Bulgaria, June 20–25, 2005* (S. Dimitrova, ed.), pp. 232–246, Sofia, Bulgaria: Heron Press, 2005.
17. F. Arickx, J. Broeckhove, P. Hellinckx, V. Vasilevsky, and A. Nesterov, "A three-cluster microscopic model for the ^5H nucleus," in *Proc. of the 24 International Workshop on Nuclear Theory, Rila Mountains, Bulgaria, June 20–25, 2005* (S. Dimitrova, ed.), pp. 217–231, Sofia, Bulgaria: Heron Press, 2005.
18. V. Vasilevsky, F. Arickx, J. Broeckhove, and V. Romanov, "Theoretical analysis of resonance states in ^4H, ^4He and ^4Li above three-cluster threshold," *Ukr. J. Phys.*, vol. **49**, no. 11, pp. 1053–1059, 2004.
19. Z. Papp and W. Plessas, "Coulomb-Sturmian separable expansion approach: Three-body Faddeev calculations for Coulomb-like interactions," *Phys. Rev. C*, vol. 54, pp. 50–56, July 1996.
20. G. Filippov and Y. Lashko, "Structure of light neutron-rich nuclei and nuclear reactions involving these nuclei," *El. Chast. Atom. Yadra*, vol. 36, no. 6, pp. 1373–1424, 2005.
21. G. F. Filippov, Y. A. Lashko, S. V. Korennov, and K. Katō, "$^6He+^6He$ clustering of ^{12}Be in a microscopic algebraic approach," *Few-Body Systems*, vol. 34, pp. 209–235, 2004.
22. S. Korennov, G. Filippov, Y. Lashko, and K. Katō, "Microscopic Hamiltonian of ^9Li and ^{10}Be and the $SU(3)$ basis," *Progr. Theor. Phys. Suppl.*, vol. 146, pp. 579–580, 2002.
23. J. M. Bang, A. I. Mazur, A. M. Shirokov, Y. F. Smirnov, and S. A. Zaytsev, "P-matrix and J-matrix approaches: Coulomb asymptotics in the harmonic oscillator representation of scattering theory," *Ann. Phys.*, vol. 280, pp. 299–335, Mar. 2000.
24. V. S. Vasilevsky and F. Arickx, "Algebraic model for quantum scattering: Reformulation, analysis, and numerical strategies," *Phys. Rev.*, vol. **A55**, pp. 265–286, 1997.
25. W. Vanroose, J. Broeckhove, and F. Arickx, "Modified J-matrix method for scattering," *Phys. Rev. Lett.*, vol. 88, p. 10404, Jan. 2002.
26. J. Broeckhove, F. Arickx, W. Vanroose, and V. S. Vasilevsky, "The modified J-matrix method for short range potentials," *J. Phys. A Math. Gen.*, vol. 37, pp. 7769–7781, Aug. 2004.
27. E. J. Heller and H. A. Yamani, "New L^2 approach to quantum scattering: Theory," *Phys. Rev.*, vol. **A9**, pp. 1201–1208, 1974.
28. Y. I. Nechaev and Y. F. Smirnov, "Solution of the scattering problem in the oscillator representation," *Sov. J. Nucl. Phys.*, vol. **35**, pp. 808–811, 1982.
29. H. A. Yamani and L. Fishman, "J-matrix method: Extensions to arbitrary angular momentum and to Coulomb scattering," *J. Math. Phys.*, vol. **16**, pp. 410–420, 1975.

30. R. G. Newton, *Scattering Theory of Waves and Particles*. New-York: McGraw-Hill, 1966.
31. P. M. Morse and H. Feshbach, *Methods of Theoretical Physics*. New-York: McGraw-Hill Book Co., Inc., 1953.
32. M. Abramowitz and A. Stegun, *Handbook of Mathematical Functions*. New-York: Dover Publications, Inc., 1972.
33. G. F. Filippov and I. P. Okhrimenko, "Use of an oscillator basis for solving continuum problems," *Sov. J. Nucl. Phys.*, vol. **32**, pp. 480–484, 1981.
34. F. Calogero, *Variable Phase Approach to Potential Scattering*. New-York and London: Academic Press, 1967.
35. V. V. Babikov, *Phase Function Method in Quantum Mechanics*. Moscow: Nauka, 1976.

Part III
Applications in Atomic Physics

The J-Matrix Method: A Universal Approach to Description of Ionization of Atoms

V.A. Knyr, S.A. Zaytsev, Yu.V. Popov and A. Lahmam-Bennani

Abstract A version of the J-matrix method for solving numerically the three-body Faddeev-Merkuriev differential equations is proposed. This method allows to obtain the correlated three charged particle continuum wave function in which two-body subdomains are considered correctly. This function is used for calculations of the fully resolved absolute differential cross sections for double ionization of helium by electron impact. Results are rather close to the experiment.

1 Introduction

The most difficult problem of a three-charged-particle scattering theory is the correct description of states with free fragments. This problem becomes rather crucial in the case of numerical calculations. On the one hand, one must use convenient and accurate algorithms to obtain a reliable result. On the other hand, the asymptotic behavior of such states is rather complicated to circumscribe it properly within any numerical scheme.

In this chapter we consider a theoretical description of the electron impact double ionization experiments on helium which were carried out in the so-called dipolar coplanar geometry [1, 2]. Such experiments are characterized by small momentum transfer and large incident and scattered electron energies, $E_i \approx E_s \approx$ 5–8 keV, while the energies of the emitted electrons are of the order of a few eV. These kinematical conditions allow to restrict ourselves to the first Born approximation (FBA) and to describe the fast electron as a plane wave. The five-fold differential cross section (5DCS) of the $\mathrm{He}(e, 3e)\mathrm{He}^{++}$ reaction is given by (atomic units are used $m_e = e = \hbar = 1$)

$$\sigma^{(5)} \equiv \frac{d^5\sigma}{d\Omega_s dE_1\, d\Omega_1 dE_2\, d\Omega_2} = \frac{4 p_s k_1 k_2}{p_i Q^4} \times \\ \times \left| \left\langle \Psi^{(-)}(\mathbf{k}_1, \mathbf{k}_2) \right| \exp(i\mathbf{Q}\mathbf{r}_1) + \exp(i\mathbf{Q}\mathbf{r}_2) - 2 \left| \Psi_0 \right\rangle \right|^2, \tag{1}$$

V.A. Knyr
Pacific National University, Khabarovsk, Russia
e-mail: knyr@fizika.khstu.ru

A.D. Alhaidari et al. (eds.), *The J-Matrix Method*, 137–143.

where $(E_i, \mathbf{p}_i)$, $(E_s, \mathbf{p}_s)$, $(E_1, \mathbf{k}_1)$ and $(E_2, \mathbf{k}_2)$ are the energies and momenta of, respectively, the incident (fast), the scattered (fast), and the both ejected slow electrons; $\mathbf{Q} = \mathbf{p}_i - \mathbf{p}_s$ is the transferred momentum; $\Psi_0(\mathbf{r_1}, \mathbf{r_2})$ is the ground state helium wave function, and $\Psi^{(-)}(\mathbf{k}_1, \mathbf{k}_2; \mathbf{r_1}, \mathbf{r_2})$ is the final state wave function.

The simplest wave function $\Psi^{(-)}$ is the so-called 3C function [3] which possesses the correct asymptotic behavior at large mutual distances between the outgoing particles

$$\Psi^{-}_{3C}(\mathbf{k}_1, \mathbf{k}_2; \mathbf{r}_1, \mathbf{r}_2) = e^{-i\mathbf{k}_{12}\mathbf{r}_{12}}\varphi^{-}_1(\mathbf{k}_1, \mathbf{r}_1)\varphi^{-}_2(\mathbf{k}_2, \mathbf{r}_2)\varphi^{-}_{12}(\mathbf{k}_{12}, \mathbf{r}_{12}), \tag{2}$$

and φ^- is the corresponding outgoing Coulomb wave. Calculations of the 5DCS with the use of this function yield fully satisfactory results at $E_1 = E_2 = 10$ eV for a wide choice of probing initial wave functions [4–6]. Nevertheless, this final state wave function is considered to be rather crude. Different approaches to improving the 3C function [7–9] are not abundant, as well as the attempts to improve the asymptotic behavior itself [10–12]. The numerical schemes become rather complicated in these cases, and the advantages are not obvious.

Among others, we would like to mention the method of hyperspherical harmonics [13], the coupled – channel method (see, for example, [14]), and the method of pseudostates [15] which are more preferable for numerical calculations.

The method of hyperspherical harmonics is a convenient tool for describing a truly three-particle scattering process. However, in atomic physics the binary channels can not be neglected, and they are hardly described in terms of the hyperspherical basis.

The coupled – channel method is a more or less effective one when studying the scattering of a third particle by a bound pair at the energies below a pair-breakup threshold. This method ignores the channels involving the breakup of the pair and fails to describe the truly three-body scattering and exchange processes.

The method of pseudostates can be regarded as an improved coupled – channel method. It makes it possible to efficiently take into account the three-body breakup channels when considering the elastic and inelastic scattering of a particle by a bound pair. For energies above the three-body breakup threshold the method of pseudostates fails to adequately describe the features of reactions because it gives rise to spurious resonances at energies close to the energies of pseudostates.

A combination of the coupled – channel method and the method of pseudostates is presently well known as the convergent close-coupling method (CCC) [16]. Calculations of the 5DCS (1) were performed within this numerical approach [1], however they showed the necessity of introducing large fitting factors to compare the theory and the experiment on an absolute scale.

A new method for solving a three-body problem was formulated in [17], based on the J-matrix formalism. It permits a combined description of both discrete and continuous spectra of quantum systems. This method improves the coupled – channel method by accurately accounting for the whole two particle spectrum. Later on, this approach was extended to the case of long-range Coulomb potentials [18]. Our method is close to that described in the papers [19, 20] and is based on the Faddeev–Merkuriev integral equations [21]. These equations are solved in terms of Coulomb–Sturmian separable expansions (see, for instance, [22] and references therein), which is equivalent to the J-matrix method.

Below a brief theoretical description of the J-matrix method is given for the most difficult case of the (e,3e) scattering.

2 Theory

The Hamiltonian of a three-body system is of the form

$$H = H_0 + \sum_{\alpha=1}^{3} V_\alpha^C(x_\alpha), \tag{3}$$

where H_0 is the kinetic energy operator

$$H_0 = -\Delta_{x_\alpha} - \Delta_{y_\alpha}, \tag{4}$$

and

$$V_\alpha^C(x_\alpha) = \frac{Z_\alpha}{x_\alpha}. \tag{5}$$

The couple $(\mathbf{x}_\alpha, \mathbf{y}_\alpha)$ stands for the set of the Jacobi coordinates [21]

$$\begin{aligned} \mathbf{x}_\alpha &= \tau_\alpha(\mathbf{r}_\beta - \mathbf{r}_\gamma), \\ \mathbf{y}_\alpha &= \mu_\alpha\left(\mathbf{r}_\alpha - \frac{m_\beta \mathbf{r}_\beta + m_\gamma \mathbf{r}_\gamma}{m_\beta + m_\gamma}\right), \end{aligned} \tag{6}$$

where m_i are the particle masses and

$$\tau_\alpha = \sqrt{2\frac{m_\beta m_\gamma}{m_\beta + m_\gamma}}, \qquad \mu_\alpha = \sqrt{2m_\alpha\left(1 - \frac{m_\alpha}{m_1 + m_2 + m_3}\right)}. \tag{7}$$

The corresponding conjugate momenta are

$$\mathbf{k}_\alpha = \frac{1}{\tau_\alpha}\frac{m_\gamma \mathbf{q}_\beta - m_\beta \mathbf{q}_\gamma}{m_\beta + m_\gamma}, \qquad \mathbf{p}_\alpha = \frac{1}{\mu_\alpha}\frac{(m_\beta + m_\gamma)\mathbf{q}_\alpha - m_\alpha(\mathbf{q}_\beta + \mathbf{q}_\gamma)}{m_1 + m_2 + m_3}, \tag{8}$$

where $\mathbf{q}_\alpha$ is the momentum of a particle with the radius-vector $\mathbf{r}_\alpha$ ($\mathbf{q}_1 + \mathbf{q}_2 + \mathbf{q}_3 = 0$).

The interaction V_α can be split into the short- and long-range parts, $V_\alpha^{(s)}$ and $V_\alpha^{(l)}$ respectively [21],

$$V_\alpha^{(s)}(x_\alpha, y_\alpha) = V_\alpha(x_\alpha)\zeta_\alpha(x_\alpha, y_\alpha), \; V_\alpha^{(l)}(x_\alpha, y_\alpha) = V_\alpha(x_\alpha)[1 - \zeta_\alpha(x_\alpha, y_\alpha)] \tag{9}$$

with a "separation" function of the form [23]

$$\zeta(x_\alpha, y_\alpha) = 2/\{1 + \exp[(x_\alpha/x_0)^\nu/(1 + y_\alpha/y_0)]\} \tag{10}$$

where $\nu > 2$. Thus, the function $V_\alpha^{(s)}$ decreases rather rapidly in the truly three-body asymptotic domain Ω_0 and coincides with the initial potential in the two-body asymptotic domain Ω_α ($x_\alpha \ll y_\alpha$).

In the general case of different particles, the total wave function is represented as a sum of three components ψ_α. For the system (e, e, He^{++}), particles 1 and 2 (electrons) are identical, and the function $\Psi^{(-)}$ reduces to the sum of two components ψ_1 and ψ_2 [23]. They are related to each other as follows: $\psi_2 = gP_{12}\psi_1$ ($g = +1$ and $g = -1$ for singlet and triplet spin states, respectively, P_{12} is the permutation operator). Taking into account the spatial symmetry of the total wave function, we obtain a single equation for the component ψ_1 [23]:

$$[H_0 + V_1(x_1) + V_3(x_3) + V_2^{(l)}(x_2) - E]\psi_1(\mathbf{x}_1, \mathbf{y}_1) = -gV_1^{(s)}P_{12}\psi_1(\mathbf{x}_1, \mathbf{y}_1). \tag{11}$$

The wave function $\Psi^{(-)}$ of the system (e, e, He^{++}) can be presented by the following expansion

$$\Psi^{(-)} = \frac{2}{\pi}\frac{1}{p_0 k_0}\frac{(1+gP_{12})}{\sqrt{2}} \sum_{L\,\ell_0\,\lambda_0, m_0\,\mu_0} (\ell_0\, m_0\, \lambda_0\, \mu_0|L\, M)\times$$
$$\times\, i^{\ell_0+\lambda_0}\, e^{-i(\sigma_{\ell_0}+\sigma_{\lambda_0})} Y^*_{\ell_0\, m_0}(\widehat{k_0}) Y^*_{\lambda_0\,\mu_0}(\widehat{p_0})\, \psi^{L\,M}_{\ell_0\,\lambda_0}. \tag{12}$$

The spatial part $\psi^{L\,M}_{\ell_0\,\lambda_0}(\mathbf{x}, \mathbf{y})$ of the component ψ_1 ($\mathbf{x} \equiv \mathbf{x}_1$, $\mathbf{y} \equiv \mathbf{y}_1$) in (12) can be decomposed in turn into the set on the bispherical functions and the square-integrable Coulomb-Sturmian functions

$$\psi^{L\,M}_{\ell_0\,\lambda_0} = \sum_{\ell,\lambda,n,\nu} C^{L\,(\ell\,\lambda)}_{n\,\nu}(E)\; |n\,\ell\,\nu\,\lambda;\; LM\rangle\,, \tag{13}$$

$$|n\,\ell\,\nu\,\lambda;\; LM\rangle = \frac{\phi_n^\ell(x)\,\phi_\nu^\lambda(y)}{xy}\,\mathcal{Y}^{LM}_{\ell\,\lambda}(\widehat{x}, \widehat{y}), \tag{14}$$

$$\mathcal{Y}^{LM}_{\ell\,\lambda}(\widehat{x}, \widehat{y}) = \sum_{m\,\mu}(\ell\, m\, \lambda\, \mu|L\, M)\, Y_{\ell\, m}(\widehat{x})\, Y_{\lambda\,\mu}(\widehat{y}). \tag{15}$$

The Laguerre basis functions ϕ_n^ℓ [24] are used in (13), (14):

$$\phi_n^\ell(x) = \left[(n+1)_{(2\ell+1)}\right]^{-1/2}(2bx)^{\ell+1}e^{-bx}L_n^{2\ell+1}(2bx), \tag{16}$$

with b being the basis parameter, a suitable choice of which affects the rate of convergence of the numerical results.

The coefficients $C_{n\,\nu}^{L\,(\ell\,\lambda)}$ in turn can be decomposed into the set

$$C_{n\,\nu}^{L\,(\ell\,\lambda)}(E) = \sum_j \mathcal{S}_{n\ell}^{(j)}(Z_1)\, a_\nu^{\ell\,\lambda}(i\kappa_j) + \frac{2}{\pi}\int_0^\infty dk\, \mathcal{S}_{n\ell}(k, Z_1)\, a_\nu^{\ell\,\lambda}(k), \tag{17}$$

by functions $\mathcal{S}_{n\ell}^{(j)}(Z_1)$ and $\mathcal{S}_{n\ell}(k, Z_1)$. They are the matrix elements between the sine-like spectral functions of the Hamiltonian

$$h_x = -\Delta_x + \frac{Z_1}{x} \tag{18}$$

describing the subsystem (2, 3), and the basis functions (16). Note, that $Z_1 = -2\sqrt{2}$ in the limit $m_3 \to \infty$. More precise definitions may be found in [18].

In an earlier work (see Eq. (33) of [18]), the system of algebraic equations for the coefficients $C_{n\,\nu}^{L\,(\ell\,\lambda)}$ was obtained for (e,2e) reactions. In the case of double ionization the analogous system is expressed as

$$C_{n\,\nu}^{L\,(\ell\,\lambda)}(E) = \delta_{(\ell\,\lambda)(\ell_0\,\lambda_0)}\, \mathcal{S}_{n\ell_0}(k_0, Z_1)\, \mathcal{S}_{\nu\lambda_0}(p_0, Z_{11}) + \sum_{n',\,\nu',\,n''\,\nu''=0}^{N-1}$$
$$\left(\sum_{\ell'',\,\lambda''} \left[\int dk\, \mathcal{S}_{n\ell}(k, Z_1)\, \mathcal{S}_{n'\,\ell}(k, Z_1)\, \mathcal{G}_{\nu\,\nu'}^{\lambda(-)}(p, Z_{11}) \right] V_{n'\,\nu',\,n''\,\nu''}^{L(\ell\,\lambda)(\ell''\,\lambda'')}\, C_{n''\,\nu''}^{L\,(\ell''\,\lambda'')}(E). \right) \tag{19}$$

In (19) $\int dk$ means the summation over the discrete index j and integration over the continuous index (see (17)) within the domain $k^2 + p^2 = E$. Note, that the potential Z_{11}/y describes the Coulomb interaction of the particle 1 with the center-of-mass of the subsystem (2, 3), and $Z_{11} = -\sqrt{2}$ in the limit $m_3 \to \infty$.

3 Results

The helium ground-state wave function Ψ_0 is obtained as a result of diagonalization of the matrix (3) which was calculated in the basis (16). Here we set $\ell_{max} = 3$ and $n_{max} = \nu_{max} = 15$. Choosing the basis parameter $b_0 = 1.193$ yields the value $E_0 = -2.903256$ for the ground-state energy.

Calculating the final-state wave function $\Psi^{(-)}$ (12), we restrict ourselves to the maximum value of the total orbital angular momentum, $L_{max} = 2$, and $\ell,\ \lambda \le 3$. The parameters x_0, y_0 and ν of the "separating" function (10) are as follows: $x_0 = 3$, $y_0 = 20$, $\nu = 2.1$. The number $N = 20$ of the Laguerre basis functions was used.

The results for the 5DCS, $\sigma^{(5)}$, are presented in Fig. 1. (thick solid lines) in comparison with the 3C calculations [4] (dashed lines). In the experiment, the in-plane angle θ_1 of one slow electron was fixed, while the in-plane angle θ_2 of the other slow electron was varied. The energy of the scattered electron was $E_s = 5500\,\mathrm{eV}$,

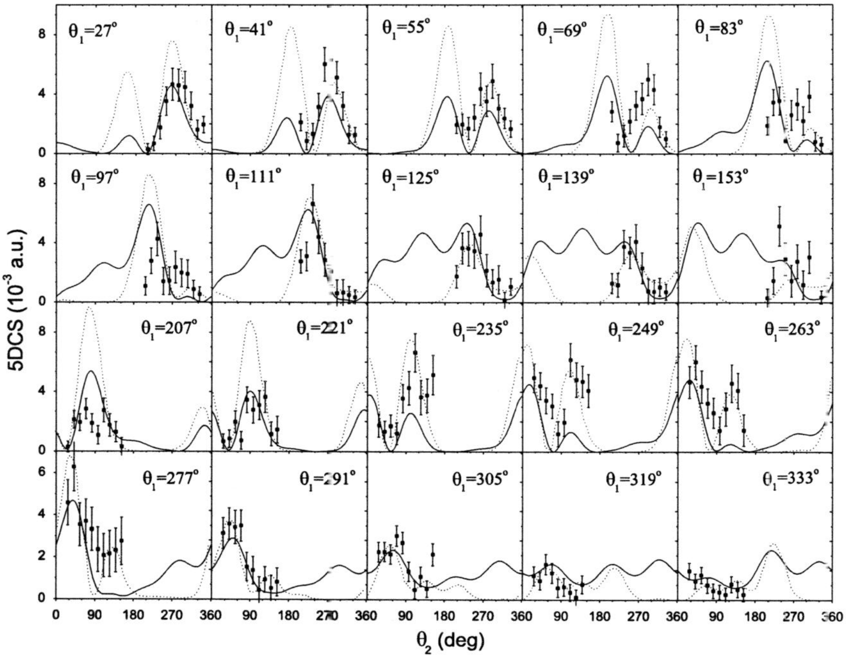

Fig. 1 The fully resolved five-fold differential cross section 5DCS of the electron-impact double ionization reaction He$(e, 3e)$He^{++}. The energy of the scattered electron is $E_s = 5500$ eV, and the energies of the slow ejected electrons are $E_1 = E_2 = 10$ eV. The scattering angle of the fast incident electron is fixed $\theta_s = 0.45°$, and the angles of the ejected electrons are θ_1 and θ_2, where one angle is fixed while the other angle varies relative to the incident electron direction. All electron velocities are disposed in the same plane. The absolute measurements are borrowed from [1]. Thick solid line: present results; dashed line: results from the 3C calculations

and its in-plane angle $\theta_s = 0.45°$ was also fixed. The energies of the slow electrons were chosen as $E_1 = E_2 = 10$ eV. It is seen that our results quite satisfactorily agree with the experimental distributions both in shape and in magnitude. The latter observation favorably distinguish our calculations from the previous results [1, 25] obtained within the analogous numerical scheme, but with the use of pseudostates.

In summary, we can conclude that the proposed numerical scheme and the performed calculations demonstrate the importance of accounting for the whole two-body spectrum. The method of pseudostates, which replaces the continuum by a finite number of states with positive energies, seems to be confronted with difficulties, especially when the resulting final-state wave function is applied to the calculation of (e,3e) reactions.

Acknowledgments This work is partially supported by the scientific program "Far East – 2008" of the Russian Foundation for Basic Research (regional grant 08-02-98501).

References

1. A. Kheifets, I. Bray, A. Lahmam-Bennani, A. Duguet, and I. Taouil, *J. Phys. B* **32**, 5047 (1999).
2. A. Lahmam-Bennani, I. Taouil, A. Duguet, M. Lecas, L. Avaldi, and J. Berakdar, *Phys. Rev. A* **59**, 3548 (1999).
3. H. Klar, *Z. Phys. D* **16**, 231 (1990).
4. S. Jones, D. H. Madison, *Phys. Rev. Lett.* **91**, 073201 (2003).
5. L. U. Ancarani, T. Montagnese, and C. Dal Cappello, *Phys. Rev. A* **70**, 012711 (2004).
6. O. Chuluunbaatar, I. V. Puzynin, P. S. Vinitsky, Yu. V. Popov, K. A. Kouzakov, and C. Dal Cappello, *Phys. Rev. A* **74**, 014703 (2006).
7. J. Berakdar, *Phys. Rev. A* **53**, 2314 (1996); ibid **54**, 1480 (1996).
8. G. Gasaneo, F. D. Colavecchia, and C. R. Garibotti, *Phys. Rev. A* **55**, 2809 (1997).
9. G. Gasaneo, F. D. Colavecchia, *Nucl. Instrum. Methods. Phys. Res. B* **192**, 150 (2002).
10. A. S. Kadyrov, A. M. Mukhamedzhanov, A. T. Steblovics, I. Bray, and F. Pirlepesov, *Phys. Rev. A* **68**, 022703 (2003).
11. A. M. Mukhamedzhanov, A. S. Kadyrov, and F. Pirlepesov, *Phys. Rev. A* **73**, 012713 (2006).
12. A. S. Kadyrov, A. M. Mukhamedzhanov, A. T. Steblovics, and I. Bray, *Phys. Rev. A* **70**, 062703 (2004).
13. L. M. Delves, *Nucl.Phys.* **9**, 391 (1959); F. T. Smith, *Phys.Rev.* **140**, 1058 (1960).
14. H. S. W. Massey, *Rev. Mod. Phys.* **28**, 199 (1956).
15. P. G. Burke, D. F. Callaher, and S. Geltman. *J. Phys. B* **2**, 1142 (1969).
16. I. Bray, *Phys. Rev. A* **49**, 1066 (1994); D. V. Fursa, I. Bray, *J. Phys. B* **30**, 5895 (1997).
17. V. A. Knyr, L. Ya. Stotland, *Phys. Atomic Nuclei* **59**, 575 (1996).
18. S. A. Zaytsev, V. A. Knyr, Yu. V. Popov, and A. Lahmam-Bennani, *Phys. Rev. A* **75**, 022718 (2007).
19. Z. Papp, C. -Y. Hu, Z. T. Hlousek, B. Konya, and S. L. Yakovlev, *Phys. Rev. A* **63**, 062721 (2001).
20. B. Konya, G. Levai, and Z. Papp, *Phys. Rev. C* **61**, 034302 (2000).
21. L. D. Faddeev, S. P. Merkuriev. *Quantum Scattering Theory for Several Particle Systems*, (Kluwer Academic Publishers, Dordrecht, 1993).
22. Z. Papp, *Phys. Rev. C* **55**, 1080 (1997).
23. A. A. Kvitsinsky, A. Wu, and C.-Yu Hu, *J. Phys. B* **28**, 275 (1995).
24. E. J. Heller, H. A. Yamani, *Phys. Rev. A* **9**, 1201 (1974); H. A. Yamani, L. Fishman, *J. Math. Phys.* **16**, 410 (1975).
25. A. Kheifets, I. Bray, *Phys. Rev. A* **69**, 050701(R) (2004).

J-Matrix Green's Operators and Solving Faddeev Integral Equations for Coulombic Systems

Z. Papp

Abstract The two-body Coulomb Hamiltonian, in the Coulomb-Sturmian basis, has an infinite symmetric tridiagonal (Jacobi) matrix structure. This allows us to construct the Green's operator in terms of ${}_2F_1$ hypergeometric function, which can be evaluated by a continued fraction. Using this two-body Coulomb Green's matrix, we developed an approximation method for solving Faddeev-type integral equations of the three-body Coulomb problem. The corresponding three-body Green's operators are calculated as a convolution integral of the two-body Coulomb Green's operators. As examples, the electron-hydrogen scattering and the resonances of the $e - Ps$ system are presented.

1 Introduction

Hamiltonians with infinite symmetric tridiagonal, i.e. Jacobi-matrix or J-matrix form have been studied extensively in the context of the L^2 approach to quantum scattering theory [1]. The Hamiltonian is represented on an appropriate L^2 basis, which is chosen in such a way that the asymptotic part of the Hamiltonian, i.e. the kinetic energy or the kinetic energy plus the Coulomb potential possesses a J-matrix form. This results in a three-term recursion relation, which can be solved analytically, yielding the expansion coefficients of both a "sine-like" $\widetilde{S}(r)$ and the "cosine-like" solution $\widetilde{C}(r)$. The J-matrix solutions $\widetilde{S}(r)$ and $\widetilde{C}(r)$ are then used to obtain the exact solution to a model scattering problem defined by approximating the potential V by its projection V^N onto the finite subspace spanned by the first N terms of the L^2 basis. Also the Green's function of a J-matrix Hamiltonian can be constructed from the coefficients of the solutions $\widetilde{S}(r)$ and $\widetilde{C}(r)$ [2].

At about the same time as the L^2 approach to quantum scattering theory was developed, a conceptually similar but technically different method was proposed [3]. This method is the potential separable expansion approach, which is based on the representation of the potential operator on a finite subset of an L^2 basis. This

Z. Papp
Department of Physics and Astronomy, California State University, Long Beach, CA 90840, USA
e-mail: zpapp@csulb.edu

A.D. Alhaidari et al. (eds.), *The J-Matrix Method*, 145–171.

turns the Lippmann–Schwinger equation into a matrix equation, which contains the matrix elements of the Green's operator as well. If the basis is chosen in a clever way, the matrix elements of the Green's operator can be evaluated analytically. This has a great value for practical applications. The concept has been elaborated in several different forms. In Refs. [3, 4] harmonic oscillator functions were used, which allowed the representation of the free Green's operator, and thus the corresponding approximation method could handle problems with free asymptotics. In Refs. [5–8], the Coulomb–Sturmian basis was applied, which allowed the inclusion of the long-range Coulomb potential into the Green's operator, and thus led to the exact treatment of the Coulomb asymptotics. This latter approach has also been extended to solve the three-body Coulomb problem in the Faddeev approach. So far good results have been achieved for bound- [9], resonant- [11–14], and below-breakup scattering-state problems both with repulsive [10] and attractive [15] Coulomb forces.

On the level of physical assumptions the L^2 approach to scattering and the method of separable expansion of the potential on an L^2 basis are equivalent in the same sense as the Schrödinger equation is equivalent to the Lippmann–Schwinger equation. The approximation of the potential V by its projection V^N is a kind of separable expansion. In the practical calculation of the matrix elements of Green's operators in the adopted basis we rely on three-term recursion relations, which results from the J-matrix structure of the underlying asymptotic Hamiltonian. In the separable expansion approach integral equations are solved and the boundary conditions are thereby automatically incorporated. This is especially useful when the boundary conditions are not known well enough, like in the three-body Coulomb problem.

The aim of the present work is to sketch the Coulomb-Sturmian separable expansion method. We show how the Coulomb-Sturmian matrix elements of the Coulomb Green's operator can be determined by making use of the J-matrix structure of the Coulomb Hamiltonian. Then, we show how the separable expansion approach works for two-body and three-body systems, i.e. how to solve Lippmann–Schwinger and Faddeev-type integral equations.

2 Coulomb-Sturmian Matrix Elements of the Coulomb Green's Operator

2.1 The Coulomb-Sturmian Basis

The radial Coulomb Hamiltonian is given by

$$h_l^{\mathrm{C}} = -\frac{\hbar^2}{2m}\left(\frac{\mathrm{d}^2}{\mathrm{d}r^2} - \frac{l(l+1)}{r^2}\right) + \frac{Z}{r}\,, \tag{1}$$

where m is the reduced mass, l is the angular momentum and Z is the strength of the Coulomb potential. The Coulomb-Sturmian (CS) functions [16] are the solutions of the Sturm–Liouville problem of the Coulomb Hamiltonian

$$\left(-\frac{\mathrm{d}^2}{\mathrm{d}r^2}+\frac{l(l+1)}{r^2}-\frac{2b(n+l+1)}{r}+b^2\right)\langle r|nl;b\rangle=0, \tag{2}$$

where b is a parameter and n is the radial quantum number. In configuration space the CS functions are given by

$$\langle r|nl;b\rangle=\sqrt{\frac{n!}{(n+2l+1)!}}\exp(-br)(2br)^{l+1}L_n^{(2l+1)}(2br)\,, \tag{3}$$

where L denotes the Laguerre polynomials [17], in momentum space they also have a nice analytic form

$$\langle p|nl;b\rangle=\frac{2^{l+3/2}l!(n+l+1)\sqrt{n!}}{\sqrt{\pi}(n+2l+1)!}\frac{(2bp)^{l+1}}{(p^2+b^2)^{2l+2}}G_n^{(l+1)}\left(\frac{p^2-b^2}{p^2+b^2}\right), \tag{4}$$

where G stands for the Gegenbauer polynomials [17]. By defining the functions $\langle r|\widetilde{nl;b}\rangle\equiv\langle r|nl;b\rangle/r$ the orthogonality and completeness properties of the Coulomb-Sturmian functions take the forms

$$\langle\widetilde{n'l;b}|nl;b\rangle=\langle n'l;b|\widetilde{nl;b}\rangle=\delta_{nn'} \tag{5}$$

and

$$\mathbf{1}=\lim_{N\to\infty}\sum_{n=0}^{N}|\widetilde{nl;b}\rangle\langle nl;b|=\lim_{N\to\infty}\sum_{n=0}^{N}|nl;b\rangle\langle\widetilde{nl;b}|. \tag{6}$$

2.2 The Coulomb Green's Operator

We denote the Coulomb Green's operator as

$$g_l^{\mathrm{C}}(z)=(z-h_l^{\mathrm{C}})^{-1} \tag{7}$$

and our aim is to determine its CS matrix elements $\underline{g}^{\mathrm{C}}_{nn'}=\langle\widetilde{nl;b}|g_l^{\mathrm{C}}|\widetilde{n'l;b}\rangle$. Starting with the defining equation

$$(z-h_l^{\mathrm{C}})g_l^{\mathrm{C}}=\mathbf{1}, \tag{8}$$

applying it to the ket $|\widetilde{jl;b}\rangle$

$$(z-h_l^{\mathrm{C}})g_l^{\mathrm{C}}|\widetilde{jl;b}\rangle=|\widetilde{jl;b}\rangle, \tag{9}$$

inserting the completeness relation between $z - h_l^{\mathrm{C}}$ and g_l^{C}, and, finally, by multiplying from the left by the bra $\langle i'; b|$ we get

$$\sum_{i'=0}^{\infty} \langle il; b|(z - h_l^{\mathrm{C}})|i'l; b\rangle \langle \widetilde{i'l; b}|g_l^{\mathrm{C}}|\widetilde{jl; b}\rangle = \delta_{ij}. \tag{10}$$

Using Eqs. (2) and (5) we can easily verify that the matrix $\underline{J}_{ii'}^{\mathrm{C}} = \langle il; b|(z - h_l^{\mathrm{C}}) |i'l; b\rangle$ possesses an infinite symmetric tridiagonal (Jacobi) structure, i.e. all elements are zero, except for the following

$$J_{ii}^{\mathrm{C}} = 2(i + l + 1)(k^2 - b^2)\frac{\hbar^2}{4mb} - Z, \tag{11}$$

$$J_{ii-1}^{\mathrm{C}} = -[i(i + 2l + 1)]^{1/2}(k^2 + b^2)\frac{\hbar^2}{4mb} \tag{12}$$

and

$$J_{ii+1}^{\mathrm{C}} = -[(i + 1)(i + 2l + 2)]^{1/2}(k^2 + b^2)\frac{\hbar^2}{4mb}, \tag{13}$$

where $k = (2mz/\hbar^2)^{1/2}$. So, the infinite sum in Eq. (10) reduces only to three terms and we arrive at a three-term recursion relation for the matrix elements $g_{ij}^{\mathrm{C}} = \langle \widetilde{il; b}|g_l^{\mathrm{C}}|\widetilde{jl; b}\rangle$:

$$J_{ii-1}^{\mathrm{C}} g_{i-1j}^{\mathrm{C}} + J_{ii}^{\mathrm{C}} g_{ij}^{\mathrm{C}} + J_{ii+1}^{\mathrm{C}} g_{i+1j}^{\mathrm{C}} = \delta_{ij}. \tag{14}$$

For the $i = j = 0$ case Eq. (10) takes the form

$$J_{01}^{\mathrm{C}} g_{10}^{\mathrm{C}} + J_{00}^{\mathrm{C}} g_{00}^{\mathrm{C}} = 1. \tag{15}$$

This recursion relation is readily computable if g_{00}^{C} is known. Then we can calculate g_{10}^{C} from Eq. (15), and then g_{i0}^{C} using Eq. (14). By interchanging the indices in Eq. (14) we get another recursion relation, which provides us with all the g_{ij}^{C} elements, starting form g_{i0}^{C}. The g_{00}^{C} matrix element can be given in a closed analytic form [5, 8]

$$\underline{g}_{00}^{C} = \frac{4mb}{\hbar^2}\frac{1}{(b - \mathrm{i}k)^2}\frac{1}{l + \mathrm{i}\gamma + 1}\, {}_2F_1\left(-l + \mathrm{i}\gamma, 1; l + \mathrm{i}\gamma + 2, \left(\frac{b + \mathrm{i}k}{b - \mathrm{i}k}\right)^2\right), \tag{16}$$

where $\gamma = Zm/(\hbar^2 k)$ and ${}_2F_1$ is the hypergeometric function. For those cases where the first or the second index of ${}_2F_1$ is equal to unity, there exists a continued fraction representation [18], which is very efficient in practical calculations. It should also be noted that if $\gamma = 0$, i.e. for the free Green's operator, the ${}_2F_1$ hypergeometric function reduces to a simple polynomial of order l.

It turns out, however, that the forward evaluation of the recursion relation converges only for positive energies. Therefore, we construct another method which uses the recursion relation in a different way and does not need g^{C}_{00} as a starting value. From the theory of special matrices we know that the inverse of a Jacobi-matrix has the following structure [19]

$$g^{\mathrm{C}}_{ij} = \begin{cases} p_i q_j, & \text{if} \quad i \le j \\ p_j q_i, & \text{if} \quad j \le i \end{cases}, \tag{17}$$

with some p and q arrays. Therefore, for $j \le N$ and $i \le N$ we can write Eq. (10) as

$$\begin{aligned} \delta_{ij} &= \sum_{i'=0}^{N} (J^{\mathrm{C}}_{ii'} g^{\mathrm{C}}_{i'j} - \delta_{iN}\, J^{\mathrm{C}}_{iN+1}\, g^{\mathrm{C}}_{jN+1}) \\ &= \sum_{i'=0}^{N} (J^{\mathrm{C}}_{ii'} - \delta_{iN}\, \delta_{i'N}\, J^{\mathrm{C}}_{iN+1}\, p_{N+1}/p_N) g^{\mathrm{C}}_{i'j} \\ &= \sum_{i'=0}^{N} (J^{\mathrm{C}}_{ii'} - \delta_{iN}\, \delta_{i'N}\, J^{\mathrm{C}}_{iN+1}\, g^{\mathrm{C}}_{0N+1}/g^{\mathrm{C}}_{0N}) g^{\mathrm{C}}_{i'j}, \end{aligned} \tag{18}$$

i.e. the inverse of the $N \times N$ $(g^{\mathrm{C}})^{(N)}$ Green's matrix reads

$$((g^{\mathrm{C}})^{(N)})^{-1} = J^{\mathrm{C}(N)}_{ij} - \delta_{jN}\, \delta_{iN}\, J^{\mathrm{C}}_{NN+1}\, g^{\mathrm{C}}_{0N+1}/g^{\mathrm{C}}_{0N}. \tag{19}$$

In other words, the inverse of the N-th leading submatrix of the $\infty \times \infty$ Green's matrix is given by the elements of the $N \times N$ Jacobi-matrix and the ratio of two consecutive, yet unknown, elements of g^{C}_{0i}. The only fact we know about g^{C}_{0i} is that they satisfy a homogeneous three-term recursion relation. We will show in the following that this is sufficient to determine $g^{\mathrm{C}}_{0N+1}/g^{\mathrm{C}}_{0N}$.

A three-term recursion relation is defined by

$$X_{n+1} = b_n X_n + a_n X_{n-1} \qquad n = 1, 2, 3, \ldots, \tag{20}$$

where a_n, b_n are complex numbers and $a_n \neq 0$. This recursion has two independent solutions. The $\{x_n\}$ nontrivial (i.e.$\neq \{0\}$) solution is said to be minimal [18, 20] if there exists another solution $\{y_n\}$ such that

$$\lim_{n\to\infty} x_n/y_n = 0. \tag{21}$$

The solution $\{y_n\}$ is called dominant. The minimal solution is unique, apart from a multiplicative constant.

The existence of the minimal solution is intimately related to the convergence of a continued fraction constructed from the coefficients of the recursion relation. According to Pincherle's theorem [18, 20] the following statements hold:
A: the three-term recursion (20) has a minimal solution if and only if the continued fraction

$$\frac{a_1}{b_1 + \frac{a_2}{b_2 + \frac{a_3}{b_3 + \cdots}}}, \tag{22}$$

converges,
B: if $\{x_n\}$ is a minimal solution then for $N = 0, 1, 2, \ldots,$

$$\frac{x_{N+1}}{x_N} = -\frac{a_{1+N}}{b_{1+N} + \frac{a_{2+N}}{b_{2+N} + \frac{a_{3+N}}{b_{3+N} + \cdots}}}. \tag{23}$$

The second statement asserts that the ratio of two successive elements of the minimal solution is determined by a continued fraction.

In Refs. [21, 22] we have shown that for negative energies g^{C}_{0i} is the minimal solution of Eq. (14). This is related to the square-integrability of the bound-state wave function. Therefore, for $\Re(z) < 0$

$$\underline{g}^{\mathrm{C}(N)} = [\underline{J}^{\mathrm{C}} + \delta_{jN}\,\delta_{iN}\,J^{\mathrm{C}}_{N\,N+1}\;C]^{-1}, \tag{24}$$

where

$$C = -\frac{u_N}{d_N + \frac{u_{N+1}}{d_{N+1} + \frac{u_{N+2}}{d_{N+2} + \cdots}}}, \tag{25}$$

with coefficients

$$u_i = -J^{\mathrm{C}}_{i,i-1}/J^{\mathrm{C}}_{i,i+1}, \quad d_i = -J^{\mathrm{C}}_{i,i}/J^{\mathrm{C}}_{i,i+1}. \tag{26}$$

The scattering states are not square integrable. Therefore, at positive energies, g^{C}_{i0} is not a minimal solution of Eq. (14), and consequently the continued fraction C does not converge. On the other hand Eq. (24) yields the Green's matrix on the $\Re(z) < 0$ domain of the complex z-plane and the Green's operator is an analytic function of z. So, values for another domain can be gained by analytic continuation of Eq. (24), i.e. by analytic continuation of the continued fraction C.

In our case the coefficients u_i and d_i possess the limit properties

$$u \equiv \lim_{i\to\infty} u_i = -1 \tag{27}$$

and

$$d \equiv \lim_{i\to\infty} d_i = 2(k^2 - b^2)/(k^2 + b^2). \tag{28}$$

So, the continued fraction appears as

$$C = -\cfrac{u_N}{d_N + \cfrac{u_{N+1}}{d_{N+1} + \cdots + \cfrac{u}{d + \cfrac{u}{d + \cdots}}}}. \tag{29}$$

The tail

$$w = \cfrac{u}{d + \cfrac{u}{d + \cdots}} \tag{30}$$

satisfies the implicit relation

$$w = \frac{u}{d + w}, \tag{31}$$

which has solution

$$w_\pm = (b \pm \mathrm{i}k)^2/(b^2 + k^2). \tag{32}$$

By substituting the tail of the continued fraction by its explicit analytical form $w_\pm$, we can speed up the convergence and, more importantly, turn a non-convergent continued fraction into a convergent one [18]. In fact, by using $w_\pm$ instead of the non-converging tail, we perform an analytic continuation. In Ref. [21] we have shown that w_+ provides an analytic continuation of the Green's matrix to the physical, while w_- to the unphysical Riemann-sheet. This way Eq. (25) together with Eq. (24) provide the CS basis representation of the Coulomb Green's operator on the whole complex energy plane.

In a recent work [23] we used an equivalent form for Eq. (24)

$$(\underline{g}^{\mathrm{C}(N)})^{-1} = [\underline{J}^{\mathrm{C}} - \delta_{jN}\, \delta_{iN}\, (J^{\mathrm{C}}_{NN+1})^2\, C_{N+1}], \tag{33}$$

where

$$C_N^{-1} = J^{\mathrm{C}}_{N,N} - \cfrac{(J^{\mathrm{C}}_{N,N+1})^2}{J^{\mathrm{C}}_{N+1,N+1} - \cfrac{(J^{\mathrm{C}}_{N+1,N+2})^2}{J^{\mathrm{C}}_{N+2,N+2} - \cfrac{(J^{\mathrm{C}}_{N+2,N+3})^2}{\ddots}}}. \tag{34}$$

In our Coulomb case, this continued fraction looks like the continued fraction of the ratio of two ${}_2F_1$ hypergeometric functions with consecutive indexes

$$C_N = -\frac{4m/\hbar^2 b}{(b-\mathrm{i}k)^2\,(N+l+1+\mathrm{i}\gamma)}\,\frac{{}_2F_1\left(-l+\mathrm{i}\gamma, N+1; N+l+2+\mathrm{i}\gamma; \left(\frac{b-\mathrm{i}k}{b-\mathrm{i}k}\right)^2\right)}{{}_2F_1\left(-l+\mathrm{i}\gamma, N; N+l+1+\mathrm{i}\gamma; \left(\frac{b+\mathrm{i}k}{b-\mathrm{i}k}\right)^2\right)}. \tag{35}$$

This ratio of ${}_2F_1$ hypergeometric functions can again be represented by a continued fraction [18], which is again easily computable on the whole complex k plane.

For solving scattering equations we need the overlap between the radial Coulomb function $\varphi_l^C(k,r)$ and the CS functions $\langle\widetilde{nl}|\varphi_l^C\rangle$. They can also be calculated by using the same three-term recursion with the starting value [6]

$$\langle\widetilde{0l}|\varphi_l^C\rangle = \exp\left(2\gamma\arctan(k/b)\right)\left(\frac{2\pi\gamma}{\exp(2\pi\gamma)-1}\right)^{1/2}\left(\frac{2k/b}{1+k^2/b^2}\right)^{l+1}$$
$$\times\prod_{i=1}^{l}\left(\frac{\gamma^2+i^2}{i(i+1/2)}\right)^{1/2}. \tag{36}$$

3 CS Separable Expansion Method for Solving Lippmann–Schwinger Integral Equations

In this section we recapitulate the solution of the Lippmann–Schwinger equation by using the Coulomb-Sturmian separable expansion method. The details are found in Refs. [5–8]. We suppose that the total Hamiltonian h_l can be split into two terms

$$h_l = h_l^{\mathrm{C}} + v_l, \tag{37}$$

where v_l is an asymptotically irrelevant short-range potential and h_l^{C} is given by Eq. (1). The Green's operator of h_l is defined by $g_l(z) = (z-h_l)^{-1}$, and it is connected to $g_l^{\mathrm{C}}(z)$ via the resolvent relation

$$g_l(z) = g_l^{\mathrm{C}}(z) + g_l^{\mathrm{C}}(z)v_l g_l(z). \tag{38}$$

The scattering wave function $|\psi_l^{(+)}\rangle$ satisfies the inhomogeneous Lippmann–Schwinger equation

$$|\psi_l^{(+)}\rangle = |\varphi_l^{\mathrm{C}}\rangle + g_l^{\mathrm{C}}(E+\mathrm{i}0)v_l|\psi_l^{(+)}\rangle, \tag{39}$$

while the bound- and resonant-state wave function satisfies the homogeneous Lippmann–Schwinger equation

$$|\psi_l\rangle = g_l^{\mathrm{C}}(E) v_l |\psi_l\rangle \tag{40}$$

at negative real and complex E energies, respectively.

We solve these equations in a unified way by approximating only the short range potential v_l. For this purpose we approximate the unit operator in a double basis

$$\mathbf{1} = \lim_{N\to\infty} \mathbf{1}_N, \tag{41}$$

where

$$\mathbf{1}_N = \sum_{n,m=0}^{N} |\widetilde{nl; b_g}\rangle (\underline{O}^{-1})_{nm} \langle ml; b_v| \tag{42}$$

with $(\underline{O})_{mn} = (\langle ml; b_v|\widetilde{nl; b_g}\rangle)$. It obviously follows from the completeness relation Eq. (6) that this double sum is also a possible expression for unity. If $b_v = b_g$, Eq. (42) falls back to the usual completeness relation. By adopting different values for b_v and b_g, we can achieve a faster convergence in the separable approximation.

In Eqs. (39) and (40) the term $v_l|\psi_l\rangle$ is square integrable, and belongs to the Hilbert space. Therefore, it can be approximated by

$$v_l|\psi_l\rangle \approx \mathbf{1}_N v_l |\psi_l\rangle \approx \mathbf{1}_N v_l \mathbf{1}_N |\psi_l\rangle = v_l^N |\psi_l\rangle = \sum_{n,n'}^{N} |\widetilde{nl; b_g}\rangle\, \underline{v}_{nn'}\, \langle \widetilde{n'l; b_g}|\psi_l\rangle \tag{43}$$

with finite N, where

$$\underline{v}_{nn'} = \sum_{m,m'=0}^{N} (\underline{O}^{-1})_{nm} \langle ml; b_v| v_l |m'l; b_v\rangle (\underline{O}^{-1})_{nm'}. \tag{44}$$

In general, the matrix elements $\langle ml; b_v|v_l|m'l; b_v\rangle$ have to be calculated numerically. Note that the potential is sandwiched between CS states with parameter b_v and the potential operator becomes a linear combination of CS ket-bra operators $|\widetilde{nl; b_g}\rangle\langle\widetilde{n'l; b_g}|$. This approximation is called separable expansion because the operator v_l^N, e.g. in coordinate representation, appears in the form

$$\langle r|v_l^N|r'\rangle = \sum_{n,n'=0}^{N} \langle r|\widetilde{nl; b_g}\rangle\, \underline{v}_{nn'}\, \langle\widetilde{n'l; b_g}|r'\rangle, \tag{45}$$

i.e. the dependence on r and r' is separated. This type of separable expansion scheme with two different b parameters was originally proposed in Ref. [24]. It follows from Schwinger-type variational principles. Then it was extended for Coulomb-like potentials [25]. Previously [5–8], we adopted a simpler separable expansion with $b_v = b_g$ and the O^{-1} matrix was replaced by a diagonal matrix containing some smoothing factors σ. We found, however, that the new scheme provides much faster convergence.

With this separable potential Eqs. (39) and (40) become

$$|\psi_l^{(+)}\rangle = |\varphi_l^{\mathrm{C}}\rangle + \sum_{n,n'=0}^{N} g_l^{\mathrm{C}}(E+\mathrm{i}0)|\widetilde{nl;b_g}\rangle\, \underline{v}_{nn'}\, \langle\widetilde{n'l;b_g}|\psi_l^{(+)}\rangle, \tag{46}$$

and

$$|\psi_l\rangle = \sum_{n,n'=0}^{N} g_l^{\mathrm{C}}(E)|\widetilde{nl;b_g}\rangle\, \underline{v}_{nn'}\, \langle\widetilde{n'l;b_g}|\psi_l\rangle, \tag{47}$$

respectively. To derive equations for the coefficients $\underline{\psi}_l^{(+)} = \langle\widetilde{n'l;b_g}|\psi_l^{(+)}\rangle$ and $\underline{\psi}_l = \langle\widetilde{n'l;b_g}|\psi_l\rangle$, we have to act with states $\langle\widetilde{n''l;b_g}|$ from the left. Then, the following inhomogeneous and homogeneous algebraic equations are obtained for scattering and bound-state problems, respectively:

$$[(\underline{g}_l^{\mathrm{C}}(E+\mathrm{i}0))^{-1} - \underline{v}_l]\underline{\psi}_l^{(+)} = (\underline{g}_l^{\mathrm{C}}(E+\mathrm{i}0))^{-1}\underline{\varphi}_l^{\mathrm{C}}, \tag{48}$$

and

$$[(\underline{g}_l^{\mathrm{C}}(E))^{-1} - \underline{v}_l]\underline{\psi}_l = 0. \tag{49}$$

The homogeneous equation (49) is solvable if and only if

$$\det[(\underline{g}_l^{\mathrm{C}}(E))^{-1} - \underline{v}_l] = 0 \tag{50}$$

holds, which is an implicit nonlinear equation for the bound- and resonant-state energies. As far as the scattering states are concerned the solution of (48) provides the overlap $\langle\widetilde{nl;b_g}|\psi_l\rangle$. From this quantity any scattering information can be inferred, for example the Coulomb-modified scattering amplitude reads [26]

$$a_l = \langle\varphi_l^{\mathrm{C}(-)}|v_l|\psi_l^{(+)}\rangle = \underline{\varphi}_l^{\mathrm{C}(-)}\underline{v}_l\,\underline{\psi}_l^{(+)}, \tag{51}$$

which is related to the Coulomb-modified short-range phase shift δ_l through

$$a_l = \frac{1}{k}\exp(\mathrm{i}(2\eta_l + \delta_l))\sin\delta_l, \tag{52}$$

where $\eta_l = \arg\Gamma(l + \mathrm{i}\gamma + 1)$ is the Coulomb phase shift.

To show the performance of this separable expansion scheme we have calculated the deuteron bound-state energy and $p-p$ Coulomb-modified nuclear phase shifts. The short range potential was taken in the Malfliet–Tjon form

$$v^s_{l=0} = v_0 \exp(-\beta_0 r)/r + v_1 \exp(-\beta_1 r)/r, \tag{53}$$

with $v_0 = -626.885\,\text{MeV}$, $\beta_0 = 1.55\,\text{fm}$, $v_1 = 1438.720\,\text{MeV}$, $\beta_1 = 3.11\,\text{fm}$. This potential is one of the simplest S-wave models for the nucleon–nucleon interaction. In the expansion the values $b_g = 3.8\,\text{fm}^{-1}$ and $b_v = 2.5\,\text{fm}^{-1}$ were adopted. Table 1 shows the convergence of the binding energy of the $p-n$ system, as well as the $p-p$ scattering phase shift for $l = 0$ at various energies. The expansions converge almost equally fast over the whole spectrum and provide extremely accurate results. To reach 6-digits accuracy the method needs 10–13 basis states. We have observed similar results over a wide range of values for b_v and b_g.

The fast and uniform convergence is due to the fact that in this approach only the short range potential operator is approximated, the asymptotically important h^C_l term is treated exactly. Although we are working with finite matrices the solution is defined over the whole Hilbert space, as in the L^2 approach. The wave functions are not a simple linear combinations of basis functions, but rather, as Eqs. (46) and (47) indicate, linear combinations of functions $\langle r|g^C_l(E)|\widetilde{nl;b_g}\rangle$, which have been shown to possess the correct Coulomb asymptotic behavior [7].

Another advantage of this method is its versatility and flexibility. It can be applied for calculating bound, resonant, and scattering states with free and Coulomb asymptotics. Since the potential enters into the formalism through its CS matrix elements, the method is applicable to practically any type of potentials: the potential can be real or complex, local or nonlocal, coordinate or momentum dependent, etc. The extension to multichannel problems is also straightforward.

Note also that the Green's matrix of the total Hamiltonian, which is equivalent to the complete solution of the physical system, can be constructed as the solution of Eq. (38),

Table 1 Convergence of the binding energy of the deuteron and the $p-p$ Coulomb-modified nuclear phase shift $\delta_0(E)$ (in radians) at different energies with respect to the number of basis states N The values are stable to 6 digits beyond 13 basis states

N	ϵ_{p-n}	$\delta_0(E)$			
		$E = 0.1$ MeV	$E = 1$ MeV	$E = 10$ MeV	$E = 100$ MeV
2	−1.500168	−0.151364	−0.797903	1.453638	0.253576
3	−2.094504	−0.122391	−0.710521	1.479076	0.351795
4	−2.170223	−0.120468	−0.705228	1.477592	0.403522
5	−2.227395	−0.119136	−0.701377	1.477278	0.402637
6	−2.227976	−0.119182	−0.701573	1.479584	0.404552
7	−2.230356	−0.119165	−0.701565	1.480309	0.407128
8	−2.230646	−0.119165	−0.701549	1.480743	0.407224
9	−2.230647	−0.119164	−0.701523	1.480921	0.407345
10	−2.230692	−0.119164	−0.701519	1.480945	0.407467
11	−2.230684	−0.119162	−0.701508	1.480958	0.407487
12	−2.230688	−0.119162	−0.701507	1.480957	0.407488
13	−2.230687	−0.119162	−0.701505	1.480957	0.407497
14	−2.230687	−0.119162	−0.701504	1.480957	0.407499
15	−2.230687	−0.119162	−0.701504	1.480957	0.407499

$$\underline{g}_l(z) = [(\underline{g}_l^{\mathrm{C}}(z))^{-1} - \underline{v}_l]^{-1}. \tag{54}$$

This is very useful as practically every physical quantity can be obtained from the Green's operator. For example, the bound-state wave function can also be calculated as the residuum of the Green's operator

$$|\psi_l(E_i)\rangle\langle\psi_l(E_i)| = \frac{1}{2\pi \mathrm{i}} \oint_C \underline{g}_l(z)dz, \tag{55}$$

where C encircles the bound-state pole E_i in a counterclockwise direction without incorporating other poles. For scattering states

$$|\psi_l(E)\rangle\langle\psi_l(E)| = -\frac{1}{2\pi \mathrm{i}}(\underline{g}_l(E + \mathrm{i}0) - \underline{g}_l(E - \mathrm{i}0)) \tag{56}$$

is valid. Also any function of the selfadjoint operator h can be calculated from its Green's operator

$$f(h) = \frac{1}{2\pi i} \oint_C dz\, f(z)\, g(z), \tag{57}$$

where C encircles the spectrum of h in a counterclockwise direction and f is analytic on the encircled domain. All these will be used in the solution of three-body problems.

4 CS Separable Expansion Approach for Solving Faddeev-Type Integral Equations

The Faddeev integral equations [27] are the fundamental equations of quantum mechanical three-body systems. They possess connected kernels and are therefore Fredholm-type integral equations of second kind. This ensures the existence and the uniqueness of the solution. Originally, the Faddeev equations were derived for short range interactions and if we simply introduce a Coulomb-like potential they become singular. The necessary modification has been proposed by Merkuriev [27,28]. In Merkuriev's approach the Coulomb interactions were split into short-range and long-range parts. The long-range parts were included into the "free" Green's operators and the Faddeev procedure was performed only with the short-range potentials. The corresponding modified Faddeev, or Faddeev–Merkuriev equations are mathematically well-behaved. They have all the nice properties of the original Faddeev equations. In particular, they possess compact kernels for all energies, even in the case of attractive Coulombic interactions. Merkuriev's approach also incorporates, as a special case, the approach proposed by Noble [29], which is applicable only when all the Coulomb interactions are repulsive.

The extension of the Coulomb-Sturmian separable expansion method for resolving Faddeev-type integral equations was described first in Ref. [9]. Then, it was extended for solving three-body scattering problems with repulsive [10] and attractive [15] Coulomb interactions. The basic concept is a "three-potential" picture, where the S matrix is decomposed into three terms. The decomposition naturally generates the set of Faddeev-type equations, which, however, contain rather complicated Green's operators. These three-body Coulomb Green's operators are calculated by independent Lippmann–Schwinger-type integral equations, whose kernels contain the channel-distorted Coulomb Green's operator, which can be calculated as a contour integral of two-body Coulomb Green's operators. The details of this approach can be found in Refs. [10, 15].

Here we present the formalism on the example of the electron-hydrogen ($e-H = e^- - (e^- + p^+)$) scattering and the resonances of the electron-positronium ($e - Ps = e^- - (e^- + e^+)$) system. Resonant states are the solutions of the homogeneous Faddeev equations at complex energies; $E = E_r - \mathrm{i}\Gamma/2$, where E_r and Γ are the location and width of the resonance, respectively. The $e - Ps$ system is very similar to the $e-H$ system; the proton is replaced by a positron. In subsection A we describe briefly the Faddeev–Merkuriev integral equations, in subsection B we present the solution method, and finally show some results.

4.1 The Faddeev–Merkuriev Integral Equations

In $e - H$ or $e - Ps$ systems the two electrons are identical particles. Let us denote them by 1 and 2, and the proton or positron by 3. The Hamiltonian is given by

$$H = H^0 + v_1^C + v_2^C + v_3^C, \tag{58}$$

where H^0 is the three-body kinetic energy operator and v_α^C denotes the Coulomb interaction of each subsystem $\alpha = 1, 2, 3$. We use the usual configuration-space Jacobi coordinates x_α and y_α, where x_α is the distance between the pair (β, γ) and y_α is the distance between the center of mass of the pair (β, γ) and the particle α. Thus, the potential v_α^C, the interaction of the pair (β, γ), appears as $v_\alpha^C(x_\alpha)$. The Hamiltonian Eq. (58) is defined in the three-body Hilbert space. So, the two-body potential operators are formally embedded in the three-body Hilbert space,

$$v_\alpha^C = v_\alpha^C(x_\alpha)\mathbf{1}_{y_\alpha}, \tag{59}$$

where $\mathbf{1}_{y_\alpha}$ is a unit operator in the two-body Hilbert space associated with the y_α coordinate.

The role of a Coulomb potential in a three-body system is twofold. It acts like a long-range potential since it modifies the asymptotic motion. On the other hand, it also acts like a short-range potential, since it strongly correlates the particles and may even support two-body bound states. Merkuriev introduced a separation of the three-body configuration space into different asymptotic regions [27]. The two-body asymptotic region Ω_α is defined as a part of the three-body configuration space where the conditions

$$|x_\alpha| < x_{\alpha_0}(1 + |y_\alpha|/y_{\alpha_0})^{1/\nu}, \tag{60}$$

with parameters $x_{\alpha_0} > 0$, $y_{\alpha_0} > 0$ and $\nu > 2$, are satisfied. Merkuriev proposed to split the Coulomb interaction in the three-body configuration space into short-range and long-range terms

$$v_\alpha^C = v_\alpha^{(s)} + v_\alpha^{(l)}, \tag{61}$$

where the superscripts s and l indicate the short- and long-range attributes, respectively. The splitting is carried out with the help of a splitting function ζ_α,

$$v_\alpha^{(s)}(x_\alpha, y_\alpha) = v_\alpha^C(x_\alpha)\zeta_\alpha(x_\alpha, y_\alpha), \tag{62a}$$

$$v_\alpha^{(l)}(x_\alpha, y_\alpha) = v_\alpha^C(x_\alpha)\left[1 - \zeta_\alpha(x_\alpha, y_\alpha)\right]. \tag{62b}$$

The function ζ_α vanishes asymptotically within the three-body sector, where $x_\alpha \sim y_\alpha \to \infty$, and approaches 1 in the two-body asymptotic region Ω_α, where $x_\alpha << y_\alpha \to \infty$. Consequently in the three-body sector $v_\alpha^{(s)}$ vanishes and $v_\alpha^{(l)}$ approaches v_α^C. In practice, the functional form

$$\zeta_\alpha(x_\alpha, y_\alpha) = 2/\left\{1 + \exp\left[(x_\alpha/x_{\alpha_0})^\nu/(1 + y_\alpha/y_{\alpha_0})\right]\right\} \tag{63}$$

is used. Typical shapes for $v^{(s)}$ and $v^{(l)}$ are shown in Figs. 1 and 2, respectively.

In the Hamiltonian Eq. (58) the Coulomb potential v_3^C, the interaction between the two electrons, is repulsive, and does not support bound states. Consequently, there are no two-body channels associated with this fragmentation. Therefore, the entire v_3^C can be considered as long-range potential. Then, the long-range Hamiltonian is defined as

$$H^{(l)} = H^0 + v_1^{(l)} + v_2^{(l)} + v_3^C, \tag{64}$$

and the three-body Hamiltonian takes the form

$$H = H^{(l)} + v_1^{(s)} + v_2^{(s)}. \tag{65}$$

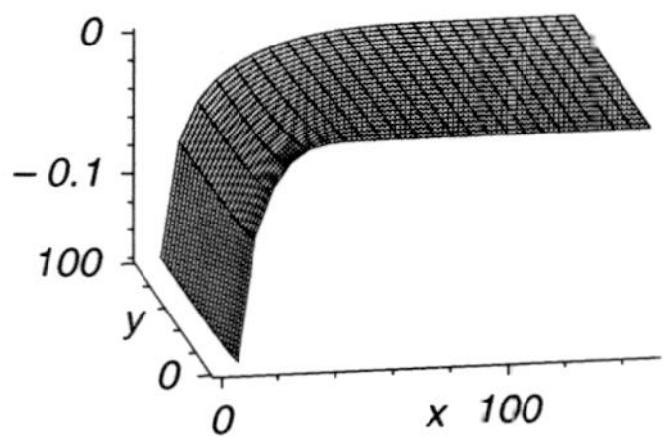

Fig. 1 Typical shape of $v^{(s)}$ for attractive Coulomb potentials

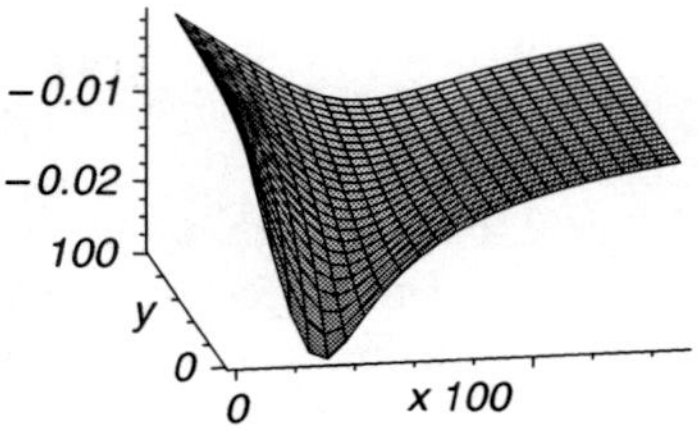

Fig. 2 Typical shape of $v^{(l)}$ for attractive Coulomb potentials

This Hamiltonian looks like an ordinary three-body Hamiltonian with two short-range interactions. We solve the Schrödinger equation

$$H|\Psi\rangle = E|\Psi\rangle \tag{66}$$

by using the Faddeev method. We split the wave function into two components

$$|\Psi\rangle = |\psi_1\rangle + |\psi_2\rangle, \tag{67}$$

and for the components arrive at a set of two-component integral equations

$$|\psi_1\rangle = |\Phi_1^{(l)}\rangle + G_1^{(l)} v_1^{(s)} |\psi_2\rangle \tag{68a}$$

$$|\psi_2\rangle = G_2^{(l)} v_2^{(s)} |\psi_1\rangle, \tag{68b}$$

where $G_\alpha^{(l)}$ is the resolvent operator of the channel Coulomb Hamiltonian

$$H_\alpha^{(l)} = H^{(l)} + v_\alpha^{(s)} \tag{69}$$

and the inhomogeneous term $|\Phi_1^{(l)}\rangle$ is an eigenstate of $H_1^{(l)}$. The differential equation form of Eq. (68) are given by

$$(E - H_1^{(l)})|\psi_1\rangle = v_1^{(s)}|\psi_2\rangle \tag{70a}$$

$$(E - H_2^{(l)})|\psi_2\rangle = v_2^{(s)}|\psi_1\rangle. \tag{70b}$$

By adding these two equations and taking into account Eqs. (69) and (67) we get back the Schrödinger equation.

To shed some light on the motivation of the Faddeev procedure and of the Merkuriev splitting of the potentials let us examine the spectral properties of the asymptotic Hamiltonian

$$H_1^{(l)} = H^{(l)} + v_1^{(s)} = H^0 + v_1^C + v_2^{(l)} + v_3^C. \tag{71}$$

It is obvious that it supports infinitely many two-body channels associated with the bound states of the attractive Coulomb potential v_1^C. The potential v_3^C is repulsive, does not have bound states, and there are no two-body channels associated with fragmentation 3. The three-body potential $v_2^{(l)}$ is attractive. It is a valley along a parabola-like curve which gets shallower and shallower, and finally disappears as y_2 goes to infinity (see Fig. 2). So, $v_2^{(l)}(x_2, y_2)$ does not support two-body bound states either in the subsystem x_2 if $y_2 \to \infty$. Consequently, there are no two-body channels associated with fragmentations 2. The asymptotic Hamiltonian $H_1^{(l)}$ has only type-1 two-body channels, where the particle 1 is at infinity and particles 2 and 3 form a bound state. If either particle 2 or 3 is at infinity, no bound states are allowed in the respective subsytem. So, the corresponding $G_1^{(l)}$ Green's operator, acting on the $v_1^{(s)}|\psi_2\rangle$ term in (68a), will generate only type-1 two-body channels in $|\psi_1\rangle$. Similar analysis is valid also for $|\psi_2\rangle$. Thus, the Faddeev–Merkuriev procedure results in a separation of the three-body wave function into components in such a way that each component has only one type of two-body channel. This is the main advantage of the original Faddeev equations and, as the above analysis shows, this property remains valid also for attractive Coulomb potentials if the Merkuriev splitting is adopted.

In $e^-e^-p^+$ and $e^-e^-e^+$ systems the particles 1 and 2, the two electrons, are identical and indistinguishable. Therefore, the Faddeev components $|\psi_1\rangle$ and $|\psi_2\rangle$, in their own natural Jacobi coordinates, have to have the same functional forms

$$\langle x_1 y_1|\psi_1\rangle = \langle x_2 y_2|\psi_2\rangle = \langle xy|\psi\rangle. \tag{72}$$

On the other hand, by interchanging the two electrons we have

$$\mathcal{P}|\psi_1\rangle = p|\psi_2\rangle, \tag{73}$$

where $\mathcal{P}$ is the operator for the permutation of indices 1 and 2, and $p = \pm 1$ is the eigenvalue of $\mathcal{P}$. Building this information into the formalism we arrive at the integral equation

$$|\psi\rangle = |\Phi_1^{(l)}\rangle + G_1^{(l)} v_1^{(s)} p\mathcal{P}|\psi\rangle, \tag{74}$$

which by itself determines $|\psi\rangle$. We notice that so far no approximation has been made, and even though this integral equation has only one component, it gives a full account of the asymptotic and symmetry properties of the system.

4.2 Coulomb-Sturmian Separable Expansion Approach for Faddeev-Type Equations

Since the three-body Hilbert space is a direct product of two-body Hilbert spaces, an appropriate basis is the bipolar basis, which can be defined as the angular-momentum-coupled direct product of the two-body bases,

$$|n\nu l\lambda; b_x b_y\rangle_\alpha = |nl; b_x\rangle_\alpha \otimes |\nu\lambda; b_y\rangle_\alpha, \quad (n, \nu = 0, 1, 2, \ldots), \tag{75}$$

where $|nl;b_x\rangle_\alpha$ and $|\nu\lambda;b_y\rangle_\alpha$ are associated with the coordinates x_α and y_α, respectively. With this basis the completeness relation takes the form (with angular momentum summation implicitly included)

$$\mathbf{1} = \lim_{N\to\infty} \sum_{n,\nu=0}^{N} |\widetilde{n\nu l\lambda; b_x b_y}\rangle_\alpha \, {}_\alpha\langle n\nu l\lambda; b_x b_y| = \lim_{N\to\infty} \mathbf{1}_\alpha^N, \tag{76}$$

where $\langle xy|\widetilde{n\nu l\lambda; b_x b_y}\rangle = \langle xy|n\nu l\lambda; b_x b_y\rangle/(xy)$.

Similarly to the two-body case, $v_1^{(s)} p\mathcal{P}|\psi\rangle$ is square integrable. Therefore, we can make the following approximation to the integral equation (74)

$$|\psi\rangle = |\Phi_1^{(l)}\rangle + G_1^{(l)} \mathbf{1}_1^N v_1^{(s)} p\mathcal{P} \mathbf{1}_1^N |\psi\rangle, \tag{77}$$

i.e. the operator $v_1^{(s)} p\mathcal{P}$ is approximated formally in the three-body Hilbert space by a separable form, viz.

$$\begin{aligned} v_1^{(s)} p\mathcal{P} &= \lim_{N\to\infty} \mathbf{1}_1^N v_1^{(s)} p\mathcal{P} \mathbf{1}_1^N \\ &\approx \mathbf{1}_1^N v_1^{(s)} p\mathcal{P} \mathbf{1}_1^N \approx \sum_{n,\nu,n',\nu'=0}^{N} |\widetilde{n\nu l\lambda, b_x b_y}\rangle_1 \, \underline{v}_1^{(s)} \, {}_1\langle \widetilde{n'\nu' l'\lambda'; b_x b_y}|, \end{aligned} \tag{78}$$

where $\underline{v}_1^{(s)} = {}_1\langle n\nu l\lambda; b_x b_y|v_1^{(s)} p\mathcal{P}|n'\nu' l'\lambda'; b_x b_y\rangle_1$. Utilizing the properties of the exchange operator $\mathcal{P}$ these matrix elements can be written in the form $\underline{v}_1^{(s)} = p \times (-)^{l'} \, {}_1\langle n\nu l\lambda; b_x b_y|v_1^{(s)}|n'\nu' l'\lambda'; b_x b_y\rangle_2$, and can be evaluated numerically by using the transformation of the Jacobi coordinates [31]. The completeness of the CS basis guarantees the convergence of the expansion with increasing N and angular momentum channels.

Now, by applying the bra $\langle \widetilde{n''\nu'' l''\lambda''; b_x b_y}|$ on Eq. (77) from the left, the solution of the inhomogeneous Faddeev–Merkuriev equation turns into the solution of a matrix equation for the component vector $\underline{\psi} = {}_1\langle \widetilde{n\nu l\lambda; b_x b_y}|\psi\rangle$

$$\underline{\psi} = \underline{\Phi}_1^{(l)} + \underline{G}_1^{(l)} \underline{v}_1^{(s)} \underline{\psi}, \tag{79}$$

where

$$\underline{\Phi}_1^{(l)} = {}_1\langle \widetilde{n\nu l\lambda; b_x b_y}|\Phi_1^{(l)}\rangle \tag{80}$$

and

$$\underline{G}_1^{(l)} = {}_1\langle \widetilde{n\nu l\lambda; b_x b_y}|G_1^{(l)}|\widetilde{n'\nu' l'\lambda'; b_x b_y}\rangle_1. \tag{81}$$

The formal solution of Eq. (79) is given by

$$\underline{\psi} = [(\underline{G}_1^{(l)})^{-1} - \underline{v}_1^{(s)}]^{-1} (\underline{G}_1^{(l)})^{-1} \underline{\Phi}_1^{(l)}. \tag{82}$$

For bound and resonant states we need to solve the homogeneous Faddeev equation

$$\underline{\psi} = \underline{G}_1^{(l)} \underline{v}_1^{(s)} \underline{\psi}, \tag{83}$$

which is solvable if and only if

$$\det[(\underline{G}_1^{(l)})^{-1} - \underline{v}_1^{(s)}] = 0. \tag{84}$$

Unfortunately, neither $\underline{G}_1^{(l)}$ nor $\underline{\Phi}_1^{(l)}$ is known. They are related to the asymptotic Hamiltonian $H_1^{(l)}$, which is still a complicated three-body Coulomb Hamiltonian. As we showed before it has only type-1 two-body channels. For such systems a single Lippmann–Schwinger equation provides a unique solution [32]. The approximation scheme for $\underline{G}_1^{(l)}$ and $\underline{\Phi}_1^{(l)}$ is presented in Ref. [15]. It is based on the Lippmann–Schwinger equation for $G_1^{(l)}$, proposed by Merkuriev [27],

$$G_1^{(l)}(z) = G_1^{as}(z) + G_1^{as}(z) V_1^{as} G_1^{(l)}(z), \tag{85}$$

where G_1^{as} and V_1^{as} are the asymptotic channel Green's operator and potential, respectively. A similar equation is valid for $|\Phi_1^{(l)}\rangle$

$$|\Phi_1^{(l)}\rangle = |\Phi_1^{as}\rangle + G_1^{as}(z) V_1^{as} |\Phi_1^{(l)}\rangle. \tag{86}$$

Merkuriev constructed G_1^{as} in the different asymptotic regions of the three-body configuration space and showed that the kernel of Eqs. (85) and (86) is completely continuous (compact) [27]. Therefore, V_1^{as} can also be approximated by separable form

$$\begin{aligned} V_1^{as} &= \lim_{N\to\infty} \mathbf{1}_1^N V_1^{as} \mathbf{1}_1^N \approx \mathbf{1}_1^N V_1^{as} \mathbf{1}_1^N \\ &\approx \sum_{n,\nu,n',\nu'=0}^{N} |\widetilde{n\nu l\lambda; b_x b_y}\rangle_1 \; \underline{V}_1^{as} \; {}_1\langle \widetilde{n'\nu' l'\lambda'; b_x b_y}|, \end{aligned} \tag{87}$$

where $\underline{V}_1^{as} = {}_1\langle n\nu l\lambda; b_x b_y | V_1^{as} | n'\nu' l'\lambda'; b_x b_y\rangle_1$. The solutions of Eqs. (85) and (86) can be expressed formally as

$$(\underline{G}_1^{(l)})^{-1} = (\underline{G}_1^{as})^{-1} - \underline{V}_1^{as} \tag{88}$$

and

$$\underline{\Phi}_1^{(l)} = [(\underline{G}_1^{as})^{-1} - \underline{V}_1^{as}]^{-1} (\underline{G}_1^{as})^{-1} \underline{\Phi}_1^{as}, \tag{89}$$

respectively, where

$$\underline{G}_1^{as} = {}_1\langle n\nu l\lambda; b_x b_y | G_1^{as} | n'\nu' l'\lambda'; b_x b_y\rangle_1, \tag{90}$$

$$\underline{V}_1^{as} = {}_1\langle n\nu l\lambda; b_x b_y | V_1^{as} | n'\nu' l'\lambda'; b_x b_y\rangle_1 \tag{91}$$

and

$$\underline{\Phi}_1^{as} = {}_1\langle \widetilde{n\nu l\lambda; b_x b_y} | \Phi_1^{as}\rangle. \tag{92}$$

The matrix elements (90)–(92) have to be calculated for a finite number of square-integrable CS states. In the integrals for the matrix elements the CS functions, as a function of x_1, decay exponentially for large x_1. So the domain of integration is confined to Ω_1, where x_1 is either finite or $x_1 << y_1$ as $x_1 \to \infty$. In this region, as Merkuriev showed [28], G_1^{as} coincides with the channel-distorted Coulomb Green's operator

$$G_1^{as}(x_1, y_1, x_1', y_1', z) = \widetilde{G}_1(x_1, y_1, x_1', y_1', z), \tag{93}$$

where $\widetilde{G}_1$ is the resolvent of the Hamiltonian

$$\widetilde{H}_1 = H^0 + v_1^C. \tag{94}$$

Therefore, in calculating the matrix elements (90) G_1^{as} can be replaced by $\widetilde{G}_1$. Similarly, in calculating (91) V_1^{as} can also be replaced by

$$U_1 = v_2^{(l)} + v_3^C, \tag{95}$$

and consequently (88) becomes

$$(\underline{G}_1^{(l)})^{-1} = (\underline{\widetilde{G}}_1)^{-1} - \underline{U}_1, \tag{96}$$

where

$$\underline{\widetilde{G}}_1 = {}_1\langle \widetilde{n\nu l\lambda; b_x b_y} | \widetilde{G}_1 | \widetilde{n'\nu' l'\lambda'; b_x b_y}\rangle_1 \tag{97}$$

and

$$\underline{U}_1 = {}_1\langle n\nu l\lambda; b_x b_y | U_1 | n'\nu' l'\lambda'; b_x b_y\rangle_1. \tag{98}$$

The $\underline{U}_1$ matrix elements can again be evaluated numerically.

Also Φ_1^{as}, in Ω_1, equates to $\widetilde{\Phi}_1$, the eigenstate of $\widetilde{H}_1$, and Eq. (89) becomes

$$\underline{\Phi}_1^{(l)} = [(\underline{\widetilde{G}}_1)^{-1} - \underline{U}^1]^{-1}(\underline{\widetilde{G}}_1)^{-1}\underline{\widetilde{\Phi}}_1, \tag{99}$$

where $\underline{\widetilde{\Phi}}_1 = {}_1\langle \widetilde{n\nu l\lambda; b_x b_y} | \widetilde{\Phi}_1\rangle$.

4.3 The Matrix Elements of $\widetilde{G}_1$

The three-particle free Hamiltonian can be written as a sum of two-particle free Hamiltonians

$$H^0 = h^0_{x_1} + h^0_{y_1}. \tag{100}$$

Thus the Hamiltonian $\widetilde{H}_1$ of Eq. (94) appears as a sum of two two-body Hamiltonians acting on different coordinates

$$\widetilde{H}_1 = h_{x_1} + h_{y_1}, \tag{101}$$

with $h_{x_1} = h^0_{x_1} + v^C_1(x_1)$ and $h_{y_1} = h^0_{y_1}$, which, of course, commute. Therefore $|\widetilde{\Phi}_1\rangle$, the eigenstate of $\widetilde{H}_1$, in CS representation, is given by

$$_1\langle \widetilde{n\nu l\lambda; b_x b_y} | \widetilde{\Phi}_1\rangle = {}_1\langle \widetilde{nl; b_x} | \phi_1\rangle \times {}_1\langle \widetilde{\nu\lambda; b_y} | \chi_1\rangle. \tag{102}$$

Here $|\phi_1\rangle$ is a bound eigenstate of h_{x_1}: $h_{x_1}|\phi_1\rangle = \epsilon_1|\phi_1\rangle$, with ϵ_1 being the bound-state energy of the 2–3 pair. The state $|\chi_1\rangle$ is a scattering eigenstate of h_{y_1} at energy $E - \epsilon_1$. The CS matrix elements $_1\langle \widetilde{nl; b_x} | \phi_1\rangle$ and $_1\langle \widetilde{\nu\lambda; b_y} | \chi_1\rangle$ are known from the two-body case.

The most crucial point in this procedure is the calculation of the matrix elements $\underline{\widetilde{G}}_1$. The Green's operator $\widetilde{G}_1$ is a resolvent of the sum of two commuting Hamiltonians h_{x_1} and h_{y_1}, and is a function of a selfadjoint operator h_{y_1}. Thus, according to Eq. (57), the three-body Green's operator $\widetilde{G}_1$ can be written as a convolution integral of two-body Green's operators, i.e.

$$\widetilde{G}_1(z) = \frac{1}{2\pi i}\oint_C dz'\, g_{x_1}(z - z')\, g_{y_1}(z'), \tag{103}$$

where $g_{x_1}(z) = (z - h_{x_1})^{-1}$ and $g_{y_1}(z) = (z - h_{y_1})^{-1}$. The contour C should be taken counterclockwise around the singularities of g_{y_1} so that g_{x_1} might be analytic on the domain encircled by C.

In the time-independent scattering theory, the Green's operator has a branch-cut singularity at scattering energies In our formalism $\widetilde{G}_1(E)$ should be understood as $\widetilde{G}_1(E) = \lim_{\varepsilon\to 0}\widetilde{G}_1(E + \mathrm{i}\varepsilon)$, with $\varepsilon > 0$. We limit our study for energies below the three-body breakup threshold, so $\Re(E) < 0$. To examine the analytic structure of the integrand in Eq. (103) let us take ε as finite. By doing so, the singularities of g_{x_1} and g_{y_1} become well separated. In fact, g_{y_1} is a free Green's operator with a branch-cut on the $[0, \infty)$ interval, while $g_{x_1}(E + \mathrm{i}\varepsilon - z')$ is a Coulomb Green's operator, which, as function of z', has a branch-cut on the $(-\infty, E + \mathrm{i}\varepsilon]$ interval and infinitely many poles accumulated at $E + \mathrm{i}\varepsilon$. Now, the branch cut of g_{y_1} can easily be encircled so that the singularities of g_{x_1} lie outside the encircled domain (Fig. 3). However, this would not be the case in the $\varepsilon \to 0$ limit. Therefore the contour C is

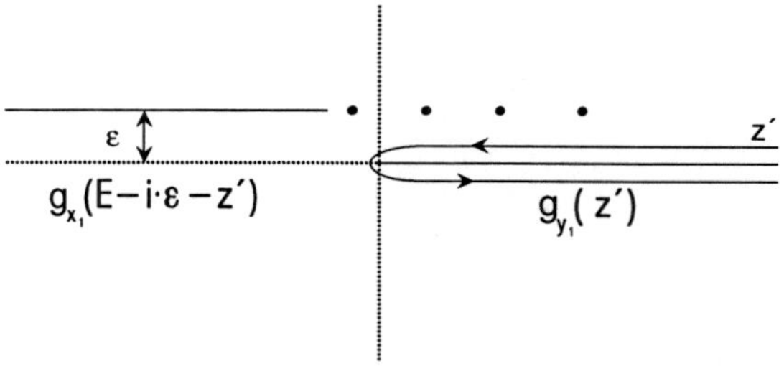

Fig. 3 Analytic structure of $g_{x_1}(E + \mathrm{i}\varepsilon - z')g_{y_1}(z')$ as a function of z', $\varepsilon > 0$. The operator $g_{y_1}(z')$ has a branch-cut on the $[0, \infty)$ interval, while $g_{x_1}(E + \mathrm{i}\varepsilon - z')$ has a branch-cut on the $(-\infty, E + \mathrm{i}\varepsilon]$ interval and infinitely many poles accumulated at $E + \mathrm{i}\varepsilon$ (denoted by dots). The contour C encircles the branch-cut of g_{y_1} and avoids the singularities of g_{x_1}. In the $\varepsilon \to 0$ limit the singularities of $g_{x_1}(E + \mathrm{i}\varepsilon - z')$ would penetrate into the area covered by C

deformed analytically in such a way that the upper part descends into the unphysical Riemann sheet of g_{y_1}, while the lower part of C is detoured away from the cut (Fig. 4). Now, even in the $\varepsilon \to 0$ limit, the contour avoids the singularities of g_{x_1} (Fig. 5). Thus, the mathematical conditions for the contour integral representation of $\widetilde{G}_1$ in Eq. (103) are met also for scattering-state energies. To calculate resonances we need $\widetilde{G}_1(E + \mathrm{i}\epsilon)$ at complex energies with $\epsilon < 0$. With this contour, this is also possible (Fig. 6). The contour still encircles the singularities of g_{y_1} without penetrating into the singularities of g_{x_1}.

In the three-potential formalism [10, 15] the S matrix can be decomposed into three terms. The first one describes a single channel Coulomb scattering, the second one is a multichannel two-body-type scattering due to the potential U, and the third one is a genuine three-body scattering. In our $e^- + H$ case the target is neutral and therefore the first term is absent. For the on-shell T matrix we have

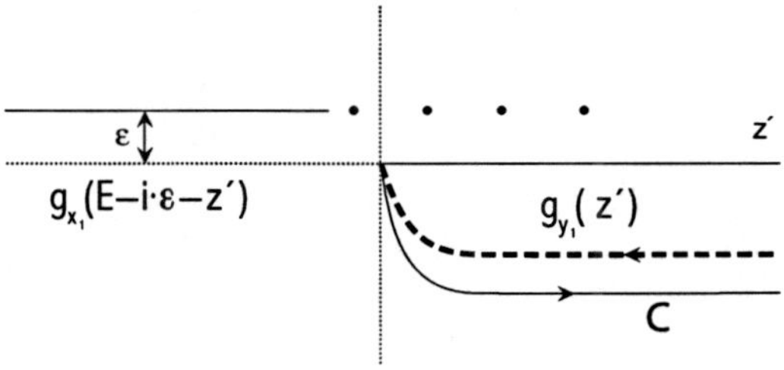

Fig. 4 The contour of Fig. 3 is deformed analytically such that a part of it goes on the unphysical Riemann-sheet of g_{y_1} (drawn by broken line) and the other part detoured away from the cut

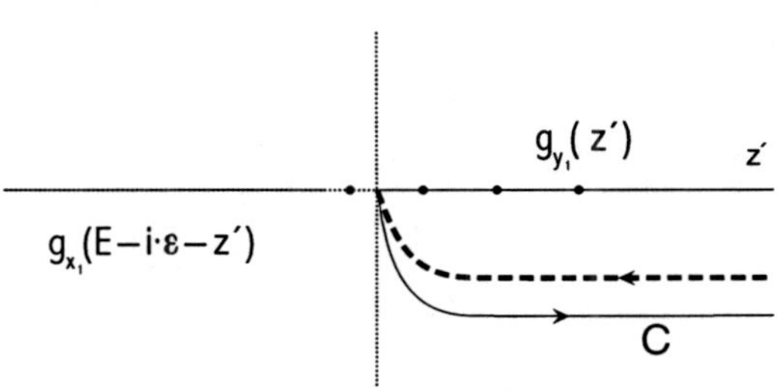

Fig. 5 In the $\varepsilon \to 0$ limit the singularities of g_{x_1} are outside the area encircled by C

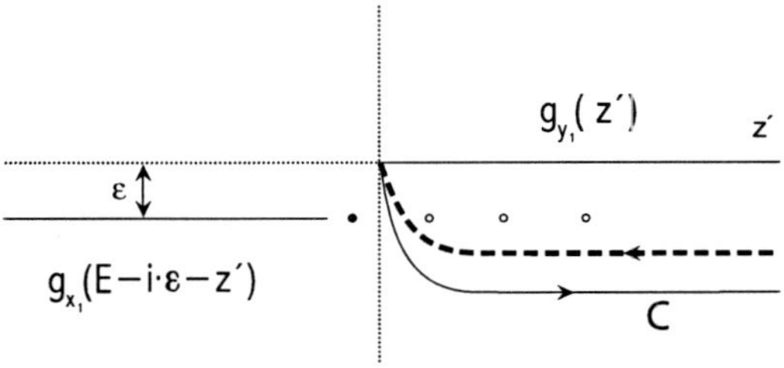

Fig. 6 Even in the $\varepsilon < 0$ case, which is needed to calculate resonances, the singularities of g_{x_1} do not penetrate into the area covered by C

$$T_{fi} = \sqrt{\frac{\mu_f \mu_i}{k_f k_i}} \left(\langle \widetilde{\Phi}_{1f}^{(-)} | U_1 | \Phi_{1i}^{(l)(+)} \rangle + \langle \Phi_{1f}^{(l)(-)} | v_1^{(s)} | \psi_{2i}^{(+)} \rangle \right), \tag{104}$$

where i and f refer to the initial and the final states, respectively, μ is the channel reduced mass and k is the channel wave number. Given the solutions $\underline{\psi}$ and $\underline{\Phi}^{(l)}$ and the matrix elements $\underline{U}_1$ and $\underline{v}_1^{(s)}$, the T matrix elements can easily be evaluated, and then all scattering information can be inferred.

4.4 Application to the $e - H$ Elastic Scattering

We present here S-wave scattering results. More detailed calculations, including S and P wave partial cross sections up to the $n = 4$ thresholds, can be found in Ref. [30]. In the separable expansion we take up to 9 bipolar angular momentum

Table 2 Singlet ($^1S^e$, $p = +1$) and triplet ($^3S^e$, $p = -1$) phase shifts of elastic S-wave $e^- + H$ scattering

k	Ref. [33]	Ref. [34]	Ref. [35]	Ref. [36]	Ref. [37]	This work
$^1S^e$, $p = +1$						
0.1	2.553	2.550	2.553	2.555	2.553	2.552
0.2	2.0673	2.062	2.066	2.066	2.065	2.064
0.3	1.6964	1.691	1.695	1.695	1.694	1.693
0.4	1.4146	1.410	1.414	1.415	1.415	1.412
0.5	1.202	1.196	1.202	1.200	1.200	1.197
0.6	1.041	1.035	1.040	1.041	1.040	1.037
0.7	0.930	0.925	0.930	0.930	0.930	0.927
0.8	0.886		0.887	0.887	0.885	0.884
$^3S^e$, p $= -1$						
0.1	2.9388	2.939	2.938	2.939	2.939	2.938
0.2	2.7171	2.717	2.717	2.717	2.717	2.717
0.3	2.4996	2.500	2.500	2.500	2.499	2.499
0.4	2.2938	2.294	2.294	2.294	2.294	2.294
0.5	2.1046	2.105	2.104	2.104	2.105	2.104
0.6	1.9329	1.933	1.933	1.933	1.933	1.932
0.7	1.7797	1.780	1.780	1.780	1.779	1.779
0.8	1.643		1.645	1.644	1.641	1.643

channels with up to $N = 36$ CS functions. We use the parameter values $\nu = 2.1$, $x_0 = 3$, $y_0 = 20$ and $b = 0.6$. Table 2 shows elastic phase shifts at several values of electron momenta k_1. Our results, which were achieved by using finite proton mass, agree very well with variational calculations of Ref. [33], *R*-matrix calculations of Ref. [34], finite-element method of Ref. [35], as well as with the results of direct numerical solution of the Schrödinger equation of Ref. [36], where infinite proton mass was adopted. We also compare our calculation with the differential equation solution of the modified Faddeev equations [37]. We can observe perfect agreement with all the previous calculations.

4.5 Resonant States in the $e - Ps$ Three-Body System

We present here *S*-wave resonances of the $e - Ps$ three-body system. The details are given in Refs. [11–14]. The calculations were carried out with three different sets of parameters: $x_0 = 18$ and $y_0 = 25$, $x_0 = 25$ and $y_0 = 50$, $x_0 = 5$ and $y_0 = 1000$, and $\nu = 2.1$. In all cases we found that the calculations converged well and that the results are insensitive to the choice of parameter b over a broad interval between $b = 0.2$ and $b = 0.6$. The basis includes up to $N = 25$ CS states and angular momentum channels up to $l = \lambda = 10$. With that basis set we achieved $5 - 6$ digits stability for the significant digits. We located all the previously known *S*-wave resonances reported in an early calculation by Ho [38], and observed excellent agreements for the position of resonances in all cases. However, as far as the widths are concerned, our results, in some cases, were several order of magnitude smaller, which indicates a several order of magnitude longer lifetime for the resonances. An independent recent calculation by Li and Shakeshaft confirmed our results [39].

Besides the resonant states known before, we found about 50–60 new resonances attached to thresholds. Figures 7 and 8 show the resonances attached to the $n = 1$ and $n = 2$ thresholds, respectively. It is obvious that these resonances have the same origin. We conjectured that they may come from Efimov effect. Indeed, Efimov pointed out that if the two-body subsystem has a bound state at zero energy the three-body system has infinitely many bound states [40]. They become resonant states if they are embedded into the continuum associated with another deeper two-body bound state.

In atomic three-body systems the attractive Coulomb potential supports infinitely many two-body bound states accumulating at zero energy. So, we have not only one but infinitely many two-body bound states in the close vicinity of zero energy. It is quite natural that they produce infinitely many resonant states in three-body systems attached to two-body thresholds.

This finding is quite remarkable. The resonances of the $e - Ps$ system has been studied by various methods, yet no previous study reported the accumulation of resonances above thresholds. To our knowledge, during the approximation scheme, no other method can retain one of the most important property of a three-body atomic

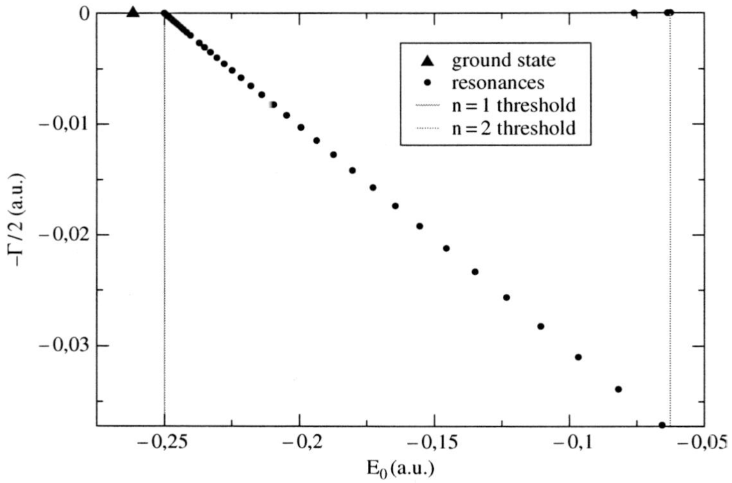

Fig. 7 ^{1}S resonances in the $e - Ps$ system starting from the $n = 1$ threshold

system, that the attractive Coulomb potentials in two two-body subsystems support infinitely many bound states. In our approach the two-body Coulomb potential is treated exactly by the exact analytical evaluation of the Green's operator. This might be the reason why we see the effect of infinitely many zero energy two body bound states in three-body systems.

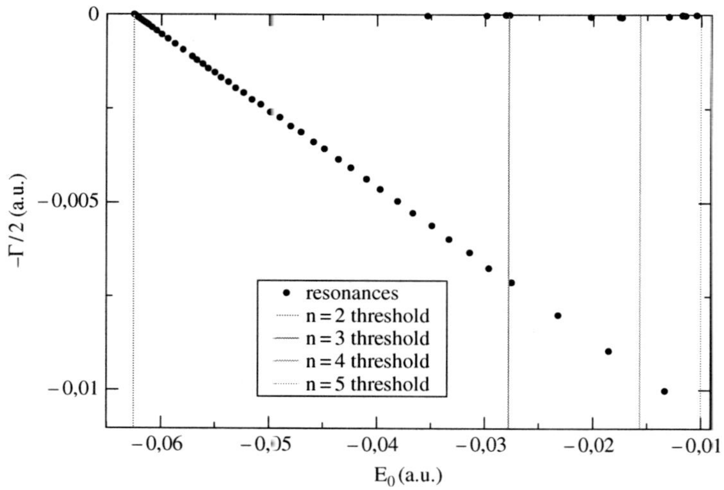

Fig. 8 ^{1}S resonances in the $e - Ps$ system starting from the $n = 2$ threshold

5 Summary and Outlook

In this work we gave a summary of a quantum mechanical approximation scheme which is based on the separable expansion of the potential in the Coulomb-Sturmian basis. Only the short-range potentials, which do not modify the nature of the asymptotic motion, are approximated and the asymptotically relevant terms are treated exactly by an analytical evaluation of the Green's operator. The method is applicable for solving two-body Lippmann–Schwinger and three-body Faddeev equations with short-range plus Coulomb potential. The Coulomb potential can be either attractive or repulsive.

The method can handle practically all kinds of short-range potentials. This is especially advantageous when we study the possible nonlocality of the nucleon-nucleon potential by calculating the bound-state properties of the triton and ^3He [41] and $p-d$ scattering observables at energy below the breakup threshold [42]. The method works also for three-quark problems [43]. We realized that from a mathematical point of view the Coulomb and confining potentials play a rather similar role in the three-body equations; both are long range potentials and modify the character of the asymptotic motion. All these applications prove that this is a practical and useful method with a broad range of applicability in atomic, nuclear and particle physics. We believe that it can be extended to solve even more challenging problems, like Coulomb scattering above the three-body breakup threshold.

At the heart of the whole procedure we find the J-matrix structure of the two-body Coulomb Hamiltonian. This allows an efficient and analytic evaluation of the two-body Coulomb Green's operators. The Green's operator of the total two-body problem can be calculated by solving a Lippmann–Schwinger equation. With the help of the Faddeev–Merkuriev integral equations the treatment of three-body Coulombic systems can be formulated in such a way that after a series of approximations for the short-range type terms the Green's operator appears as a convolution integral of two-body Green's operators. The convolution integral can be continued analytically to all energies of physical interest.

Although approaches utilizing the J-matrix structure of the Hamiltonian already have rich history and a broad range of applications the future looks even more interesting. One possible future direction is to extend the method to Hamiltonians which have band-matrix structure. A band matrix can always be considered as a block tridiagonal matrix. Then the Green's matrix analogous to Eq. (33) reads

$$G^{(N)} = (J_{i,j}^{(N)} - \delta_{i,N}\delta_{j,N} J_{N,N+1} C_{N+1} J_{N+1,N})^{-1}, \tag{105}$$

where $J_{i,j}^{(N)}$ is the upper left $N \times N$ block J-matrix and C_{N+1} is a matrix continued fraction

$$C_{N+1} = \Big(J_{N+1,N+1} - J_{N+1,N+2}\big(J_{N+2,N+2} - J_{N+2,N+3}(J_{N+3,N+3} - \cdots)^{-1} \\ \times J_{N+3,N+2}\big)^{-1} \times J_{N+2,N+1}\Big)^{-1} \tag{106}$$

built from $m \times m$ block matrices In Ref. [44] we have shown that the nonrelativistic Coulomb Hamiltonian plus polynomial potentials, in the CS basis, possess a band-matrix structure. We studied the Coulomb plus linear confinement potential, which is a common model for quark confinement and is related to the Stark effect as well. This Hamiltonian is pentadiagonal in the CS basis. We also studied the Coulomb plus quadratic confinement in two dimensions, which is related to the Zeeman effect, and whose Hamiltonian is septadiagonal. A pentadiagonal band matrix can be treated as a block-Jacobi matrix with 2×2, while a septadiagonal with 3×3 blocks. The exact knowledge of these Green's matrices may facilitate the use of integral equations to describe quantum processes in external fields. The ability of calculating the Green's matrices of band-matrix Hamiltonian is very promising and may considerably broaden the scope of the J-matrix method.

Acknowledgments This work has been supported by the Research Corporation. I am thankful to my collaborators, especially to W. Plessas, P. Doleschall, B. Kónya, J. Darai, J. Zs. Mezei, Z. T. Hlousek, C-.Y. Hu, and S. L. Yakovlev.

References

1. E. J. Heller and H. A. Yamani, Phys. Rev. A **9**, 1201 (1974); ibid. **9**, 1209 (1974); H. A. Yamani and L. Fishman, J. Math. Phys. **16**, 410 (1975).
2. E. J. Heller, Phys. Rev. A **12**, 1222 (1975).
3. J. Révai, JINR Preprint E4-9429, Dubna, (1975).
4. F. A. Gareev, M. Ch. Gizzatkulov and J. Révai, Nucl. Phys. **A 286**, 512 (1977); E. Truhlik, Nucl. Phys. **A 296**, 134 (1978); F. A. Gareev, S. N. Ershov, J. Révai, J. Bang and B. S. Nillsson, Phys. Scripta **19**, 509 (1979); B. Gyarmati, A. T. Kruppa and J. Révai, Nucl. Phys. **A 326**, 119 (1979); B. Gyarmati and A. T. Kruppa, Nucl. Phys. A **378**, 407 (1982); B. Gyarmati and A. T. Kruppa, Z. Papp and G. Wolf, Nucl. Phys. A **417**, 393 (1984); A. T. Kruppa and Z. Papp, Comp. Phys. Comm. **36**, 59 (1985); J. Révai, M. Sotona and J. Žofka, J. Phys. G: Nucl. Phys. **11**, 745 (1985); K. F. Pál, J. Phys. A: Math. Gen. **18**, 1665 (1985).
5. Z. Papp, J. Phys. A **20**, 153 (1987).
6. Z. Papp, Phys. Rev. C **38**, 2457 (1988).
7. Z. Papp, Phys. Rev. A **46**, 4437 (1992).
8. Z. Papp, Comp. Phys. Comm. **70**, 426 (1992); ibid. **70**, 435 (1992).
9. Z. Papp and W. Plessas, Phys. Rev. C, **54**, 50 (1996); Z. Papp, Few-Body Syst., **24**, 263 (1998).
10. Z. Papp, Phys. Rev. C, **55**, 1080 (1997).
11. Z. Papp, J. Darai, C-.Y. Hu, Z. T. Hlousek, B. Kónya and S. L. Yakovlev, Phys. Rev. A **65**, 032725 (2002).
12. Z. Papp, J. Darai, A. Nishimura, Z. T. Hlousek, C-.Y. Hu and S. L. Yakovlev, Phys. Lett. A **304**, 36–42 (2002).
13. Z. Papp, J. Darai, J. Zs. Mezei, Z. T. Hlousek and C-.Y. Hu, Phys. Rev. Lett. **94**, 243201 (2005).
14. J. Zs. Mezei and Z. Papp, Phys. Rev. A **73**, 030701(R) (2006).
15. Z. Papp, C.-Y. Hu, Z. T. Hlousek, B. Kónya and S. L. Yakovlev, Phys. Rev. A **63**, 062721 (2001).
16. M. Rotenberg, Ann. Phys. (N.Y.) **19**, 262 (1962); M. Rotenberg, Adv. At. Mol. Phys. **6**, 233 (1970).
17. M. Abramowitz and I. Stegun, *Handbook of Mathematical Functions* (Dover, New York, 1970).

18. L. Lorentzen and H. Waadeland, *Continued Fractions with Applications* (Noth-Holland, Amsterdam, 1992), p. 296.
19. P. Rózsa, *Linear Algebra and Its Applications*, (in Hungarian) Műszaki Könyvkiadó, Budapest (1976).
20. W. B. Jones and W. J. Thron, *Continued Fractions: Analytic Theory and Applications* (Addison-Wesley, Reading, 1980).
21. B. Kónya, G. Lévai and Z. Papp, J. Math. Phys. **38**, 4832 (1997).
22. B. Kónya, G. Lévai and Z. Papp, Phys. Rev. C **61**, 034302 (2000).
23. F. Demir, Z. T. Hlousek and Z. Papp, Phys. Rev. A **74**, 014701 (2006).
24. S.K. Adhikari and L. Tomio, Phys. Rev. C, **36**, 1275 (1987).
25. J. Darai, B. Gyarmati, B. Kónya and Z. Papp, Phys. Rev. C **63**, 057001 (2001).
26. R. G. Newton, *Scattering Theory of Waves and Particles* (Springer, New York, 1982).
27. L. D. Faddeev and S. P. Merkuriev, *Quantum Scattering Theory for Several Particle Systems*, (Kluwer, Dordrecht, 1993).
28. S. P. Merkuriev, Ann. Phys. NY, **130**, 395 (1980).
29. J. Noble, Phys. Rev. **161**, 945 (1967).
30. Z. Papp, C.-Y. Hu, Phys. Rev. A **66**, 052714 (2002).
31. R. Balian and E. Brézin, Nuovo Cim. B **2**, 403 (1969).
32. W. Sandhas, *Few-Body Nuclear Physics*, (IAEA Vienna), 3 (1978).
33. C. Schwartz, Phys. Rev. **124**, 553 (1961).
34. T. Scholz, P. Scott and P. G. Burke, J. Phys. B: At. Mol. Opt. Phys. **21**, L139 (1988).
35. J. Botero and J. Shertzer, Phys. Rev. A **46**, R1155 (1992).
36. Y. D. Wang and J. Callaway, Phys. Rev. A **48**, 2058 (1993); Phys. Rev. A **50**, 2327 (1994).
37. A. A. Kvitsinsky, A. Wu and C.-Y. Hu, J. Phys. B: At. Mol. Opt. Phys. **28** 275 (1995).
38. Y. K. Ho, Phys. Lett., **102A**, 348 (1984).
39. T. Li and R. Shakeshaft, Phys. Rev. A, **71**, 052505 (2005).
40. V. Efimov, Phys. Lett. **33 B**, 563 (1970).
41. P. Doleschall, I. Borbély, Z. Papp and W. Plessas, Phys. Rev. C **67**, 064005 (2003).
42. P. Doleschall and Z. Papp, Phys. Rev. C **67**, 064005 (2003).
43. Z. Papp, Few-Body Syst. **26** 99–103 (1999); Z. Papp, A. Krassnigg and W. Plessas, Phys. Rev. C **62**, 044004 (2000).
44. E. Kelbert, A. Hyder, F. Demir, Z. T. Hlousek and Z. Papp, J. Phys. A: Math. Theor. 40, 7721 (2007).

The Use of a Complex Scaling Method to Calculate Resonance Partial Widths

H.A. Yamani and M.S. Abdelmonem

Abstract The diagonalization of a complex-scaled Hamiltonian is used to calculate the full Green's matrix elements that are valid in the complex energy plane, including the exposed portion of the energy sheet. Three approximate methods are introduced to calculate the partial widths associated with a projectile-target system having several scattering channels. Three potentials are used to check these methods. The results are compared with those obtained by other methods, including the J-matrix method.

The methods that are used to characterize resonances may be described as either direct or indirect. The direct methods [1–6] are based on the formal definition of resonances as bound-state-like isolated eigenvalues of a complex-scaled Hamiltonian, or as poles of the associated Green's function, T-matrix or S-matrix. The indirect methods [2, 7–12] extract the real resonance energy and resonance total width from the behavior of certain physical quantities, such as the cross-section, eigenphase shifts, or the density of resonance states in the immediate neighborhood of the real resonance position. The resonance energies and widths calculated by the two methods are in agreement only for narrow resonances. These are resonances having a long lifetime or a complex energy with a small imaginary part. In fact, it is difficult to extract resonance information for broad resonances by the indirect methods. This is because the effect of broad resonances on scattering quantities is diminished and simultaneously spread over a larger interval of scattering energies. The study of narrowness of a resonance and its implementation insures that it has a strong effect on the computed physical quantities, while at the same time being limited to a small interval of energies. The resonance is then term-isolated in the sense that its effect is registered on an energy interval with minimal effects from other resonances. The task of finding the resonance position and partial widths is almost invariably focused on narrow resonances. Complex scaling [3, 13–15] is perhaps the only method that locates all resonances, narrow or broad, with equal ease, provided

---

H.A. Yamani
Ministry of Commerce & Industry, P.O. Box 5729, Riyadh 11127, Saudi Arabia
e-mail: haydara@sbm.net.sa

A.D. Alhaidari et al. (eds.), *The J-Matrix Method*, 173–179.

of course that the size of the complex scaled Hamiltonian and the rotation angle are large enough. However, the utilization of this method has been restricted to the calculation of real resonance positions and total widths at the exclusion of partial widths. The computation of partial widths associated with a projectile-target system having several scattering channels is more difficult than finding the position of the complex resonance energy alone. The methods that use the "golden rule" [1, 16] require the resonance wave function itself, and the matrix element of the potential between the resonance wave function and an unperturbed continuum function of the specified channel. Methods that use the S-matrix [4, 5] find the partial widths from a calculation of the residues of the S-matrix pole at the resonance energy. Furthermore, methods that use the density of states [8] calculate the product of the S-matrix and its derivative at several energies near resonance.

In this chapter, we propose three approximation methods that essentially endow the complex scaling method with the ability to compute resonance partial widths, and without the need for information other than that which the diagonalization of the complex-scaled full Hamiltonian yields and that provided by the J-matrix quantities related to the free Hamiltonian.

We consider a projectile-target system where the projectile is spinless and structureless, while the target is composed of M internal states having threshold energies $E_1, E_2, \ldots, E_M$. The channel potentials $V^{\alpha\beta}$ $(\alpha, \beta = 1, 2, \ldots, M)$ are assumed to be of a short range. We form a complex symmetric matrix representation of the multichannel full Hamiltonian, $\tilde{H}$. We use in each channel α a set of N_α orthonormal complex-scaled square integrable functions $\left\{\phi_n^{(\alpha)}(re^{-i\theta})\right\}_{n=1}^{N_\alpha}$, where θ is the rotation angle. The diagonalization of $\tilde{H}$ results in a set of complex Harris [17] eigenvalues $\left\{\tilde{\varepsilon}_\mu\right\}_{\mu=1}^{N_C}$, where $N_C = \sum_{\alpha=1}^{M} N_\alpha$ is the dimension of each of the Hamiltonian matrices. For sufficiently large values of N_C and θ, the discrete set of Harris eigenvalues of the full Hamiltonian contains the system bound states and the exposed set of system resonances, while the remainder attempt to discretize the channel continua. The (α, β)-channel sub-matrix of the full Green's matrix may be written as

$$g_{n,m}^{(\alpha,\beta)}(E) = \sum_{\mu=1}^{N_c} \frac{\Lambda_{n,\mu}^{(\alpha)} \Lambda_{m,\mu}^{(\beta)}}{\tilde{\varepsilon}_\mu - E} \tag{1}$$

The set $\left\{\Lambda_{n,\mu}^{(\alpha)}\right\}_{n=1}^{N_\alpha}$ contains the coefficients of the eigenvector $\left|\upsilon_\mu\right\rangle$ in channel α associated with the eigenvalue $\tilde{\varepsilon}_\mu$. More explicitly, we have

$$\left|\upsilon_\mu\right\rangle = \sum_{\alpha=1}^{M} \sum_{n=1}^{N_\alpha} \Lambda_{n,\mu}^{(\alpha)} \left|\phi_n^{(\alpha)}\right\rangle \tag{2}$$

As these eigenvectors are themselves orthogonal, we also normalize them by requiring that

$$\sum_{\alpha=1}^{M} \tilde{R}_\mu^{(\alpha)} = 1 \tag{3}$$

for each $\mu = 1, 2, \ldots, N_C$, where

$$\tilde{R}_{\mu}^{(\alpha)} = \sum_{n=1}^{N_\alpha} [\Lambda_{n,\mu}^{(\alpha)}]^2 \tag{4}$$

It is clear that the finite matrix representation (1) of the Green's function does not have the same analytical structure of the exact Green's function. The representation (1) replaces the channel continua with a set of poles. We are, however, interested in the resonance energy whose position in this approximation is stable relative to the variations in the parameters defining the approximation. Thus, one of the discrete energies $\{\tilde{\varepsilon}_\mu\}$ of interest can be easily identified with the resonance energy, ε_r. We may then write

$$\varepsilon_r = E_r - i\Gamma_r/2, \tag{5}$$

where E_r is the 'real' resonance position that usually appears prominently in the indirect resonance calculation. For example, it is the energy at which the density of states curve attains a maximum [18], displaying a shape that is typically Lorentzean. Γ_r is then the total resonance width of the curve.

The identification of E_r and Γ_r is the usual extent of the utility of the complex scaling method to the study of resonance. We now outline three approximate methods to further extract the partial width of each channel from the information already obtained in the complex scaling calculation. For simplicity we limit the setup to two channels, although the results are applicable to any number of channels.

1 Method (A)

This is the crudest of the approximations. It assumes that the total width Γ_r of the resonance is the sum of the partial widths associated with the channels, and that the contribution of each channel is the weighted contribution of the channel to the resonance eigenfunction. In other words,

$$\Gamma_1 = \frac{1}{\Omega}\left|\sum_{n=0}^{N_1-1} [\Lambda_{n,r}^{(1)}]^2\right| \Gamma_r \tag{6a}$$

$$\text{and} \quad \Gamma_2 = \frac{1}{\Omega}\left|\sum_{n=0}^{N_2-1} [\Lambda_{n,r}^{(2)}]^2\right| \Gamma_r, \tag{6b}$$

$$\text{where } \Omega = \left|\sum_{n=0}^{N_1-1} [\Lambda_{n,r}^{(1)}]^2\right| + \left|\sum_{n=0}^{N_2-1} [\Lambda_{n,r}^{(2)}]^2\right|.$$

2 Method (B)

If the basis $\left\{\phi_n^{(\alpha)}\right\}$ is a J-matrix oscillator basis, we may use the results obtained in the analysis of the multi-channel J-matrix Green's function [19,20]. We utilize the definition of the partial width, Γ_α, associated with channel α as the residue of the scattering matrix $S_{\alpha\alpha}(\varepsilon)$, where ε is near the resonance energy. That is,

$$S_{\alpha\alpha}(\varepsilon) = S_{\alpha\alpha}^{\text{bg}}(\varepsilon) - \frac{i\Gamma_\alpha}{\varepsilon - \varepsilon_r}, \tag{7}$$

where $S_{\alpha\alpha}^{\text{bg}}(\varepsilon)$ is a background term that is non-singular at $\varepsilon = \varepsilon_r$. On the other hand, we know [20] that $S_{\alpha\alpha}(\varepsilon)$ is related to Green's function $G_{N_\alpha-1,N_\alpha-1}^{(+)(\alpha,\alpha)}(\varepsilon)$ as represented by the expression

$$S_{\alpha\alpha}(\varepsilon) = \tilde{S}_{\alpha\alpha}(\varepsilon) - \frac{ik_\alpha}{(c_{N_\alpha-1} + is_{N_\alpha-1})^2} G_{N_\alpha-1,N_\alpha-1}^{(+)(\alpha,\alpha)}(\varepsilon), \tag{8}$$

where s_n and c_n are the sine-like and cosine-like J-matrix Fourier coefficients of the asymptotic reference wavefunction. In this approximation, we take

$$G_{N_\alpha-1,N_\alpha-1}^{(+)(\alpha,\alpha)}(\varepsilon) \simeq g_{N_\alpha-1,N_\alpha-1}^{(\alpha,\alpha)}(\varepsilon) \tag{9}$$

The finite complex rotated Green's matrix is given by (1). Since $g_{N_\alpha-1,N_\alpha-1}^{(\alpha,\alpha)}(\varepsilon)$ behaves near ε_r as

$$\frac{\left[\Lambda_{N_\alpha-1,r}^{(\alpha)}\right]^2}{\varepsilon_r - \varepsilon}, \tag{10}$$

we conclude that

$$\Gamma_\alpha = \left| \left\{ \frac{k_\alpha}{(c_{N_\alpha-1} + is_{N_\alpha-1})^2} \left[\Lambda_{N_\alpha-1,r}^{(\alpha)}\right]^2 \right\}_{\varepsilon=\varepsilon_r} \right| \tag{11}$$

This translates after some manipulation into

$$\Gamma_\alpha = \left| \left\{ J_{N_\alpha-1,N_\alpha} T_{N_\alpha-1}(\varepsilon) \left[R_{N_\alpha}^{+}(\varepsilon) - R_{N_\alpha}^{-}(\varepsilon)\right] \left[\Lambda_{N_\alpha-1,r}^{(\alpha)}\right]^2 \right\}_{\varepsilon=\varepsilon_r} \right|, \tag{12}$$

where

$$J_{N_\alpha-1,N_\alpha} = \left(\lambda_\alpha^2/2\right) \sqrt{N_\alpha\,(N_\alpha + \ell + 1/2)} \tag{13}$$

$$\text{and} \quad R_{N_\alpha}^{\pm} = \frac{c_{N_\alpha}^{(\alpha)} \pm is_{N_\alpha}^{(\alpha)}}{c_{N_\alpha-1}^{(\alpha)} \pm is_{N_\alpha-1}^{(\alpha)}}, \ \text{and}\ T_{N_\alpha-1} = \frac{c_{N_\alpha-1}^{(\alpha)} - is_{N_\alpha-1}^{(\alpha)}}{c_{N_\alpha-1}^{(\alpha)} + is_{N_\alpha-1}^{(\alpha)}}. \tag{14}$$

Similar steps may be followed to set-up the method using the J-matrix Laguerre basis.

3 Method (C)

We use the same arguments leading to (8) as used in Method (B), with the exception that we improve our approximation (9) by using the exact relation

$$G^{(+)(\alpha,\alpha)}_{N_\alpha-1,N_\alpha-1}(\varepsilon) = g^{(\alpha,\alpha)}_{N_\alpha-1,N_\alpha-1}(\varepsilon)\Big/\Delta^{(+)}(\varepsilon), \tag{15}$$

where

$$\Delta^{(+)}(\varepsilon) = \begin{vmatrix} 1+g_{11}J_1R_1^+ & g_{12}J_2R_2^+ \\ g_{21}J_1R_1^+ & 1+g_{22}J_2R_2^+ \end{vmatrix}, \tag{16}$$

and we abbreviate $g_{\alpha\beta}$ as shown in the expression

$$g_{\alpha\beta} = g^{(\alpha,\beta)}_{N_\alpha-1,N_\beta-1}(\varepsilon_r) = \sum_{\mu\neq r} \frac{\Lambda^{(\alpha)}_{N_\alpha-1,\mu}\Lambda^{(\beta)}_{N_\beta-1,\mu}}{\tilde{\varepsilon}_\mu - \varepsilon_r} \tag{17}$$

We still assume that ε_r represents the resonance position. The result is a modified expression of (12) as derived from Method (B), which is now expressed as

$$\Gamma_\alpha = \left|\left\{J_{N_\alpha-1,N_\alpha}T_{N_\alpha-1}(\varepsilon)\left[R^+_{N_\alpha}(\varepsilon) - R^-_{N_\alpha}(\varepsilon)\right]\left[\Lambda^{(\alpha)}_{N_\alpha-1,r}\right]^2\Big/\Delta^{(+)}(\varepsilon)\right\}_{\varepsilon=\varepsilon_r}\right| \tag{18}$$

4 Examples and Results

In order to check these approximations we consider three examples of a two-channel problem, all with threshold energies $E_1 = 0.0$ a.u. and $E_1 = 0.1$ a.u. The matrix elements of the potential all have the form:

$$V^{\alpha\beta} = V_0^{\alpha\beta} r^2 e^{-r}, \quad \alpha = 1, 2.$$

with the following particulars:

$$\textbf{Potential (1): } V_0^{\alpha\beta} = \begin{pmatrix} -1.0 & -7.5 \\ -7.5 & 7.5 \end{pmatrix}, \tag{19a}$$

$$\textbf{Potential (2): } V_0^{\alpha\beta} = \begin{pmatrix} -1.0 & 2.0 \\ 2.0 & 7.5 \end{pmatrix}, \text{ and} \tag{19b}$$

$$\textbf{Potential (3): } V_0^{\alpha\beta} = \begin{pmatrix} -1.0 & -7.5 \\ -7.5 & -7.5 \end{pmatrix} \tag{19c}$$

Table 1 The sharp resonance of potential (1) that is located at $\varepsilon_r = 4.7682 - \mathrm{i}\ 7.1\mathrm{E}{-}4$. We used the Laguerre basis with parameters: $N = 30$, $|\lambda| = 5.0$ and $\theta = 0.15$ radians. The results of Noro and Taylor [18] are also shown

Method	Γ_1	Γ_2
Method (A)	3.5E–4	1.07E–3
Method (B)	6.6E–5	1.35E–3
Method (C)	6.4E–5	1.32E–3
Noro and Taylor [18]	6.1E–5	1.36E–3

The short-range potential (1) has been studied thoroughly [18]; it is known to possesses a sharp resonance, located at $\varepsilon = 4.7682 - \mathrm{i}\ 7.1\mathrm{E}{-}4$. The partial widths are calculated and shown in Table (1). They are in good agreement with the results of Noro and Taylor [18].

The short-range potential (2) has also received a thorough investigation [1]. It is known to possesses a sharp resonance, located at $\varepsilon = 3.66565{-}\mathrm{i}0.00955$. As shown in Table (2), the calculated partial widths compare well with the results of Rescigno and McCurdy [1].

The short-range potential (3) possesses a sharp resonance that is located at $\varepsilon = 2.1397 - \mathrm{i}\ 9.114\mathrm{E}{-}2$. As shown in Table (3), the partial widths are calculated and compared with the exact result using calculations derived from the J-matrix method [20]. The latter locates the complex resonance energy when $\Delta^{(+)}(\varepsilon)$ equals zero. Details of the exact J-matrix calculation have been provided previously [20].

It is clear that the proposed three methods endow the complex scaling method with the ability to compute resonance partial widths, and more importantly, without the need for information other than the results of the diagonalization of the complex-scaled full Hamiltonian and the tools provided by the J-matrix method.

Table 2 The sharp resonance of potential (2) that is located at $\varepsilon_r = 3.66565 - \mathrm{i}\ 9.555\mathrm{E}{-}3$. We used the Laguerre basis with parameters: $N = 30$, $|\lambda| = 5.0$ and $\theta = 0.2$ radians. The results of Rescigno and McCurdy [1] are also shown

Method	Γ_1	Γ_2
Method (A)	8.530E–4	1.826E–2
Method (B)	1.113E–3	1.800E–2
Method (C)	1.116E–3	1.805E–2
Rescigno and McCurdy [1]	1.114E–3	1.802E–2

Table 3 The sharp resonance of potential (3) that is located at $\varepsilon_r = 2.1397 - \mathrm{i}\ 9.114\mathrm{E}{-}2$. We used the Laguerre basis with parameters: $N = 30$, $|\lambda| = 5.0$ and $\theta = 0.2$ radians. The calculations of the J-matrix method [20] are also included

Method	Γ_1	Γ_2
Method (A)	0.1262	5.609E–2
Method (B)	0.1147	6.757E–2
Method (C)	0.1190	7.013E–2
J-matrix calculation [20]	0.1190	6.990E–2

References

1. Rescigno T N and McCurdy C W, Phys. Rev. A **34**, 1882 (1986)
2. Watson D K, Phys. Rev. A **34**, 1016 (1986)
3. Reinhardt W P, Ann. Rev. Phys. Chem. **90**, 3492 (1982)
4. Yamani H A and Abdelmonem M S, J. Phys. A **26**, L1183 (1993)
5. Yamani H A and Abdelmonem M S, J. Phys. A **27**, 5345 (1994)
6. Alhaidari A D, Int. J. Mod. Phys. A **20**, 2657 (2005)
7. Mandelshtam V A, Ravuri T R and Taylor H S, Phys. Rev. Lett. **70**, 1932 (1993)
8. Jang H W and Light J C, J. Chem. Phys. **99**, 1057 (1993)
9. Macek J, Phys. Rev. A **2**, 1101 (1970)
10. Schneider B I, Phys. Rev. A **24**, 1 (1981)
11. Smith F T, Phys. Rev. **118**, 349 (1960)
12. Bowman J M, J. Phys. Chem. **90**, 3492 (1986)
13. Junker B R, Adv. At. Mol. Phys. **18**, 207 (1982)
14. Froelich P, Davison E R and Brandas E, Phys. Rev. A **28**, 2641 (1983)
15. Csoto A, Phys. Rev. C **49**, 2244 (1994)
16. Watson D K, J. Phys. B **19**, 293 (1986)
17. Harris F E, Phys. Rev. Lett. **19**, 287 (1967)
18. Noro T and Taylor H S, J. Phys. B **13**, L377 (1980)
19. Yamani H A and Abdelmonem M S, J. Phys. A **29**, 6991 (1996)
20. Yamani H A and Abdelmonem M S, J. Phys. B **30**, 1633 (1997)

Part IV
Applications in Nuclear Physics

J-Matrix Approach to Loosely-Bound Three-Body Nuclear Systems

Yu.A. Lurie and A.M. Shirokov

Abstract We review our papers devoted to the studies of three-body loosely-bound nuclear systems within the J-matrix approach. In particular, we discuss the extension of the oscillator-basis J-matrix formalism on the case of true few-body scattering. The formalism is applied to loosely-bound ^{11}Li and ^{6}He nuclei treated as three-body systems $^{11}\text{Li} =^{9} \text{Li}+n+n$ and $^{6}\text{He} = \alpha+n+n$. The J-matrix formalism is used not only for the calculation of the continuum spectrum wave functions of ^{11}Li and ^{6}He nuclei excited in nuclear reactions and decaying via three-body channels $^{11}\text{Li} \to^{9} \text{Li}+n+n$ and $^{6}\text{He} \to \alpha+n+n$, but also for the calculation of the S-matrix poles associated with the ^{11}Li and ^{6}He ground states to improve the description of the binding energies and ground state properties. The J-matrix formalism is also used to derive a phase equivalent transformation of two-body interactions. The effect of the phase equivalent transformation on the properties of the ^{6}He nucleus is examined.

1 Introduction

Investigation of the decay properties of nuclear resonant states is an important component of the nuclear structure studies. Usually in such investigations only two-body decay channels are considered. Three-body decays are much more informative than the two-body ones, however the complete description of a three-body decay is a very complicated problem. Nevertheless in some cases, one can use the so-called '*democratic decay approximation*' for the successful analysis of the experimental data or for obtaining reliable theoretical predictions.

Generally, a three-body continuum spectrum wave function in the asymptotic region is a superposition of the components describing two- and three-body decay

Yu.A. Lurie
The College of Judea and Samaria, Ariel 44837, Israel
e-mail: ylurie@ycariel.yosh.ac.il

Reprinted from: Loosely-bound three-body nuclear systems in the J-matrix approach, Yu.A. Lurie and A.M. Shirokov, Annals of Physics, Vol. 312:22, pp. 284–318, Elsevier, 2004, with permission from Elsevier.

A.D. Alhaidari et al. (eds.), *The J-Matrix Method*, 183–217.

channels [1]. The democratic decay approximation accounts for the three-body channel only; two-body channels associated with the appearance of bound two-body subsystems and '*non-democratic*' subdivision of the system of the type $(2+1)$ are not allowed for in this approximation. Therefore, the approximation is valid for the study of a three-body system only in the case when all two-body decay channels are closed and the only open channel is a three-body one; in other words, this is the case when none of the two-body subsystems has a bound state.

Generally speaking, if the three-body channel is the only open one, then the system can have two types of asymptotics [1]. One of the asymptotics corresponds to the situation when one of the particles is scattered by another and the third particle is a spectator. From the general physical point of view, this asymptotics is supposed to be of little importance for a nuclear system excited in some nuclear reaction and decaying via a three-body channel. The alternative type of the asymptotics is a superposition of ingoing and outgoing six-dimensional spherical waves, it corresponds to the situation when the decaying system emits (or/and absorbs) three particles from some point in space. Allowing for the three-body asymptotics of this type only is a quintessence of the democratic decay approximation; sometimes this approximation is also referred to as *true three-body scattering* or *3 → 3 scattering*.

Nuclear structure studies in the framework of the democratic decay approximation have been started in 70th by R. I. Jibuti and collaborators. The review of the results obtained by the Tbilisi group can be found in Refs. [2,3]. Recently a considerable progress has been made in both theoretical and experimental studies of democratic decays (see reviews in Refs. [4–7]).

The generalization of the J-matrix formalism on the case of $3 \to 3$ scattering (and on a more general case of $N \to N$ scattering) was suggested in Ref. [8]. First it was successfully applied in the study of monopole excitations in ^{12}C nucleus in the cluster model $^{12}\text{C} = \alpha + \alpha + \alpha$ in Ref. [9]. Later by means of this approach we studied loosely-bound neutron-excess ^{11}Li and ^{6}He nuclei in the cluster models $^{11}\text{Li} =^9 \text{Li} + n + n$ and $^{6}\text{He} = \alpha + n + n$ [10–14] and $\Lambda\Lambda$ hypernuclei [15]. A similar approach ('algebraic model') is developed for few-cluster nuclear systems by Kiev–Antwerp group [16, 17]. In this contribution, we review the results of our studies of ^{11}Li and ^{6}He nuclei.

Exotic neutron-excess nuclei like ^{11}Li and ^{6}He are extensively studied now both theoretically [4–7, 10–14, 17–37] and experimentally [38–48]. The two-neutron separation energy in both these nuclei is very small compared with the neutron separation energy in the ^{9}Li and ^{4}He clusters; as a result the wave function of the pair of neutrons decreases slowly with distance and the rms radius of the two-neutron distribution is large compared with the rms radius of the ^{9}Li or ^{4}He core (the so-called *two-neutron halo*). We note that ^{11}Li and ^{6}He are so-called *Borromean nuclei*, i.e. none of the two-body subsystems in the cluster decompositions $^{11}\text{Li} =^9 \text{Li} + n + n$ and $^{6}\text{He} = \alpha + n + n$ has a bound state. Therefore the democratic decay approximation is adequate for the study of these nuclei at least at small enough excitation energies (less than the excitation energy of the lowest excited state in the ^{9}Li and ^{4}He clusters).

We use the J-matrix $3 \to 3$ scattering formalism for the construction of the three-body continuum spectrum wave functions of ^{11}Li and ^{6}He nuclei excited in nuclear reactions. However, probably more interesting is the application of this formalism to the study of the ground states of these nuclei treated as three-body systems $^9\text{Li} + n + n$ and $\alpha + n + n$. The idea is the following. By the J-matrix formalism we calculate the $3 \rightarrow 3$ scattering S-matrix. This calculation may be extended on the complex momentum plane. Hence we can locate numerically the S-matrix poles associated with the bound states. The S-matrix pole calculations improve the variational results for binding energies obtained by the pure diagonalization of the truncated Hamiltonian matrix. Knowing the binding energies, we calculate the bound state wave functions by means of the J-matrix formalism.

The first calculation of the S-matrix poles within the J-matrix formalism was performed in Ref. [49] where resonances in α–α scattering were investigated. The S-matrix poles in He atom and H^- ion were calculated in Ref. [50, 51] in the model suggested by Broad and Reinhardt [52]. The resonance energies and widths were reproduced with a high accuracy. The S-matrix poles associated with the bound states were also calculated in Ref. [50, 51]. The binding energies were reproduced with a very high accuracy; we note also that a very large number of bound states was obtained by means of the S-matrix pole calculations: it was not only much larger than the number of bound states obtained variationally, but even much larger than the rank of the truncated Hamiltonian matrix. The S-matrix pole calculation of binding energies was also used in Refs. [10–15]. This approach not only provides an essential improvement in calculations of binding energies, but also makes it possible to calculate the ground state wave function with the correct asymptotics. It is very important for halo nuclei like ^{11}Li and ^{6}He due to the slow decrease of the wave function in the asymptotic region that cannot be reproduced by a superposition of a finite number of oscillator functions obtained in variational calculation. As a result, the rms radius, electromagnetic transition probabilities and other observables are improved essentially.

We note here that the S-matrix poles associated with bound states, play a crucial role in the J-matrix inverse scattering theory [53, 54]. This formalism was utilized for the construction of a novel high-quality nucleon-nucleon interaction [54–56]; the deuteron properties (binding energy, rms radius, quadrupole moment) provided by this interaction, are obtained in the J-matrix method by the numerical calculation of the S-matrix pole.

In this contribution, we also discuss a phase equivalent transformation based on the J-matrix formalism suggested in Ref. [13]. This transformation was used to obtain families of phase equivalent n–α potentials used in the ^{6}He calculation [14].

We start the discussion with a brief sketch of the $N \rightarrow N$ scattering J-matrix formalism suggested in Ref. [8].

2 *J*-Matrix Formalism with Hyperspherical Oscillator Basis

The wave function Ψ of a system of A particles is generally a function of A coordinates of individual particles $\boldsymbol{r}_i$. The center of mass motion can be separated and eliminated; as a result we can treat the wave function Ψ as a function of $A-1$ Jacobi coordinates $\boldsymbol{\xi}_i$. Within the democratic decay approximation, it is natural to employ the hyperspherical harmonics formalism (see, e.g., [3, 57]). The hyperspherical harmonics are now widely used in various nuclear applications including the studies of continuum states of nuclear systems [4–7, 9–17, 20–24, 31, 32, 35, 37, 58]. The $3A-3$ independent variables $\boldsymbol{\xi}_i$ are rearranged in this formalism: the so-called hyperradius $\rho = \sqrt{\sum_{i=1}^{A-1} \xi_i^2}$ is introduced and all the rest $3A - 4$ variables are the angles on the $(3A-3)$-dimensional sphere. The wave function is searched for in the form

$$\Psi = \sum_{\Gamma} d_\Gamma \Psi_\Gamma, \tag{1}$$

where

$$\Psi_\Gamma \equiv \Psi_{K\gamma} = \Phi_{K\gamma}(\rho)\mathscr{Y}_{K\gamma}(\Omega), \tag{2}$$

Ω is the set of angles on the $(3A-3)$-dimensional sphere, K is the so-called hypermomentum [physically K is the angular momentum in the $(3A-3)$-dimensional space; $K \geq L$ where L is the orbital angular momentum] and γ stands for all the rest quantum numbers labeling the hyperspherical function $\mathscr{Y}_{K\gamma}(\Omega)$, the multi-index $\Gamma = \{K, \gamma\}$. Various analytic expressions may be found for the hyperspherical functions $\mathscr{Y}_{K\gamma}(\Omega)$ in textbooks (see, e.g., [3]).

In the two-body case, ρ is proportional to the distance between the particles $r = |\boldsymbol{r}_1 - \boldsymbol{r}_2|$ and $\mathscr{Y}_{K\gamma}(\Omega)$ becomes the usual spherical function $Y_{LM}(\Omega)$ with K and γ playing the role of the angular momentum L and its projection M, respectively. Therefore Eq. (2) is a generalization on the A-body ($A \geq 3$) case of the conventional representation of the wave function used in the study of spherically symmetric two-body systems. However, contrary to the case of two-body systems with central interaction which impose the spherical symmetry, the hypermomentum K is not an integral of motion: the two-body interactions in the A-body system mix the states with different values of the hypermomentum K and we have a set of coupled equations for the hyperradial functions $\Phi_{K\gamma}(\rho)$. This set of equations has the same structure as the set of equations describing two-body systems with non-central forces; from the point of view of scattering theory, this set of coupled equations is formally equivalent to the one describing multichannel scattering in the system with the hyperspherical channels Γ.

We introduce the hyperspherical oscillator basis

$$|\kappa\Gamma\rangle \equiv |\kappa K\gamma\rangle = \mathscr{R}_{\kappa K}(\rho)\, \mathscr{Y}_{K\gamma}(\Omega), \tag{3}$$

where the hyperradial oscillator function

$$\mathscr{R}_{\kappa K}(\rho) \equiv \mathscr{R}_\kappa^{\mathscr{L}}(\rho) = \rho^{-(3A-4)/2}\, \mathfrak{r}_{\kappa K}(\rho), \tag{4}$$

$$\mathfrak{r}_{\kappa K}(\rho) \equiv \mathfrak{r}_\kappa^{\mathscr{L}}(\rho) = (-1)^\kappa \sqrt{\frac{2\kappa!}{\Gamma(\kappa + \mathscr{L} + 3/2)}}\; \rho^{\mathscr{L}+1}\, e^{-\rho^2/2}\, L_\kappa^{\mathscr{L}+\frac{1}{2}}(\rho^2), \tag{5}$$

$L_n^\alpha(x)$ is the associated Laguerre polynomial and

$$\mathscr{L} = K + \frac{3A-6}{2}. \tag{6}$$

In the two-body case ($A = 2$), $\mathscr{L}$ is equivalent to the orbital angular momentum L, $\mathscr{L} = L$, and Eqs. (3)–(6) define the conventional oscillator basis. In the many-body case ($A \geq 3$), Eqs. (3)–(6) define the A-body oscillator functions, i.e. the eigenfunctions of the A-body Schrödinger equation with the potential energy

$$U = \frac{\omega^2}{2} \sum_{i=1}^{A} m_i \, (\boldsymbol{r}_i - \boldsymbol{R})^2 = \frac{\hbar\omega}{2} \, \rho^2, \tag{7}$$

where m_i is the mass of the ith particle, $\boldsymbol{R}$ is the center-of-mass coordinate and ω is the parameter of the oscillator basis. The corresponding eigenenergy is

$$E_{\kappa K} = \left(N + \frac{3A-3}{2} \right) \hbar\omega, \tag{8}$$

where the oscillator quanta

$$N = 2\kappa + K = 2\kappa + \mathscr{L} - \frac{3A-6}{2}. \tag{9}$$

The hyperspherical oscillator basis (3) is orthonormalized:

$$\langle \kappa K \gamma | \kappa' K' \gamma' \rangle = \delta_{\kappa\kappa'} \delta_{\Gamma\Gamma'}, \tag{10}$$

$$\int_0^\infty \mathfrak{r}^*_{\kappa K}(\rho) \mathfrak{r}_{\kappa' K}(\rho) \, \mathrm{d}\rho = \delta_{\kappa\kappa'}. \tag{11}$$

The wave function Ψ is expanded in the hyperspherical oscillator function series,

$$\Psi = \sum_{\kappa, K, \gamma} \langle \kappa K \gamma | \Psi \rangle \, | \kappa K \gamma \rangle, \tag{12}$$

and the Schrödinger equation takes the form

$$\sum_{\kappa', K', \gamma'} \langle \kappa K \gamma | H - E | \kappa' K' \gamma' \rangle \, \langle \kappa' K' \gamma' | \Psi \rangle = 0, \tag{13}$$

where E is the energy, the Hamiltonian $H = T + V$, the potential energy V is usually a superposition of two-body interactions V_{ij}, $V = \sum_{i<j} V_{ij}$, and T is the A-body kinetic energy operator. Generally, the Hamiltonian matrix $\langle \kappa K \gamma | H | \kappa' K' \gamma' \rangle$

is infinite. However, within the J-matrix formalism we truncate the potential energy matrix; it is most natural to define the truncation boundary using the oscillator quanta, i.e. we suppose that

$$\langle \kappa K \gamma | V | \kappa' K' \gamma' \rangle = 0 \quad \text{if} \quad 2\kappa + K > \tilde{N} \quad \text{or} \quad 2\kappa' + K' > \tilde{N}. \tag{14}$$

The kinetic energy matrix $\langle \kappa K \gamma | T | \kappa' K' \gamma' \rangle$ is tridiagonal,

$$\langle \kappa K \gamma | T | \kappa' K' \gamma' \rangle = \frac{\hbar\omega}{2} \, \delta_{KK'} \, \delta_{\gamma\gamma'} \left[-\sqrt{(\kappa+1)\left(\kappa + \mathcal{L} + \frac{3}{2}\right)} \, \delta_{\kappa+1,\kappa'} + \left(2\kappa + \mathcal{L} + \frac{3}{2}\right) \delta_{\kappa\kappa'} - \sqrt{\kappa\left(\kappa + \mathcal{L} + \frac{1}{2}\right)} \, \delta_{\kappa-1,\kappa'} \right]. \tag{15}$$

It is seen from Eqs. (13)–(15) that the A-body hyperspherical J-matrix formalism is very close to the conventional J-matrix formalism with oscillator basis used in multichannel two-body problems. The main difference is that $\mathcal{L}$ entering Eq. (15) is half-integer if the number of particles A is odd [see Eq. (6) and note that K is always integer], in particular, $\mathcal{L}$ is half-integer in the three-body case discussed in this contribution. Hence the standard regular and irregular two-body J-matrix oscillator-basis solutions of the free Schrödinger equation that can be found, e.g., in Ref. [59, 60], cannot be used in the three-body case. The general J-matrix oscillator-basis solutions that may be used in the case of arbitrary $\mathcal{L}$, were suggested in Ref. [8]. The regular solution is

$$S_{\kappa K}(q) = \sqrt{\frac{2\kappa!}{\Gamma(\kappa + \mathcal{L} + 3/2)}} \, q^{\mathcal{L}+1} \, e^{-q^2/2} \, L_\kappa^{\mathcal{L}+\frac{1}{2}}(q^2), \tag{16}$$

the irregular solutions are

$$C_{\kappa K}(q) = -\frac{2q}{\pi S_{0K}(q)} \, \mathbf{V.P.} \int_0^\infty \frac{S_{0K}(q') \, S_{\kappa K}(q')}{q^2 - q'^2} \, dq', \tag{17}$$

$$C^{(+)}_{\kappa K}(q) = -\frac{2q}{\pi S_{0K}(q)} \int_0^\infty \frac{S_{0K}(q') \, S_{\kappa K}(q')}{q^2 - q'^2 + i0} \, dq', \tag{18}$$

$$C^{(-)}_{\kappa K}(q) = -\frac{2q}{\pi S_{0K}(q)} \int_0^\infty \frac{S_{0K}(q') \, S_{\kappa K}(q')}{q^2 - q'^2 - i0} \, dq', \tag{19}$$

where $q = \sqrt{\frac{2E}{\hbar\omega}}$, **V. P.** in Eq. (17) indicates the principal value integral, $+i0$ and $-i0$ in Eqs. (18)–(19) show how the poles of the integrand should be treated. The solutions (16)–(19) are not independent, the following relation between them is valid:

$$C^{(\pm)}_{\kappa K}(q) = C_{\kappa K}(q) \pm i\, S_{\kappa K}(q). \tag{20}$$

However, any pair of the solutions (16)–(19) can be used to construct an arbitrary solution.

The regular solution (16) is just the hyperradial momentum-space oscillator function. The irregular solutions (17)–(19) are more complicated. They were analyzed in detail in Ref. [8]. In the general case, they can be expressed through Tricomi function. However in the case of even A, the irregular solutions can be simplified and expressed through the confluent hypergeometric function. Of course, the expressions (16)–(19) in the two-body case can be reduced to the well-known expressions given, e.g., in Refs. [59,60]. The physical meaning of the solutions (16)–(19) is clear from the following expressions [8]:

$$\sum_{\kappa=0}^{\infty} S_{\kappa K}(q)\,\mathfrak{r}_{\kappa K}(\rho) = \sqrt{q\rho}\; J_{\mathscr{L}+\frac{1}{2}}(q\rho) \underset{\rho\to\infty}{\longrightarrow} \sqrt{\frac{2}{\pi}}\, \sin\!\left(q\rho - \frac{\pi\mathscr{L}}{2}\right), \tag{21}$$

$$\sum_{\kappa=0}^{\infty} C_{\kappa K}(q)\,\mathfrak{r}_{\kappa K}(\rho) \underset{\rho\to\infty}{\longrightarrow} -\sqrt{q\rho}\; N_{\mathscr{L}+\frac{1}{2}}(q\rho) \underset{\rho\to\infty}{\longrightarrow} \sqrt{\frac{2}{\pi}}\, \cos\!\left(q\rho - \frac{\pi\mathscr{L}}{2}\right), \tag{22}$$

$$\sum_{\kappa=0}^{\infty} C^{(+)}_{\kappa K}(q)\,\mathfrak{r}_{\kappa K}(\rho) \underset{\rho\to\infty}{\longrightarrow} i\,\sqrt{q\rho}\; H^{(1)}_{\mathscr{L}+\frac{1}{2}}(q\rho) \underset{\rho\to\infty}{\longrightarrow} \sqrt{\frac{2}{\pi}}\, e^{i(q\rho-\frac{\pi\mathscr{L}}{2})}, \tag{23}$$

$$\sum_{\kappa=0}^{\infty} C^{(-)}_{\kappa K}(q)\,\mathfrak{r}_{\kappa K}(\rho) \underset{\rho\to\infty}{\longrightarrow} -i\,\sqrt{q\rho}\; H^{(2)}_{\mathscr{L}+\frac{1}{2}}(q\rho) \underset{\rho\to\infty}{\longrightarrow} \sqrt{\frac{2}{\pi}}\, e^{-i(q\rho-\frac{\pi\mathscr{L}}{2})}. \tag{24}$$

Here $J_\alpha(x)$, $N_\alpha(x)$ and $H^{(1,2)}_\alpha(x)$ are Bessel, Neumann and Hankel functions, respectively.

In the case of continuum spectrum ($E > 0$), the oscillator space wave function $\langle \kappa K \gamma | \Psi^{(\Gamma')}\rangle \equiv \langle \kappa \Gamma | \Psi^{(\Gamma')}\rangle$ in the channel Γ is of the form

$$\langle \kappa\Gamma|\Psi^{(\Gamma')}\rangle = \frac{1}{2}\left(\delta_{\Gamma\Gamma'}\, C^{(-)}_{\kappa K}(q) - C^{(+)}_{\kappa K}(q)\,[\mathbf{S}]_{\Gamma\Gamma'}\right) \tag{25}$$

in the '*external region*' $\kappa \geq \kappa_\Gamma$, where $\kappa_\Gamma = (\tilde{N} - K)/2$ is the potential energy truncation boundary in the channel Γ. It is supposed that the ingoing spherical wave is present in the channel Γ' only while the outgoing spherical waves are present in all channels; $[\mathbf{S}]_{\Gamma\Gamma'}$ entering Eq. (25) is the matrix element of the hyperspherical S-matrix. The S-matrix $[\mathbf{S}]$ can be calculated by the following formula:

$$[\mathbf{S}] = \left[\mathbf{A}^s\right]^{-1}\left[\mathbf{B}^s\right], \tag{26}$$

where the matrix elements of the matrices $[\mathbf{A}^s]$ and $[\mathbf{B}^s]$ are

$$\left[\mathbf{A}^s\right]_{\Gamma'\Gamma} = \langle \kappa_{\Gamma'}\Gamma'|\mathscr{P}|\kappa_\Gamma+1,\Gamma\rangle\, C^{(+)}_{\kappa_\Gamma+1,\,K}(q) - \delta_{\Gamma\Gamma'}\, C^{(+)}_{\kappa_\Gamma K}(q), \tag{27}$$

$$\left[\mathbf{B}^s\right]_{\Gamma'\Gamma} = \langle \kappa_{\Gamma'}\Gamma'|\mathscr{P}|\kappa_\Gamma+1,\Gamma\rangle\, C^{(-)}_{\kappa_\Gamma+1,\,K}(q) - \delta_{\Gamma\Gamma'}\, C^{(-)}_{\kappa_\Gamma K}(q); \tag{28}$$

the matrix elements of the matrix $[\mathscr{P}]$ proportional to the discrete analog of the P-matrix (see Ref. [60] for details) are

$$\langle\kappa\Gamma|\mathscr{P}|\kappa_{\Gamma'}+1,\Gamma'\rangle = \langle\kappa\Gamma|\mathfrak{P}|\kappa_{\Gamma'}\Gamma'\rangle\,\langle\kappa_{\Gamma'}\Gamma'|T|\kappa_{\Gamma'}+1,\Gamma'\rangle, \tag{29}$$

the kinetic energy matrix elements $\langle\kappa_{\Gamma'}\Gamma'|T|\kappa_{\Gamma'}+1,\Gamma'\rangle$ are given by Eq. (15), and the matrix $[\mathfrak{P}] \equiv [H-E]^{-1}$ defined in the truncated model space spanned by oscillator functions (3) with $\kappa \le \kappa_\Gamma = (\tilde{N}-K)/2$ in each channel Γ, has the matrix elements

$$\langle\kappa\Gamma|\mathfrak{P}|\kappa'\Gamma'\rangle = \sum_\lambda \frac{\langle\kappa\Gamma|\lambda\rangle\langle\lambda|\kappa'\Gamma'\rangle}{E-E_\lambda}. \tag{30}$$

In Eq. (30), E_λ are eigenenergies and $\langle\kappa\Gamma|\lambda\rangle$ are the respective eigenvectors of the truncated Hamiltonian matrix, i.e. E_λ and $\langle\kappa\Gamma|\lambda\rangle$ can be found by solving Eq. (13) supposing that $\kappa \le \kappa_\Gamma$ and $\kappa' \le \kappa_{\Gamma'}$. The results obtained by diagonalization of the truncated Hamiltonian matrix we shall refer to as *variational results.*

After performing variational calculation, we obtain the S-matrix with the help of Eqs. (26)–(30) and the oscillator space wave function $\langle\kappa\Gamma|\Psi^{(\Gamma')}\rangle$ in the external region $\kappa \ge \kappa_\Gamma$ by the formula (25). The oscillator space wave function $\langle\kappa\Gamma|\Psi^{(\Gamma')}\rangle$ in the '*internal region*' $\kappa \le \kappa_\Gamma$ can be now calculated as

$$\langle\kappa\Gamma|\Psi^{(\Gamma')}\rangle = \sum_{\Gamma''} \langle\kappa\Gamma|\mathscr{P}|\kappa_{\Gamma''}+1,\Gamma''\rangle\,\langle\kappa_{\Gamma''}+1,\Gamma''|\Psi^{(\Gamma')}\rangle, \qquad \kappa \le \kappa_\Gamma. \tag{31}$$

Equation (26) can be used in the complex momentum plane; the bound state energies are associated with the S-matrix poles that can be found by the numerical solution of the obvious equation

$$\det\left[\mathbf{A}^s\right] = 0. \tag{32}$$

The matrix $[\mathbf{A}^s]$ in Eq. (32) is the extension on the complex momentum plane of Eq. (27); for the the bound states with $E = \frac{1}{2}q^2\hbar\omega < 0$ its matrix elements are

$$\left[\mathbf{A}^s\right]_{\Gamma'\Gamma} = \langle\kappa_{\Gamma'}\Gamma'|\mathscr{P}|\kappa_\Gamma+1,\Gamma\rangle\, C^{(b)}_{\kappa_\Gamma+1,\,K}(q) - \delta_{\Gamma\Gamma'}\, C^{(b)}_{\kappa_\Gamma K}(q), \tag{33}$$

where

$$C^{(b)}_{\kappa K}(q) = C^{(+)}_{\kappa K}(i|q|). \tag{34}$$

The relation (25) for the oscillator space wave function in the external region $\kappa \ge \kappa_\Gamma$ should be replaced for bound states by

$$\langle\kappa\Gamma|\Psi\rangle = -[\mathbf{S}]_{\Gamma\Gamma}\, C^{(b)}_{\kappa K}(q); \tag{35}$$

the multipliers $[\mathbf{S}]_{\Gamma\Gamma}$ are obtained by the numerical solution of the equation

$$[\mathbf{S}]_{\Gamma\Gamma}\, C^{(b)}_{\kappa_\Gamma K}(q) = \sum_{\Gamma'} \langle \kappa_\Gamma \Gamma | \mathscr{P} | \kappa_{\Gamma'} + 1, \Gamma' \rangle \; [\mathbf{S}]_{\Gamma'\Gamma'}\, C^{(b)}_{\kappa_{\Gamma'}+1,\, K}(q) \qquad (36)$$

[we note that Eq. (32) is the condition of solvability of Eq. (36) for the bound states]. Equation (31) can be used for the calculation of the bound state oscillator space wave function $\langle \kappa\Gamma | \Psi^{(\Gamma')} \rangle$ in the internal region $\kappa \le \kappa_\Gamma$. Since the set of $[\mathbf{S}]_{\Gamma\Gamma}$ can be obtained from Eq. (36) up to a common multiplier only, the bound state wave function should be normalized numerically.

The bound state energies and wave functions obtained by numerical calculation of the S-matrix poles, will be referred to as *J-matrix results*.

In practical applications, it is useless to allow for all possible channels $\Gamma = \{K, \gamma\}$ in the external region $N \equiv 2\kappa + K > \tilde{N}$. We start the calculations allowing for the channels Γ with few minimal possible values of K in the external region only, i.e. we allow for the channels with all possible values of $K \le K_{\rm tr}$ and all possible γ values for a given K in the external region. Therefore the summation over Γ'' in (31) is restricted to these allowed channels. However all possible channels with $K \le \tilde{N}$ are allowed for in the internal region $N \equiv 2\kappa + K \le \tilde{N}$. As a next step, we increase the value of $K_{\rm tr}$ and allow for more channels Γ in the external region. The convergence of all results (binding energies and other bound state observables, transition probabilities, etc.) with $K_{\rm tr}$ is carefully examined. The convergence is usually achieved at small enough values of $K_{\rm tr}$ (much less than $\tilde{N}$). Such converged with respect to $K_{\rm tr}$ results are discussed below.

3 Application to ^{11}Li Nucleus

The ^{11}Li nucleus is studied in the three-body cluster model, i.e. ^{11}Li is regarded as a system of three interacting bodies: ^{9}Li and two neutrons. The three-body cluster model for ^{11}Li was first suggested in Ref. [18]. The ^{11}Li three-body binding energy or equivalently the ^{11}Li two-neutron separation energy is 0.295 ± 0.035 MeV [43] that is much less than the excitation energy of the lowest excited state in the cluster ^{9}Li. As it was already noted, ^{11}Li is a Borromean nuclei, i.e. none of the two-body subsystems $^{9}\mathrm{Li} + n$ and $n + n$ has a bound state.

The two-body potentials are needed for the investigation of the three-body system $^{11}\mathrm{Li} =^{9} \mathrm{Li} + n + n$. Unfortunately the information about the $n{-}^{9}$Li interaction is scarce. As a result, in Ref. [18] the $n{-}^{9}$Li potential was phenomenologically parametrized and fitted to the ^{11}Li binding energy. Therefore it is reasonable to use a somewhat simplified three-body model with simplified interactions to avoid computational difficulties; a microscopic extension of the model seems to be useless due to uncertainties in the $n{-}^{9}$Li interaction.

We employ the $n{-}n$ and $n{-}^{9}$Li potentials of Ref. [18]. The Gaussian $n{-}n$ potential

$$U(r) = -V_0\, e^{-(r/R)^2} \tag{37}$$

was fitted to the low-energy singlet ($S = 0$) s wave phase shifts, its depth $V_0 = 31$ MeV and its width $R = 1.8$ fm. The n–^{9}Li potential is of a two-Gaussian form,

$$U(r) = -V_1\, e^{-(r/R_1)^2} - V_2\, e^{-(r/R_2)^2}, \tag{38}$$

with $V_1 = 7$ MeV, $R_1 = 2.4$ fm, $V_2 = 1$ MeV, and $R_2 = 3.0$ fm. It is supposed that two valent neutrons in ^{11}Li are in a singlet spin state. The total angular momentum J in ^{11}Li results from the coupling of the spin $\frac{3}{2}$ of the ^{9}Li cluster with the relative motion orbital angular momentum L; however we do not make use of the spin-flip operators in our investigation, therefore we can disregard the spin variables and exclude them from the wave function. The ^{11}Li ground state orbital angular momentum L is supposed to be zero. We can define the channels Γ as $\Gamma = \{K, l_x, l_y, L\}$ where l_x and l_y are orbital angular momenta corresponding to the Jacobi coordinates

$$\boldsymbol{x} = \sqrt{\frac{\omega}{\hbar}\,\frac{m_1 m_2}{m_1 + m_2}}\,(\boldsymbol{r}_1 - \boldsymbol{r}_2), \tag{39}$$

$$\boldsymbol{y} = \sqrt{\frac{\omega}{\hbar}\,\frac{m_3(m_1 + m_2)}{m_1 + m_2 + m_3}}\,\left(\frac{m_1\boldsymbol{r}_1 + m_2\boldsymbol{r}_2}{m_1 + m_2} - \boldsymbol{r}_3\right); \tag{40}$$

the angular momenta l_x and l_y are coupled to the orbital angular momentum L. The wave function is given by the general formula (12). The oscillator basis (3) in our case takes the form

$$|\kappa\Gamma\rangle \equiv |\kappa K l_x l_y : LM\rangle \equiv \langle\rho|\kappa K l_x l_y : LM\rangle = \mathscr{R}_{\kappa,K}(\rho)\,\mathscr{Y}_{K\gamma}(\Omega), \tag{41}$$

where the hyperspherical functions

$$\begin{aligned}\mathscr{Y}_{K\gamma}(\Omega) &\equiv \mathscr{Y}_{K l_x l_y LM}(\Omega)\\ &= N_K^{l_x l_y}\,\cos^{l_x}\alpha\,\sin^{l_y}\alpha\,P_n^{l_y+1/2,\,l_x+1/2}(\cos 2\alpha)\\ &\quad\sum_{m_x+m_y=M}(l_x m_x l_y m_y|LM)\,Y_{l_x,m_x}(\hat{\boldsymbol{x}})\,Y_{l_y,m_y}(\hat{\boldsymbol{y}}),\end{aligned} \tag{42}$$

$\alpha = \tan^{-1}(y/x)$, Y_{lm} is the spherical function, $(l_x m_x l_y m_y|LM)$ is the Clebsch–Gordan coefficient, $P_n^{\alpha,\beta}(x)$ is Jacobi polynomial [61], $n = (K - l_x - l_y)/2$, and the normalization multiplier

$$N_K^{l_x l_y} = \sqrt{\frac{2n!\,(K+2)\,(n + l_x + l_y)!}{\Gamma(n + l_x + 3/2)\,\Gamma(n + l_y + 3/2)}}\,.$$

The basis (41) is useful for the calculation of three-body decays. However it is not convenient in the calculation of the matrix elements of two-body potentials V_{ij}. These matrix elements can be easily calculated in the basis

$$\begin{aligned} |n_x l_x n_y l_y : LM\rangle &\equiv \langle \boldsymbol{x}\,\boldsymbol{y}|n_x l_x n_y l_y : LM\rangle \\ &= \sum_{m_x+m_y=M} (l_x m_x l_y m_y|LM)\,\langle \boldsymbol{x}|n_x l_x m_x\rangle\,\langle \boldsymbol{y}|n_y l_y m_y\rangle, \end{aligned} \tag{43}$$

where $\langle \boldsymbol{x}|n_x l_x m_x\rangle$ and $\langle \boldsymbol{y}|n_y l_y m_y\rangle$ are the conventional three-dimensional oscillator functions depending on the Jacobi coordinates $\boldsymbol{x}$ and $\boldsymbol{y}$, respectively. The unitary transformations relating the basises (41) and (43),

$$|\kappa K l_x l_y : LM\rangle = \sum_{n_x,n_y} \langle n_x l_x n_y l_y|\kappa K\rangle\,|n_x l_x n_y l_y : LM\rangle, \tag{44}$$

$$|n_x l_x n_y l_y : LM\rangle = \sum_{\kappa,K} \langle n_x l_x n_y l_y|\kappa K\rangle\,|\kappa K l_x l_y : LM\rangle, \tag{45}$$

are discussed in detail in Refs. [3,57]. The transformation coefficients $\langle n_x l_x n_y l_y|\kappa K\rangle$ are non-zero if only the oscillator quanta N of the basis functions (41) and (43) are the same, i.e. when $2n_x + l_x + 2n_y + l_y = 2\kappa + K$.

We suppose that the 1st and the 2nd particles are neutrons and the 3rd particle is the ^{9}Li cluster (or the ^{4}He cluster in the next section). Therefore the Jacobi coordinate $\boldsymbol{x}$ is proportional to the distance between the valent neutrons and the Jacobi coordinate $\boldsymbol{y}$ is proportional to the distance between the cluster and the center of mass of two valent neutrons. In this case the basis (43) can be used directly for the calculation of the $n-n$ interaction matrix elements. For the calculation of the matrix elements of the $n-$cluster interaction, we renumerate the particles assigning number 1 to the cluster and numbers 2 and 3 to the valent neutrons, and use Eqs. (39)–(40) to define another set of Jacobi coordinates $\boldsymbol{x}'$ and $\boldsymbol{y}'$. The $n-$cluster interaction matrix elements can be directly calculated now in the basis (43). The unitary transformation of the wave functions associated with the switching from one set of Jacobi coordinates ($\boldsymbol{x}$ and $\boldsymbol{y}$) to another ($\boldsymbol{x}'$ and $\boldsymbol{y}'$), is discussed in detail in Refs. [3,57].

The ^{11}Li ground state energy dependence on the oscillator basis parameter $\hbar\omega$ is shown on Fig. 1 for a set of values for the truncation boundary $\tilde{N}$. The dependence is typical for variational calculations and has a minimum at $\hbar\omega \approx 6.5$ MeV (the best convergence $\hbar\omega$ value). The $\hbar\omega$ ground state energy dependence obtained in variational and J-matrix calculations are of the same type. However this dependence is much less pronounced in the J-matrix calculations (note a different scale on the left and right panels of Fig. 1). Therefore the J-matrix calculation is much less sensitive to the choice of the $\hbar\omega$ value. The J-matrix ground state energy results are better than the variational ones for any value of $\hbar\omega$ and any value of $\tilde{N}$.

This is clearly seen on Fig. 2 where we present the convergence with $\tilde{N}$ of the ^{11}Li ground state energy and rms matter radius

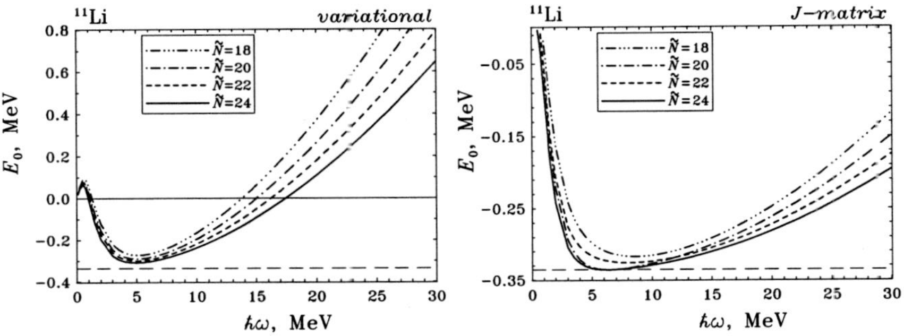

Fig. 1 The ^{11}Li ground state energy vs the oscillator basis parameter $\hbar\omega$ in the variational (*left panel*) and J-matrix (*right panel*) calculations for various values of the truncation boundary $\tilde{N}$. The horizontal dashed line depicts the convergence limit for the ground state energy

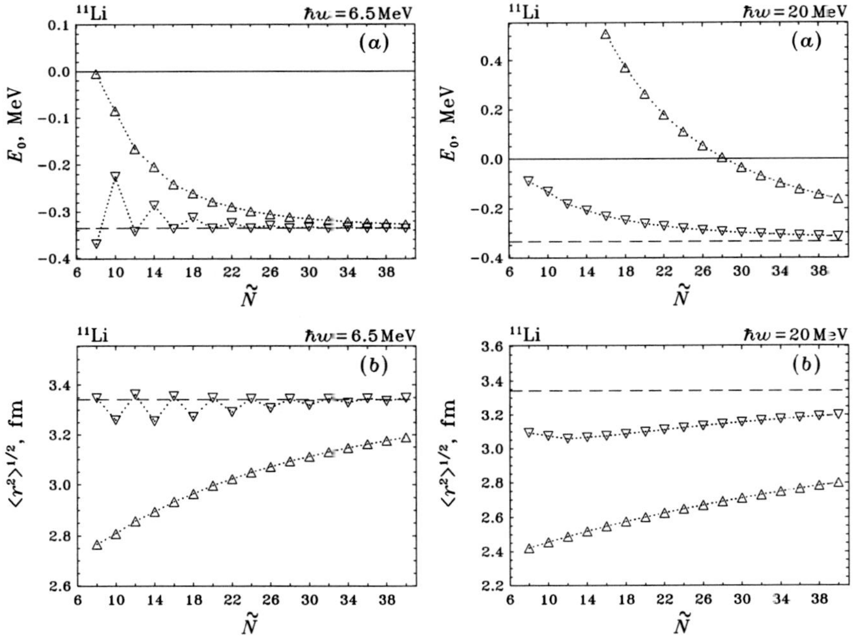

Fig. 2 Convergence with $\tilde{N}$ of the ^{11}Li three-body ground state energy (**a**) and rms matter radius (**b**) for $\hbar\omega$=6.5 MeV (*left panel*) and $\hbar\omega$=20 MeV (*right panel*). Variational and J-matrix results are shown by triangles up and triangles down, respectively. The convergence limits are shown by the dashed line

$$\langle r^2 \rangle^{1/2} = \sqrt{\frac{A-2}{A} \langle r_c^2 \rangle + \langle \rho^2 \rangle}\,, \tag{46}$$

where the ^{11}Li mass number $A = 11$ and $\langle r_c^2 \rangle^{1/2}$ is the rms radius of the ^{9}Li cluster. The variational ground state energy decreases monotonically due to the variational principal; the variational rms radius consequently monotonically increases. We note here that the variational principle is not applicable to the J-matrix calculations where the infinite oscillator basis is allowed for and the results obtained with different $\tilde{N}$ values differ by the rank of the potential energy matrix. The J-matrix results have a more intriguing dependence on $\tilde{N}$ than the variational ones. For the $\hbar\omega$ values close to the best convergence value of $\hbar\omega = 6.5$ MeV, there is a staggering of the J-matrix binding energy and rms radius with $\tilde{N}$ (Fig. 2, left panel). The amplitude of the staggering decreases with $\tilde{N}$ and the results converge rapidly. It is interesting that the binding energy and the rms radius converge to the values lying between the results obtained with two subsequent values of $\tilde{N}$. Hence we obtain not only the lower bound binding energy and rms radius estimates as in variational calculations but both lower bound and upper bound estimates for these observables. These estimates are presented in Table 1 together with experimental data and the results of calculations of other authors.

If $\hbar\omega$ differs much from the best convergence value, we have the conventional variational-type monotonic convergence of the J-matrix ground state energy and rms radius (Fig. 2, right panel). The results converge slower with $\tilde{N}$ in this case, however much faster than in the variational calculation.

The structure of the ground state wave function is the following. The total weight of the $l_x = l_y = 0$ components is 99.15%, in particular the $K = 0$ component contribution is 94.4% and the $K = 2$, $l_x = l_y = 0$ component contribution is 3.3%. This result agrees well with the results of other authors who used the three-body cluster model of ^{11}Li and the two-body interactions of the same type. In particular, the contribution of the $l_x = l_y = 0$ components was estimated in Ref. [18] as 99%. According to Ref. [20], the weight of the dominant $K = 0$, $l_x = l_y = 0$ wave function component is 95.3%, and the contribution of this component together with the next $K = 2$, $l_x = l_y = 0$

Table 1 The ^{11}Li two-neutron separation energy $E(2n)$ and rms matter radius $\langle r^2 \rangle^{1/2}$ obtained in the calculations with $\hbar\omega = 6.5$ MeV together with the results of theoretical studies of Ref. [20] and experimental data

Approximation	$E(2n)$, MeV		$\langle r^2 \rangle^{1/2}$, fm	
	$\tilde{N} = 38$	$\tilde{N} = 40$	$\tilde{N} = 38$	$\tilde{N} = 40$
Variational	0.326	0.327	3.176	3.189
J-matrix	0.335	0.336	3.336	3.348
Ref. [20]	0.3		3.32	
Experimental data	0.247 ± 0.080, Ref. [62] 0.295 ± 0.035, Ref. [43]		3.10 ± 0.17, Ref. [39] 3.53 ± 0.10, Ref. [27] 3.55 ± 0.10, Ref. [28]	

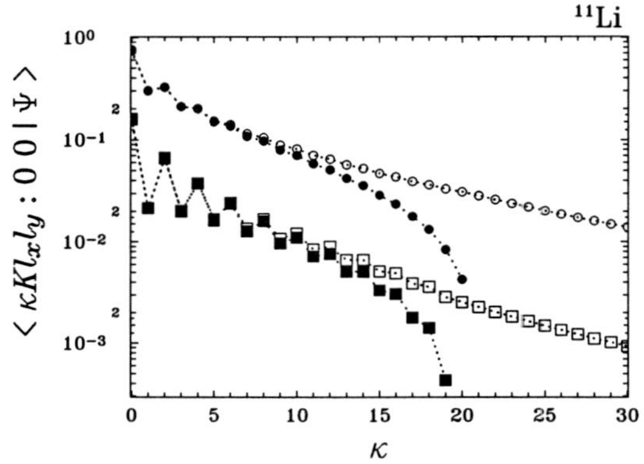

Fig. 3 The dominant $K = 0$ and $K = 2, l_x = l_y = 0$ components of the ^{11}Li ground state wave function in the oscillator representation, $\langle\kappa,\ K = 0,\ l_x = 0,\ l_y = 0 : 00\,|\,\Psi\rangle$ (circles) and $\langle\kappa,\ K = 2,\ l_x = 0,\ l_y = 0 : 00\,|\,\Psi\rangle$ (squares), obtained in the calculations with $\tilde{N} = 40$ and $\hbar\omega = 6.5$ MeV. The bold and empty symbols are the variational and the J-matrix results, respectively

component is 98.4%. The dominant components of the ^{11}Li ground state wave function in the oscillator representation, $\langle\kappa,\ K = 0,\ l_x = 0,\ l_y = 0 : 00\,|\,\Psi\rangle$ and $\langle\kappa,\ K = 2,\ l_x = 0,\ l_y = 0 : 00\,|\,\Psi\rangle$, are shown in Fig. 3. It is seen from the figure that the low-κ dominating oscillator components are well reproduced in the variational calculation. However as κ approaches the truncation boundary κ_Γ, the variational calculation underestimates $\langle\kappa,\ K = 0,\ l_x = 0,\ l_y = 0 : 00\,|\,\Psi\rangle$ and $\langle\kappa,\ K = 2,\ l_x = 0,\ l_y = 0 : 00\,|\,\Psi\rangle$ essentially.

A very interesting problem is the problem of low-energy electromagnetic transitions in ^{11}Li and other neutron-rich loosely-bound nuclei. It is supposed that the $E1$ transition strength is strongly enhanced in such nuclei at small enough energies (the so-called *soft dipole mode*). We calculate the reduced $E1$ transition probability

$$\frac{d\,\mathcal{B}(E1)}{d\,E} = \frac{1}{2J_i + 1} \sum_{J_f} \left|\langle J_f || \mathcal{M}(E1) || J_i\rangle\right|^2, \tag{47}$$

where

$$\mathcal{M}(E1\mu) = -\,\frac{Zen}{A}\sqrt{\frac{\hbar}{\omega}\,\frac{2m + M_c}{2mM_c}}\;y\,Y_{1\mu}(\hat{y}), \tag{48}$$

y is the Jacobi coordinate proportional to the distance between the ^{9}Li cluster and the center of mass of two valent neutrons, m is the nucleon mass and M_c is the ^{9}Li cluster mass, e is the electron charge, the number of protons in ^{11}Li $Z = 3$, the number of valent neutrons $n = 2$, $J_i = 3/2$ and J_f are the total angular momenta of the ground (initial) and final states, respectively. We note here that within our three-body cluster model of ^{11}Li we calculate only the so-called *cluster $E1$ transition strength* associated with the relative motion of two neutrons and the ^{9}Li cluster; the total $E1$ transition strength includes additionally excitations of nucleons forming the ^{9}Li cluster that manifest itself at higher energies. Since at low energies the only

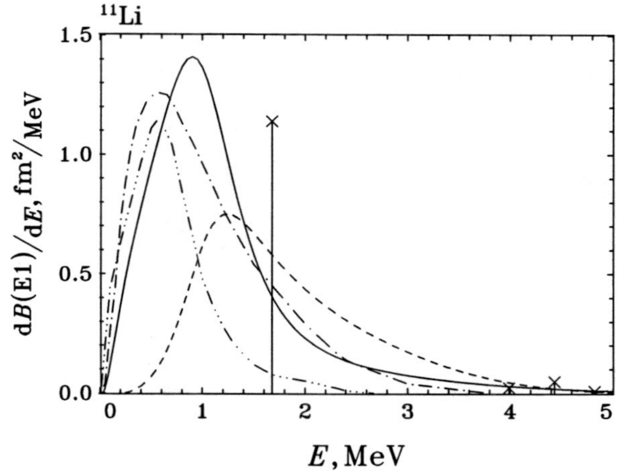

Fig. 4 Reduced $E1$ transition probability $\frac{d\mathcal{B}(E1)}{dE}$ in ^{11}Li. Vertical lines with cross at the end, dashed and solid lines were obtained in the VV, VJ and JJ approximations, respectively, with $\hbar\omega = 6.5$ MeV and $\tilde{N} = 20$ for the ground state and $\tilde{N} = 21$ for the final state. Dash-dot and dash-dot-dot lines are the calculations of Ref. [22] and the parameterizations of experimental data of Ref. [42], respectively

open decay channel is the three-body one, $^{11}\text{Li} \to {}^{9}\text{Li}+n+n$, the democratic decay approximation is well justified for the calculation of the final state wave function.

The results of the $\dfrac{d\mathcal{B}(E1)}{dE}$ calculations are presented in Fig. 4. It is interesting to compare the results obtained in the following approximations:

VV — the variational calculation of both the ground and final states;
VJ — the variational calculation of the ground state and the J-matrix calculation of the final three-body continuum state;
and
JJ — the J-matrix calculation of both the ground and final three-body continuum states.

In the VV approximation, we obtain a discrete spectrum of excited states and cannot calculate the final state wave function at an arbitrary given positive energy. As a result, the $E1$ transition strength differs from zero at the energies belonging to the discrete spectrum of final states and $\dfrac{d\mathcal{B}(E1)}{dE}$ has a form of a number of δ-function peaks shown in the figure by vertical lines with a cross at the end. The first peak at the excitation energy of approximately 2 MeV exhausts nearly 80% of the cluster energy-weighted sum rule [19]

$$\mathcal{S}_{E1}^{clust} = \int_0^\infty (E_f - E_0)\,\frac{d\mathcal{B}(E1)}{dE}\,dE_f = \frac{9}{4\pi}\,\frac{\hbar^2 e^2}{2m}\,\frac{nZ^2}{A(A-n)}, \tag{49}$$

where E_0 is the ground state energy. This is clearly seen from the Fig. 5 where we present the function

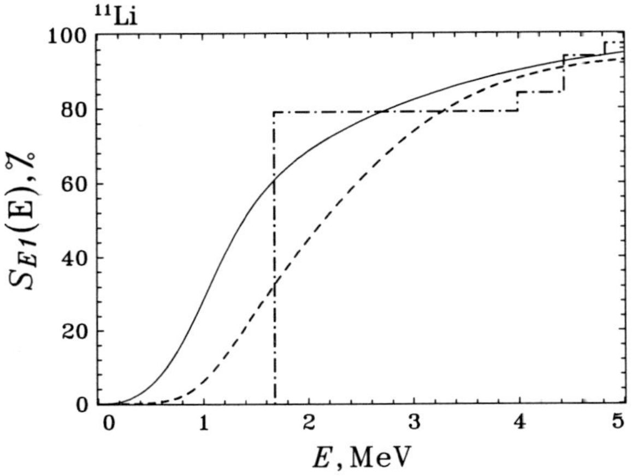

Fig. 5 Cluster $E1$ energy-weighted sum rule in ^{11}Li. Dash-dot, dashed and solid lines were obtained in the VV, VJ and JJ approximations, respectively, with $\hbar\omega = 6.5$ MeV and $\tilde{N} = 20$ for the ground state and $\tilde{N} = 21$ for the final state

$$\mathcal{S}_{E1}(E) = \frac{1}{\mathcal{S}_{E1}^{clust}} \int_0^E (E_f - E_0) \frac{d\,\mathcal{B}(E1)}{d\,E}\, dE_f. \tag{50}$$

The 2 MeV peak should be related with the soft dipole mode in ^{11}Li in the VV approach. The energy of the soft dipole mode obtained in the VV approximation differs essentially from the one of the experimental $E1$ strength low-energy maximum.

In the VJ approximation, the final energy wave function can be calculated at any given positive energy. Instead of sharp δ-peaks, we have a smooth curve of $\frac{d\,\mathcal{B}(E1)}{d\,E}$ with a maximum shifted to a lower energy that is closer to the energy of the experimental $E1$ strength maximum. This is clearly the result of allowing for the continuum spectrum effects in the final state wave function. However the external 'asymptotic' part of the model space with $2\kappa + K > \tilde{N}$ is not allowed for in the ground state wave function and hence the final state wave function components with $2\kappa + K > \tilde{N} + 2$ do not contribute to $\frac{d\,\mathcal{B}(E1)}{d\,E}$.

The contributions to $\frac{d\,\mathcal{B}(E1)}{d\,E}$ of external asymptotic part of the model space of both the ground state and the final state wave functions, are completely accounted for in the JJ approximation. These contributions are seen from Figs. 4 and 5 to be of great importance. They shift the $\frac{d\,\mathcal{B}(E1)}{d\,E}$ maximum to lower energy and change the shape of the $E1$ strength function. We note here that due to the slow decrease in the asymptotic region of the wave function of the loosely-bound state, it is needed to allow for a very large number of components in the asymptotic part of the model space in the $E1$ strength calculations. To calculate $\frac{d\,\mathcal{B}(E1)}{d\,E}$ in the low-energy region with high accuracy, we should allow for all components with the number of oscillator quanta $2\kappa + K \leq N$ where N is of the order of 1000. The convergence of

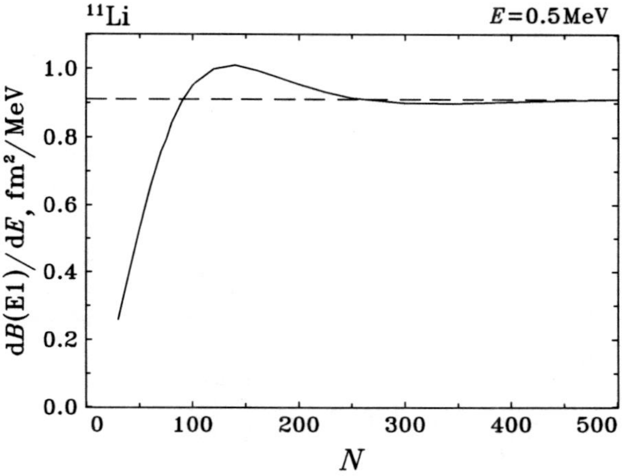

Fig. 6 The $\frac{d\mathcal{B}(E1)}{dE}$ convergence with N at the energy $E = 0.5$ MeV where all components with the number of oscillator quanta $2\kappa + K \leq N$ are allowed for in the JJ calculations with $\hbar\omega = 6.5$ MeV and $\tilde{N} = 20$ for the ground state and $\tilde{N} = 21$ for the final state

$\dfrac{d\mathcal{B}(E1)}{dE}$ with N at the energy $E = 0.5$ MeV, is shown in Fig. 6. We note here that the $N\hbar\omega$ oscillator function has the classical turning point at

$$\sqrt{\frac{\hbar}{m\omega}}\rho \approx \sqrt{\frac{2N\hbar}{m\omega}}. \tag{51}$$

Therefore supposing, say, $N = 500$, we allow for the distances up to $\sqrt{\frac{\hbar}{m\omega}}\rho \approx$ 80 fm in the $\dfrac{d\mathcal{B}(E1)}{dE}$ calculation with $\hbar\omega = 6.5$ MeV.

4 Two-Neutron Halo in ^{6}He Nucleus

The ^{6}He nucleus is studied in the three-body cluster model $^6\text{He} = \alpha + n + n$. The ^{6}He two-neutron separation energy is 0.976 MeV [63] that is much less than the excitation energy of the α particle lowest excited state and the α particle disintegration threshold. As it was already noted, ^{6}He is a Borromean nuclei, i.e. none of the two-body subsystems $\alpha + n$ and $n + n$ has a bound state.

Generally our approach to the ^{6}He nucleus is very close to the approach used in the ^{11}Li studies discussed above. However, contrary to the case of $n{-}^{11}$Li interaction, the $n{-}\alpha$ interaction is well-known. Therefore it is reasonable to use a more microscopically justified approach based on available more realistic $n{-}$cluster and $n{-}n$ potentials. In particular, it is reasonable to explicitly allow for the spin variables in our model. We couple the spins of two valent neutrons into the total spin $S = 0, 1$ (the α particle has a zero spin), the orbital angular momenta l_x and l_y are coupled to the total orbital angular momentum L, and L and S are coupled to the total angular momentum J. Hence the channel index $\Gamma = \{K, l_x, l_y, L, S, J\}$, and we introduce the following generalization of the basis functions (41):

$$|\kappa K l_x l_y(L)(S) : JM\rangle = \sum_{M_L, M_S} (LM_L SM_S|JM)\, |\kappa K l_x l_y : LM_L\rangle\, |SM_S\rangle; \quad (52)$$

the ^{6}He wave function is given by the general formula (12). In Eq. (52), $|SM_S\rangle$ is the spin component of the wave function and $|\kappa K l_x l_y : LM_L\rangle$ is given by Eqs. (41)–(42). We note here that allowing for the triplet ($S = 1$) spin states, we enlarge essentially the number of basis functions with any given number of oscillator quanta $N = 2\kappa + K$. Hence for any given $\tilde{N}$, we have the truncated Hamiltonian matrix of a much larger rank, and we are able to perform calculations with smaller $\tilde{N}$ than in the ^{11}Li case.

It would be very interesting to perform the ^{6}He studies based on modern so-called *realistic* NN potentials derived from the meson exchange theory and perfectly describing NN scattering data and deuteron observables. However the oscillator basis matrices of such potentials are extremely large and cannot be handled in calculations of lightest nuclei; the realistic potential matrix elements decrease slowly with the number of oscillator quanta N and truncation of the Hamiltonian matrix results in the slow convergence of the results. In practical applications usually *effective* interactions are used in calculations. Microscopic *ab initio* approaches (see, e.g., [64]) involve *realistic effective interactions* and *effective operators* derived from the realistic NN potentials. Unfortunately the J-matrix formalism (and other formalisms allowing for the continuum spectrum effects) is not developed still for the case of *ab initio* models based on realistic effective NN interactions. In this contribution, we are interested in the continuum spectrum effects and S-matrix pole corrections to the binding energy, and so we use *phenomenological effective potentials* for $n-n$ and $n-\alpha$ interactions. A large number of phenomenological effective $n-n$ and $n-\alpha$ potentials is available. We perform calculations with various $n-n$ and $n-\alpha$ potentials since it is not clear from the very beginning what is the best choice of these potentials; we would like also to compare our results with the results of other authors who used different combinations of these interactions.

The following effective $n-n$ potentials were employed in our studies.

The Gaussian potential (37) with $V_0 = 30.93$ MeV and $R = 1.82$ fm in the singlet ($S = 0$) state and $V_0 = 50.9$ MeV and $R = 1.65$ fm in the triplet ($S = 1$) state suggested in Ref. [65], is hereafter referred to as Gs. The potential was fitted to the s wave nucleon-nucleon phase shifts only; the interaction between the neutrons in the states with relative orbital angular momentum $l_x > 0$, is neglected.

We also make use of a bit different Gaussian potential suggested in Refs. [23,24] with $V_0 = 31.00$ MeV and $R = 1.8$ fm in the singlet state and $V_0 = 71.09$ MeV and $R = 1.4984$ fm in the triplet state. The singlet component of this potential was employed in our ^{11}Li studies. We shall refer to this potential as G.

A more realistic Minnesota $n-n$ potential of Ref. [66] will be referred to as MN. This Gaussian potential includes central, spin-orbit and tensor components.

The lowest single particle $0s_{1/2}$ is occupied in the α particle. There are two conventional approaches to the problem of the Pauli forbidden $0s_{1/2}$ state in the $n + \alpha$ system. The first approach is to add a phenomenological repulsive term to the s

wave component of the $n-\alpha$ potential. This phenomenological repulsion excludes the Pauli forbidden state in the $n+\alpha$ system and is supposed to simulate the Pauli principle effects in more complicated cluster systems. This idea was utilized in the SBB $n-\alpha$ potential suggested in Ref. [21]. This is an l-dependent Gaussian potential that includes the spin-orbit component.

Another approach is to use a deep attractive $n-\alpha$ potentials that support the Pauli forbidden $0s_{1/2}$ state in the $n+\alpha$ system. The Pauli forbidden states should be excluded in the three-body cluster system. The conventional method is to supplement the deep attractive $n-\alpha$ potential by the projecting pseudo potential (see, e.g. [26])

$$\lambda|0s_{1/2}\rangle\langle 0s_{1/2}|. \qquad (53)$$

If the parameter λ is positive and large enough, the projector (53) pushes the Pauli forbidden states to very large energies and cleans up the ground and low-lying states from the Pauli forbidden admixtures. The eigenfunction of the Pauli forbidden state supported by the deep attractive $n-\alpha$ potential should be used as $|0s_{1/2}\rangle$ (the so-called eigen-projector [26]); in this case the pseudo potential (53) does not affect the description of the scattering data provided by the initial deep attractive $n-\alpha$ potential.

The deep attractive Woods-Saxon $n-\alpha$ potential suggested in Ref. [67] will be hereafter referred to as WS. We use the WS potential parameters suggested in Ref. [27] where the radius of the potential was increased in order to fit the ^{6}He binding energy.

We use also deep attractive Majorana splitting potential suggested in Ref. [26] (hereafter referred to as MS) and deep attractive l-dependent Gaussian potential proposed by us in Ref. [13] (hereafter referred to as GP). These potentials improve the description of scattering data in high ($l > 1$) partial waves.

The convergence patterns in the ^{6}He case are very similar to the ones discussed in the ^{11}Li case. As a typical example, we present the results obtained in the Gs + SBB potential model. The ^{6}He ground state energy dependence on the oscillator parameter $\hbar\omega$ for a number of truncation boundary $\tilde{N}$ values, is depicted in Fig. 7. The $\hbar\omega$ ground state energy dependence is seen to be of the same type as in the ^{11}Li case shown in Fig. 1. The ^{6}He ground state energy and rms radius convergence with $\tilde{N}$ is presented in Fig. 8 for two $\hbar\omega$ values. As in the ^{11}Li case (see Fig. 2), there is the staggering of the J-matrix binding energy and rms radius with $\tilde{N}$ for $\hbar\omega = 10$ MeV that is close to the value providing the best convergence of the variational calculation, and there is the variational-type $\tilde{N}$ dependence of these observables for larger $\hbar\omega$ values. It is interesting that all calculations shown on the right panel of Fig. 7, result in nearly the same binding energy at $\hbar\omega \approx 15$ MeV; this $\hbar\omega$ value is clearly the best convergence value for the J-matrix calculation and it differs essentially from the best convergence $\hbar\omega$ value for the variational calculation corresponding to the minima of the curves on the left panel of Fig. 7.

The results of our calculations of the ^{6}He two-neutron separation energy and rms matter radius, are summarized in Table 2. Since the calculations using different potential models were performed with slightly different $\hbar\omega$ values, we list these

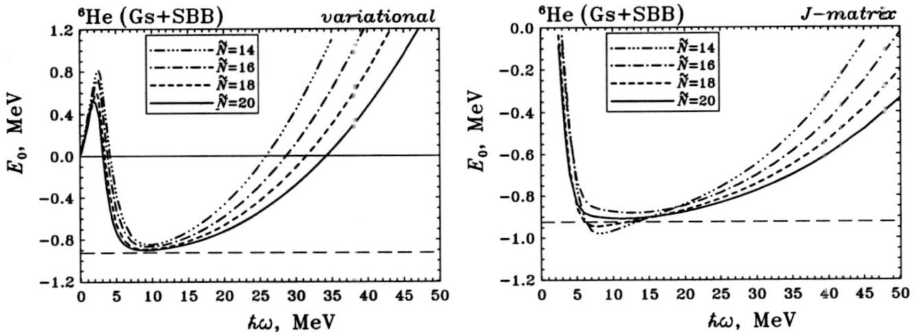

Fig. 7 The ^{6}He ground state energy in the Gs + SBB potential model vs the oscillator basis parameter $\hbar\omega$ in the variational (*left panel*) and J-matrix (*right panel*) calculations for different values of the truncation boundary $\tilde{N}$. The horizontal dashed line depicts the convergence limit for the ground state energy

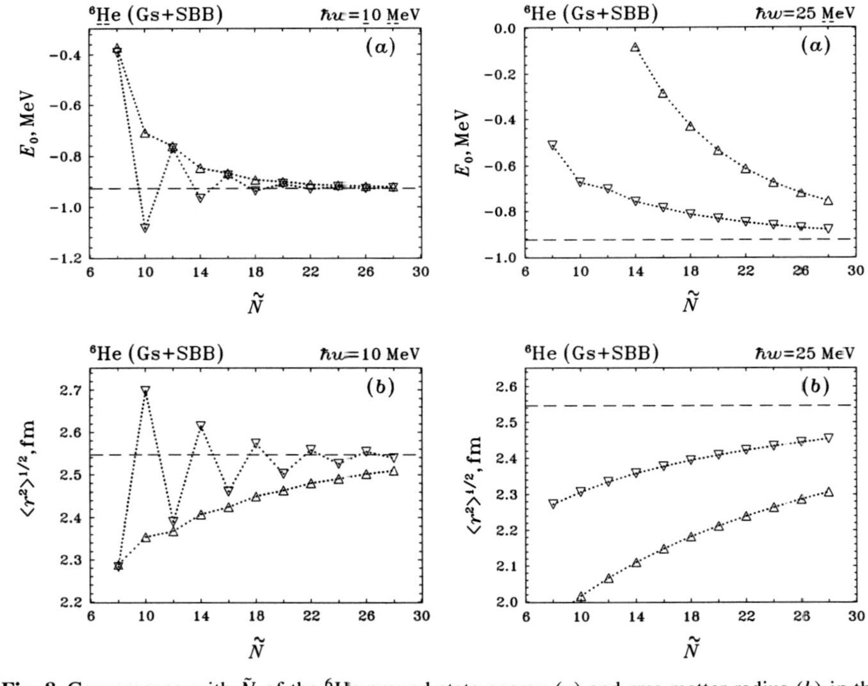

Fig. 8 Convergence with $\tilde{N}$ of the ^{6}He ground state energy (a) and rms matter radius (b) in the Gs + SBB potential model for $\hbar\omega = 10\,\text{MeV}$ (*left panel*) and $\hbar\omega = 25\,\text{MeV}$ (*right panel*). Variational and J-matrix results are shown by triangles up and triangles down, respectively. The convergence limits are shown by the dashed line

Table 2 The ^{6}He two-neutron separation energy and rms matter radius obtained in various potential models and available three-body cluster model results of other authors

Potential model	$\hbar\omega$, MeV	Approximation	$E(2n)$, MeV		$\langle r^2\rangle^{1/2}$, fm	
			$\tilde{N}=26$	$\tilde{N}=28$	$\tilde{N}=26$	$\tilde{N}=28$
	10.00	variational	0.917	0.919	2.502	2.510
Gs + SBB		J-matrix	0.927	0.923	2.554	2.539
		Ref. [21]	*"correct asymptotic value"*			
	11.90	variational	0.993	0.997	2.394	2.404
Gs + WS		J-matrix	1.003	1.008	2.444	2.461
		Ref. [27]	1.00		2.44	
G + GP	11.35	variational	0.875	0.878	2.386	2.393
		J-matrix	0.889	0.883	2.451	2.428
MN + GP	10.75	variational	0.494	0.509	2.477	2.484
		J-matrix	0.515	0.516	2.588	2.547
MN + MS	13.00	variational	0.656	0.672	2.382	2.389
		J-matrix	0.684	0.681	2.482	2.445
other potential models			Ref. [26]: 0.305; Ref. [31]: 1.00; Ref. [29]: 0.784; Ref. [30]: 0.696		Ref. [31]: 2.50	
experimental data			Ref. [63]: 0.976		Ref. [41]: 2.33 ± 0.04; Ref. [38]: 2.48 ± 0.03; Ref. [40]: 2.57 ± 0.10	

values in the corresponding rows of the Table. If for a given potential model the theoretical predictions of other authors using other approximations within the three-body cluster model are available, they are presented in the corresponding rows of the Table, too. The theoretical predictions within the three-body cluster model with other potential models and the experimental data, are presented in the Table in additional rows.

The J-matrix approach is seen from the Table to improve the variational results for both the binding energy and the rms radius in the case of any potential model. The improvement is generally more essential in the case of smaller binding energy (it is also seen comparing the ^{6}He results in Table 2 with the ^{11}Li results in Table 1). The J-matrix approach is also seen to have a faster convergence than the variational one, and again the improvement of the convergence rate is more essential in the smaller binding energy case. Our results are in good correspondence with the available results of other authors who used the same potential models.

The structure of the ^{6}He ground state wave function is presented in Table 3. We note here that within the three-body cluster model $^6\text{He} = \alpha + n + n$, the 0^+ ground state can be obtained by the coupling of the Jacobi orbital momenta l_x and l_y to the total orbital angular momentum L and subsequent coupling of L with the total neutron spin S into the total angular momentum $J = 0$, if only $L = S$ and $l_x = l_y$. We arrange the dominant components in a different manner. In the Table 3 (a) we show the total weights of all components with given $S = L$ and $l_x = l_y$, i.e. we sum the contributions of the components with different K and $N = 2\kappa + K$ values for the given $S = L$ and $l_x = l_y$ values. In the Table 3 (b) we collect the total weights of all

Table 3 Dominant component weights in the ^{6}He 0^+ ground state wave function obtained with various potential models in the J-matrix approximation with $\tilde{N} = 28$ (the respective $\hbar\omega$ values can be found in Table 2) and the results of other authors in the three-body cluster model: (**a**) — total contribution of the components with given $S = L$ and $l_x = l_y$ (summation over possible K and N values); (**b**) — total contribution of the components with given K and $S = L$ (summation over possible $l_x = l_y$ values); (**c**) — total contribution of the components with given shell model-like orbital angular momenta $l_{n_1\alpha} = l_{n_2\alpha}$

(a)

State		Weight, %						
$S=L$	$l_x=l_y$	present work				Ref. [4,21]		Ref. [26]
		Gs+SBB	Gs+WS	G+GP	MN+MS	Gs+SBB	GPT^a+SBB	RSC^b+SBB^{*c}
0	0	82.1	83.6	80.2	76.6	83.13	82.87	88.908
	2	2.46	1.90	1.64	2.07	1.77	2.31	1.035
	4	0.18	0.14	0.11	0.15	0.02	0.58	
1	1	14.3	13.4	17.6	20.1	14.54	13.96	9.692
	3	0.85	0.83	0.66	0.89	0.55	0.69	0.366
	5	0.09	0.09	0.06	0.09		0.06	

(b)

State		Weight, %					
$S=L$	K	present work				Ref. [4,21]	
		Gs + SBB	Gs + WS	G + GP	MN + MS	Gs + SBB	GPT + SBB
0	0	4.28	4.06	3.99	3.93	4.41	4.68
	2	77.2	78.9	75.5	72.4	78.93	78.10
	4	0.59	0.19	0.13	0.16	0.53	0.64
	6	1.60	1.61	1.40	1.73	1.16	1.43
	8	0.67	0.51	0.43	0.41	0.39	
1	2	14.2	13.2	17.5	20.0	13.91	13.41
	4	0.12	0.15	0.05	0.06	0.12	0.15
	6	0.82	0.80	0.63	0.85	0.53	0.67
	8	0.05	0.04	0.04	0.06	0.02	

(c)

State	Weight, %				
$l_{n_1\alpha} = l_{n_2\alpha}$	present work				Ref. [35]
	Gs + SBB	Gs + WS	G + GP	MN + MS	GPT + WS + V_3^{d}
s	8.27	7.60	7.35	7.26	7.7
p	90.4	90.9	91.5	91.6	91.0
d	0.57	0.64	0.43	0.44	
f	0.57	0.54	0.47	0.41	
g	0.07	0.07	0.07	0.05	

^{a} Gogny–Pires–de Tourreil n–n potential [68].
^{b} Reid soft-core n–n potential [69].
^{c} SBB n–α potential with modified parameters (see Ref. [26]).
^{d} Three-body $nn\alpha$ potential, see Ref. [35].

components with given $S = L$ and hypermomentum K summing the contributions with different $l_x = l_y$ values. It is also interesting to calculate the shell model type component weights which are characterized by the orbital momenta $l_{n_1\alpha}$ and $l_{n_2\alpha}$ of individual neutron motion relatively to the α core. The unitary transformation from the hyperspherical basis (52) to the shell model basis can be found in textbooks (see, e.g., [70, 71]). The ^{6}He 0^+ ground state wave function can be arranged if only $l_{n_1\alpha} = l_{n_2\alpha}$. The shell model-type component weights are listed in the Table 3 (c).

It is seen from Table 3 that different potential models result in the wave functions of nearly the same structure. For example, the component weights in different arrangements obtained with SBB potential simulating the Pauli principle by repulsive terms and with WS potential when the Pauli forbidden states are explicitly projected out, are nearly the same. Our results for component weights are in good correspondence with the results of other authors who used other approaches to the three-body cluster model and other potential models. Therefore the uncertainties of the ^{6}He wave function structure due to the uncertainties of the two-body interactions, are very small. The most essential (however not very large) difference between the dominant component weights obtained in different approaches is the difference between the results of Ref. [26] where the RSC + SBB* potential model was employed and our results and the results of Ref. [4, 21] obtained with different potential models. We note here that the ^{6}He binding energy obtained in Ref. [26] is 0.3046 MeV only, i.e. it was essentially underestimated. Most probably this is indication that the RSC + SBB* potential model is not adequate for the description of ^{6}He in the cluster model and hence it is not surprising that the component weights of Ref. [26] differ from the ones obtained with other potentials.

From the naive shell model considerations it follows that two valent neurons should occupy p states in the ^{6}He nucleus. We see from the Table 3 (c) that this situation is really utilized in the dominant component of the ^{6}He wave function. The same naive shell model considerations bring us to the $K = 2$ component dominance, and again we see from the Table 3 (b) that this is really the case. However from these naive considerations it looks strange that the dominant component corresponds to the Jacobi orbital momenta $l_x = l_y = 0$ [see the Table 3 (a)]. Nevertheless there is, of course, no contradiction between the results presented in the Table 3 (a) and the Table 3 (c), they are obtained by the summation over different quantum numbers of the same ^{6}He ground state wave functions.

An example of the reduced (cluster) $E1$ transition probability calculations in ^{6}He is shown in Fig. 9 where the results obtained with G + GP potential model in the VV, VJ and JJ approximations are presented. In the ^{6}He case, the transition strength is distributed in the VV approximation over a number of strong δ-peaks. Therefore the shape of the reduced probability $\dfrac{d\,\mathcal{B}(E1)}{d\,E}$ appears to be more complicated than in the ^{11}Li case when the effects of continuum are allowed for in the VJ and JJ approximations. As in the ^{11}Li case, we see that the external asymptotic part of the model space allowed for in the JJ approximation only, provides a very significant contribution to the electromagnetic transition probabilities.

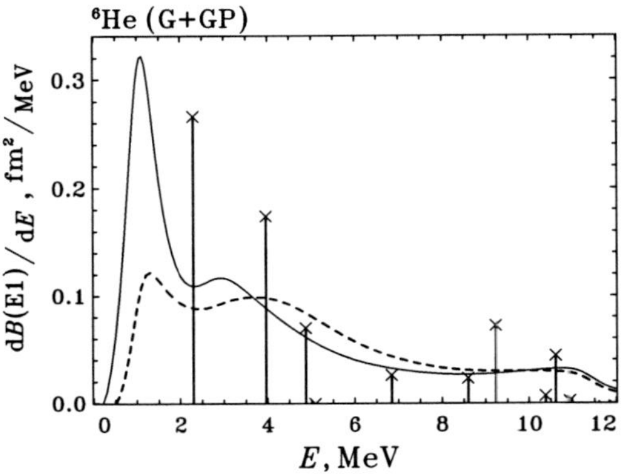

Fig. 9 Reduced $E1$ transition probability $\frac{d\mathcal{B}(E1)}{dE}$ in ^{6}He obtained in the G + GP potential model. Vertical lines with cross at the end, dashed and solid lines were obtained in the VV, VJ and JJ approximations, respectively, with $\hbar\omega = 11.35$ MeV and $\tilde{N} = 22$ for the ground state and $\tilde{N} = 23$ for the final state

The convergence of the reduced $E1$ transition probability calculations in the JJ approximation is illustrated by Fig. 10 where we show the results obtained with different truncation boundaries in the ground state calculations $\tilde{N}_{\text{g.s.}}$, the truncation boundary in the final state calculations $\tilde{N}_{\text{f.s.}} = \tilde{N}_{\text{g.s.}} + 1$. The staggering of the $\dfrac{d\mathcal{B}(E1)}{dE}$ shape as $\tilde{N}_{\text{g.s.}}$ increases, is clearly seen in the figure. For example, the weakest transition strength is obtained with $\tilde{N}_{\text{g.s.}} = 12$ and $\tilde{N}_{\text{f.s.}} = 13$ shown by the dotted line in the figure; the strongest transition strength is obtained with $\tilde{N}_{\text{g.s.}} = 14$ and $\tilde{N}_{\text{f.s.}} = 15$ shown by the dash-dot line; the transition strength obtained with $\tilde{N}_{\text{g.s.}} = 16$ and $\tilde{N}_{\text{f.s.}} = 17$ is stronger than the one obtained with $\tilde{N}_{\text{g.s.}} = 12$ and $\tilde{N}_{\text{f.s.}} = 13$ but weaker than all the rest results presented in the figure, etc. However selecting the results obtained with even $\tilde{N}_{\text{g.s.}}/2$ values only, we see that the $E1$ transition strength increases monotonically with $\tilde{N}_{\text{g.s.}}$; selecting the results obtained with odd $\tilde{N}_{\text{g.s.}}/2$ values only, we see the monotonic decrease of the $E1$ transition strength with $\tilde{N}_{\text{g.s.}}$.

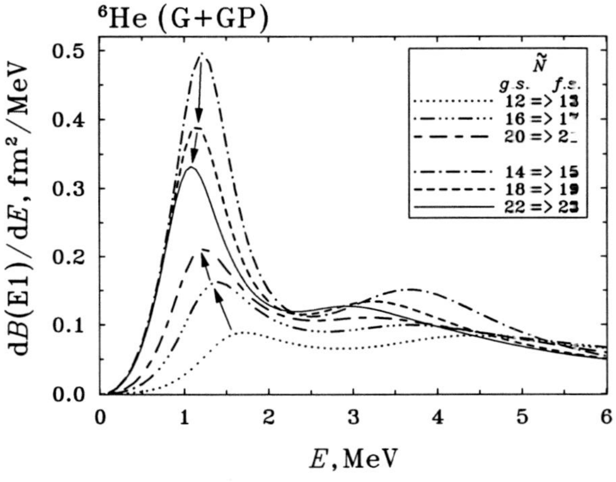

Fig. 10 Convergence of the $\frac{d\mathcal{B}(E1)}{dE}$ calculations in ^{6}He with the truncation parameter $\tilde{N}_{\text{g.s.}}$ used in the ground state calculations (the final state truncation parameter $\tilde{N}_{\text{f.s.}} = \tilde{N}_{\text{g.s.}} + 1$). The results are obtained in the JJ approximation with the G + GP potential model and $\hbar\omega = 11.35$ MeV. Arrows show the changes of $\frac{d\mathcal{B}(E1)}{dE}$ when $\tilde{N}_{\text{g.s.}}$ is increased by 4

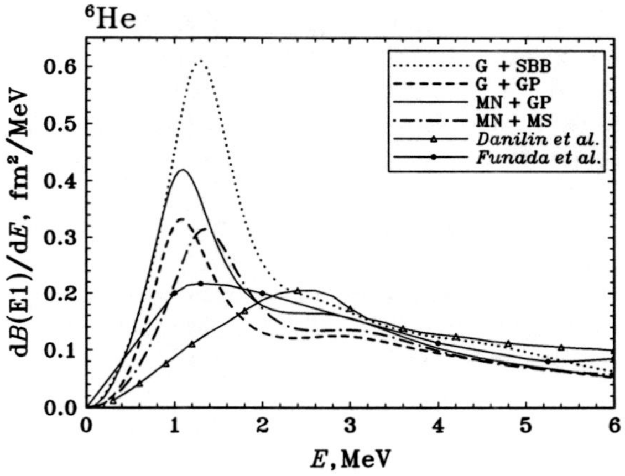

Fig. 11 Reduced $E1$ transition probability $\frac{dB(E1)}{dE}$ in ^{6}He obtained with various potential models in the JJ approximation with the ground state truncation parameter $\tilde{N}_{\text{g.s.}} = 22$ and the final state truncation parameter $\tilde{N}_{\text{f.s.}} = 23$ (the corresponding $\hbar\omega$ values can be found in Table 2) in comparison with the results of Funada et al. [25] and Danilin et al. [32]

The comparison of the reduced $E1$ transition probability results obtained with different potential models and calculations of other authors within the three-body cluster model, is shown in Fig. 11.

5 Phase Equivalent Transformation with Continuous Parameters and Three-Body Cluster System

The results presented in Table 2 show that we obtain a very good approximation for the ^{6}He two-neutron separation energy. However the exact value of the ^{6}He binding energy is not reproduced with any potential model employed. We suggest to use a phase equivalent transformation of the two-body interaction to improve the description of the ^{6}He binding. If the phase equivalent transformation depends on a continuous parameter(s) than varying the parameter we can fit the ^{6}He binding energy to the phenomenological value. We suppose this approach to be interesting for various few-body applications.

Various phase equivalent transformations have been discussed in literature. The local phase equivalent transformations (transforming a local potential into another local potential phase equivalent to the initial one) is well-known (see, e.g., Ref. [72]). In the inverse scattering theory, this transformation gives rise to the ambiguity of the potential restored from the scattering data. Recently this transformation was extended on the case of the multichannel scattering [73]. However the local phase equivalent transformation can be applied only to a two-body system that have at least one bound states and the number of continuous parameters of the transformation is equal to the number of the bound states in the system. Therefore this transformation cannot be applied to the $n + \alpha$ system that has no bound states.

Recently the so-called supersymmetry phase equivalent transformation (see, e.g., review [74]) become very popular. Using this transformation one can transform a potential with the Pauli forbidden state into exactly phase equivalent potential with additional repulsion simulating the Pauli effects. The effect of this transformation on the properties of three-body cluster systems was examined in a number of recent papers (see, e.g., Ref. [75]). The supersymmetry transformation was shown [73] to be a particular case of the local phase equivalent transformation. However the supersymmetry transformation does not have parameters and cannot be used for our purposes.

A phase equivalent transformation based on unitary transformation of the Hamiltonian, was suggested in Ref. [76]. This transformation has continuous parameters. This transformation have been used in many-body applications in Ref. [77]. However the authors of Ref. [77] used few particular cases of the transformation corresponding to particular parameter values and did not try to vary the parameters continuously.

We have developed recently [13] a phase equivalent transformation using the J-matrix formalism. Generally, our transformation is a particular case of the phase equivalent transformation of Ref. [76]. However we suppose that our phase equivalent transformation is general enough and very convenient for the use in various many-body applications utilizing any L^2 basis. In what follows, we discuss briefly the transformation of Ref. [13] and its application to the ^{6}He nucleus within the three-body cluster model [14].

We suppose that a two-body system is described by the Schrödinger equation

$$H\Psi(r) = E\Psi(r), \tag{54}$$

where the Hamiltonian $H = T + V$, T and V are kinetic and potential energy operators. Introducing a complete basis $\{\phi_\kappa(r)\}$, $\kappa = 0, 1, 2, \ldots$ of L^2 functions $\phi_\kappa(r)$, we can expand the solutions of Eq. (54) in infinite series of basis functions $|\kappa\rangle \equiv \phi_\kappa(r)$:

$$\Psi(r) = \sum_{\kappa=0}^{\infty} C_\kappa |\kappa\rangle. \tag{55}$$

The Schrödinger equation (54) takes the form of an infinite dimensional algebraic problem

$$\sum_{\kappa'=0}^{\infty} \langle\kappa|H|\kappa'\rangle\, C_{\kappa'} = EC_\kappa, \tag{56}$$

where $\langle\kappa|H|\kappa'\rangle$ are the matrix elements of the infinite dimensional Hamiltonian matrix $[H]$.

Now we define a new matrix

$$\left[H'\right] = \left[U^+\right] [H]\, [U] \tag{57}$$

with the help of the unitary matrix $[U]$ which is supposed to be of the form

$$[U] = [U_0] \oplus [I] = \left(\begin{array}{c|c} [U_0] & 0 \\ \hline 0 & [I] \end{array} \right), \tag{58}$$

where $[I]$ is the infinite dimensional unit matrix and $[U_0]$ is $N \times N$ unitary matrix. A new Hamiltonian H' is defined through its matrix $\left[H'\right]$. It is supposed that $\left[H'\right]$ is the matrix of the Hamiltonian H' in the *original* basis $\{\phi_\kappa(r)\}$.

Clearly the spectra of Hamiltonians H and H' are identical. The difference between the eigenfunctions $\Psi'(r)$ of the Hamiltonian H' and the eigenfunctions $\Psi(r)$ of the Hamiltonian H corresponding to the same energy E, is a superposition of a finite number of square integrable functions:

$$\Psi'(r) = \Psi(r) + \sum_{\kappa=0}^{N-1} \Delta C_\kappa \, \phi_\kappa(r). \tag{59}$$

The superposition of a finite number of L^2 functions cannot affect the asymptotics of scattering wave functions. Since the scattering phase shifts and the S-matrix are defined through the asymptotic behavior of the wave functions, the phase shifts associated with the functions $\Psi(r)$ and $\Psi'(r)$ are identical. In other words, the Hamiltonians H and H' are phase equivalent.

Supposing that the Hamiltonian $H' = T + V_{\mathrm{PET}}$, we introduce the potential

$$V_{\mathrm{PET}} = V + \Delta V \tag{60}$$

phase equivalent to the original potential V. The potential V_{PET} is defined through its matrix

$$[V_{\mathrm{PET}}] = [V] + \left[\Delta V\right] \tag{61}$$

in the basis $\{\phi_\kappa(r)\}$, where

$$\left[\Delta V\right] = \left[U^+\right] [H] \, [U] - [H]. \tag{62}$$

The net result of the above considerations is very simple. We introduce any complete L^2 basis $\{\phi_n(r)\}$ and calculate the Hamiltonian matrix in this basis. Next we introduce an arbitrary unitary transformation of the type (58) that affects a finite number of the basis functions. With the help of this transformation, we calculate the additional potential ΔV using Eq. (62) and obtain the phase equivalent interaction by means of Eq. (60). The additional potential ΔV is non-local, and hence the phase equivalent potential V_{PET} is non-local, too. Therefore the suggested transformation is a *non-local phase equivalent transformation*: it brings us to a non-local interaction phase equivalent to the original one.

The suggested non-local phase equivalent transformation can be easily implemented in many-body calculations utilizing any L^2 basis. One just needs to add the two-body kinetic energy matrix to the two-body interaction matrix, unitary transform the obtained matrix and subtract the kinetic energy matrix from the result to obtain the matrix of the phase equivalent two-body interaction.

The non-local phase equivalent transformation can be easily explained and understood within the J-matrix formalism. In the J-matrix formalism, the S-matrix and the phase shifts are governed by the matrix elements $\langle \kappa_\Gamma \Gamma | \mathscr{P} | \kappa_{\Gamma'} + 1, \Gamma' \rangle$ which are defined through eigenvalues E_λ and the last component $\langle \kappa_\Gamma \Gamma | \lambda \rangle$ of the eigenvectors of the truncated Hamiltonian matrix [see Eqs. (29)–(30)]. If the truncated Hamiltonian matrix is larger than the non-trivial submatrix $[U_0]$ of the unitary matrix $[U]$, than neither E_λ nor the last component $\langle \kappa_\Gamma \Gamma | \lambda \rangle$ of the eigenvectors are affected by the unitary transformation (58). Therefore neither the S-matrix nor the phase shifts are affected by the transformation. Nevertheless the wave function is seen from Eq. (59) to be subjected to changes by the transformation; in other words, the off-shell properties of the original and the transformed potentials are different. The non-local phase equivalent transformation generates the ambiguity of the interaction obtained by means of the J-matrix version of the inverse scattering theory [53, 54].

The off-shell properties of the two-body interactions play an important role in the formation of the properties of a three-body (and generally many-body) system. We applied this transformation to the NN interaction and showed that the transformation results in significant changes of the binding energies of ^{3}H and ^{4}He nuclei [78]. Therefore it is interesting to investigate the effect of the non-local phase equivalent transformation on the properties of the ^{6}He nucleus within the three-body cluster model. We note that the non-local phase equivalent transformation preserves the energies of the two-body bound states. However the properties of these states are not preserved, for example, the rms radius of the two-body bound system may be affected by the transformation [79]. The transformation can be used to fit the J-matrix inverse scattering NN potential to the deuteron properties [54] and even to the properties of heavier nuclei [55, 56]. However we have already noted that we employ effective $n-n$ interactions that are designed for the effective description of many-body nuclear systems and that are not supposed to be careful in the description of the dinucleon. We suppose that it is more natural to apply the transformation to the $n-\alpha$ interaction since the $n+\alpha$ system does not have a bound state and all the information about this interaction is extracted from the scattering data only. It is clear from the Table 3 (c) that the ^{6}He properties are dominated by the p wave component of the $n-\alpha$ interaction. Therefore we apply the non-local phase equivalent transformation to the p wave component of the $n-\alpha$ potential only.

The non-local phase equivalent transformation for the $n-\alpha$ potential p wave component is constructed in the oscillator basis with $\hbar\omega$ values used in three-body calculations with the respective potential model (see Table 2). The simplest non-local transformation involves a 2×2 unitary matrix $[U_0]$. We consider this 2×2 unitary matrix as a rotation matrix with a single continuous parameter γ:

$$[U_0] = \begin{bmatrix} \cos\gamma & -\sin\gamma \\ +\sin\gamma & \cos\gamma \end{bmatrix}. \tag{63}$$

A more complicated non-local transformation involves a 3×3 rotation matrix $[U_0]$ with two continuous parameters γ and β:

$$[U_0] = \begin{bmatrix} \cos\beta\,\cos\gamma & -\cos\beta\,\sin\gamma & \sin\beta \\ \sin\gamma & \cos\gamma & 0 \\ -\sin\beta\,\cos\gamma & \sin\beta\,\sin\gamma & \cos\beta \end{bmatrix}. \tag{64}$$

Clearly the transformation with the matrix (64) is equivalent to the transformation with the matrix (63) if the Euler angle $\beta = 0$.

We present in Fig. 12 the results of variational calculations of the ^{6}He ground and the first excited 0^+ states with $\tilde{N} = 20$ and of the lowest 1^- state with $\tilde{N} = 21$ with various n–n potentials and n–α potentials obtained by applying the phase equivalent transformation (63) to various original potentials. The γ dependence of 0^+ and 1^- state energies is very interesting. It is seen from Fig. 12 that the simplest non-local phase equivalent transformation of p wave component of the n–α potential only, can completely change the spectrum of the three-body cluster system.

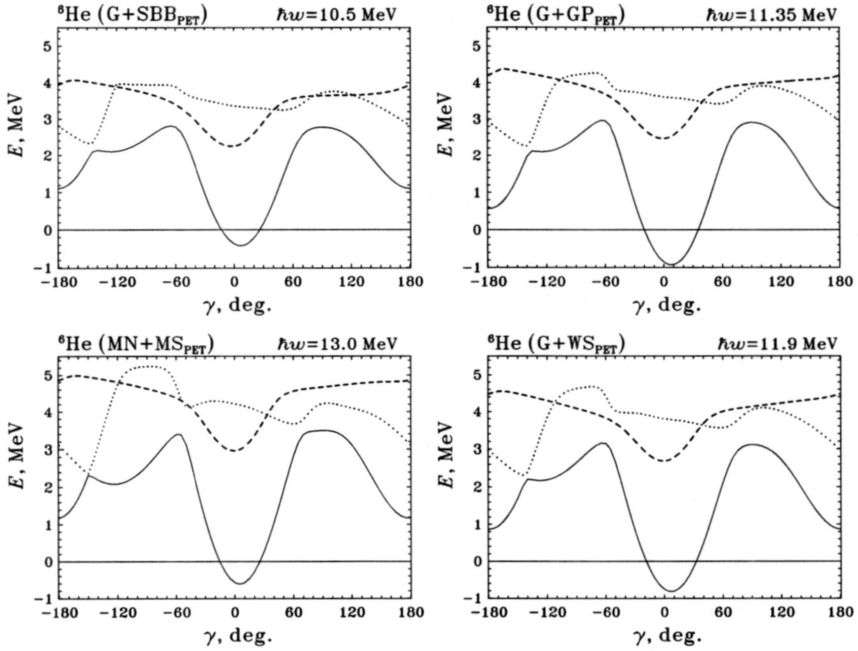

Fig. 12 The ^{6}He ground state energy (solid line) and the energies of the lowest 1^- (*dashed line*) and 0^+ (*dotted line*) excited states obtained in the variational approach with phase equivalent n–α potentials vs the transformation parameter γ of the one-parameter transformation (63). The truncation parameter $\tilde{N} = 20$ in the 0^+ state calculations and $\tilde{N} = 21$ in the 1^- state calculations

The γ dependence of the energies is seen to be very similar for all potential models under consideration. The variations of the energies of the excited 0^+ and 1^- states are seen to be smaller than the ground state energy variations. There is the 0^+ states level crossing at $\gamma \approx -150°$ for all potential model considered.

The ground state energy γ dependence passes through a minimum in the vicinity of $\gamma = 5{-}7°$. The minimum corresponds to the increase of the binding by approximately 0.04–0.07 MeV or 7%–12% depending on the potential model, that improves the results obtained with the original potentials ($\gamma = 0$) given in Table 2. The ground states may be additionally shifted down using the two-parameter transformation Eq. (64). Supposing $\gamma = 7°$ and varying the parameter β we obtain variational results shown in Fig. 13. The ground state β dependence has a minimum at $\beta \approx -2.5°$ for all potential models. This minimum corresponds to the additional binding of $0.019 \div 0.023$ MeV (about 2%–4%).

The variational results presented in Figs. 12 and 13 are interesting for understanding the general trends of variation of the ground state energy and ^{6}He spectrum when the parameters of the phase equivalent transformations (63)–(64) are varied in a wide range of values. The corrections to the ^{6}He binding due to the effect of the phase equivalent transformation are better seen in Fig. 14 where we present in a larger scale the J-matrix results for the ground state energies. The ^{6}He binding energy is seen to be very close to the empirical value in the G + GP$_{PET}$ potential model when $\gamma \approx 7°$. However the results presented in Fig. 14 were obtained with not very large value of the truncation parameter $\tilde{N} = 20$. With $\tilde{N} = 28$ and $\gamma = 7.5°$ we obtain in this potential model an excellent description of the ^{6}He binding energy $E_b = 0.952$ MeV and rms radius $\langle r^2 \rangle^{1/2} = 2.37$ fm. Using the two-parameter transformation (64) with $\gamma = 7.5°$ and $\beta = -3°$, we obtain $E_b = 0.973$ MeV and $\langle r^2 \rangle^{1/2} = 2.36$ fm, i.e. the phenomenological value $E_b = 0.976$ MeV is reproduce nearly exactly.

The correlations between the ^{6}He rms radius $\langle r^2 \rangle^{1/2}$ and the square root of the binding energy $E_b^{1/2}$, are depicted in Fig. 15. We present the J-matrix results

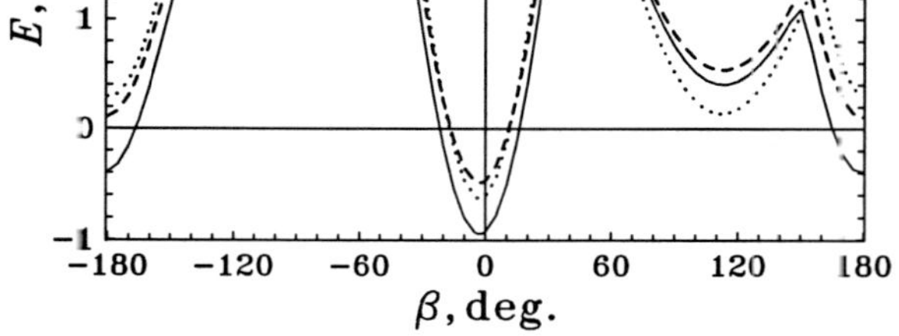

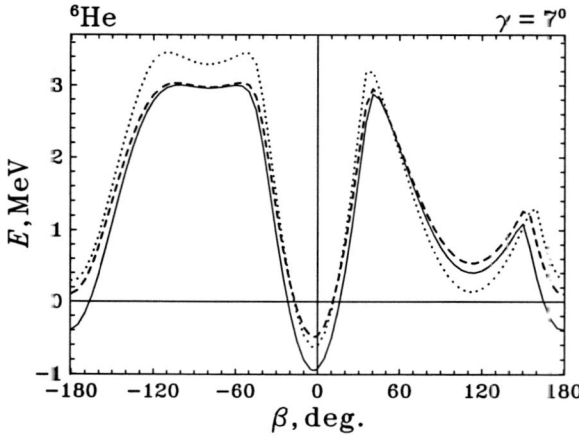

Fig. 13 The ^{6}He ground state energy vs the transformation parameter β of the phase equivalent transformation (64) with $\gamma = 7°$. Solid, dashed and dotted lines are calculations in the variational approximation with G + GP$_{PET}$, MN + GP$_{PET}$ and MN + MS$_{PET}$ potential models, respectively, with $\hbar\omega$ values given in Table 2 and $\tilde{N} = 20$

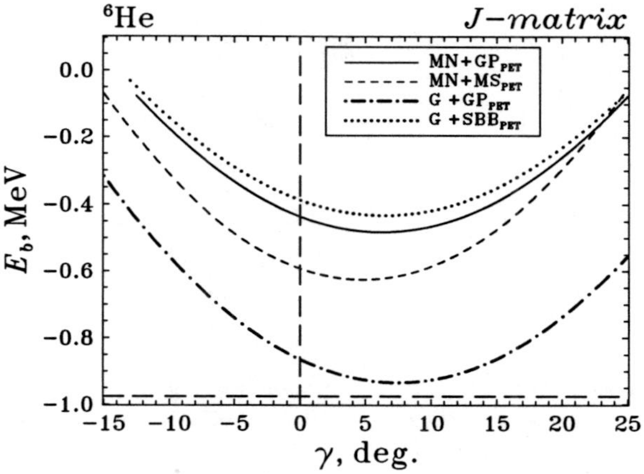

Fig. 14 The ^{6}He ground state energy obtained in the J-matrix approach with phase equivalent $n-\alpha$ potentials vs the transformation parameter γ of the one-parameter transformation (63). The truncation parameter $\tilde{N} = 20$, the $\hbar\omega$ values for each potential model can be found in Table 2. The straight dashed line is the experimental ground state energy

obtained with various potential models when the parameter γ of the phase equivalent transformation (63) varies on the interval where ^{6}He appears to be bound in the variational approximation in the given potential model. The correlations are seen to be very interesting: there are two very different $\langle r^2\rangle^{1/2}$ values that are in correspondence with the same binding energy. For $\gamma \lesssim -6°$, the rms radius $\langle r^2\rangle^{1/2}$ decreases linearly with $E_b^{1/2}$.

The effect of the phase equivalent transformation (63) on the reduced $E1$ transition probability $\dfrac{d\,\mathcal{B}(E1)}{d\,E}$ in ^{6}He, is illustrated by Fig. 16. The effect is seen to be essential. The naive expectation is that the soft dipole mode will be more enhanced if the binding energy is smaller. However the strength of the $E1$ transitions in the vicinity of the maximum of $\dfrac{d\,\mathcal{B}(E1)}{d\,E}$ does not demonstrate so simple dependence on the binding energy E_b (the corresponding E_b values are listed in the figure).

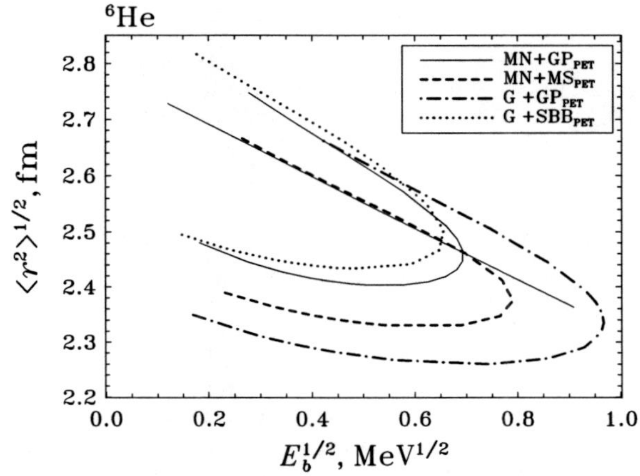

Fig. 15 Correlations between $E_b^{1/2}$ and rms radius in ^{6}He obtained by variation of the parameter γ of the phase equivalent transformation (63) in various potential models (J-matrix approximation with $\tilde{N} = 20$, the corresponding $\hbar\omega$ values can be found in Table 2). A straight solid line is added to visualize the linear correlation between $\langle r^2\rangle^{1/2}$ and $E_b^{1/2}$ on a part of the trajectory

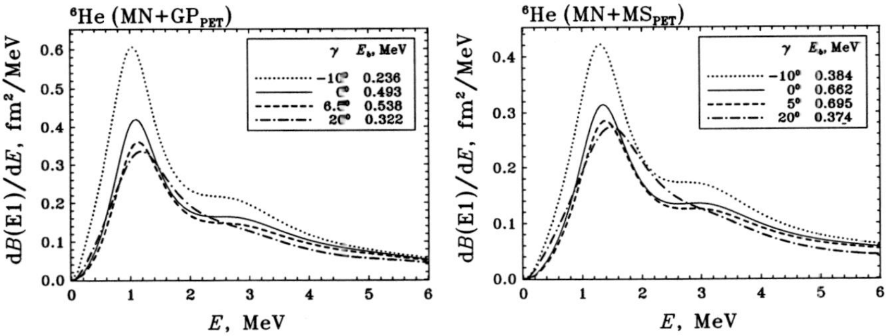

Fig. 16 Reduced $E1$ transition probability $\frac{dB(E1)}{dE}$ in ^{6}He obtained in the JJ approximation with the ground state truncation parameter $\tilde{N}_{g.s.} = 22$ and the final state truncation parameter $\tilde{N}_{f.s.} = 23$ with the MN + GP$_{PET}$ (*left panel*) and MN + MS$_{PET}$ (*right panel*) potential models (the corresponding $\hbar\omega$ values can be found in Table 2) and different values of the parameter γ of the phase equivalent transformation (63) (the corresponding binding energies are listed in the figure)

6 Conclusions

The hyperspherical J-matrix formalism makes it possible to study not only the few-body disintegration of the system but also to calculate the S-matrix poles in the A-body system. This approach appears to be a very powerful tool for calculations of the bound state properties. As a result, we obtain a unified theory that is capable to investigate in a unique approach both the discrete and continuum spectra of few-body systems.

The suggested J-matrix motivated non-local phase equivalent transformation may be used to fit binding energies of many-body systems in the case when all information about two-body interaction is extracted from the scattering data only. The transformation can be easily utilized in the studies of many-body systems with any L^2 basis.

Acknowledgments This work was supported in part by the State Program "Russian Universities", by the Russian Foundation of Basic Research Grant No 05-02-17429 and by the US DOE Grants DE-FG02-87ER40371 and DE-FC02-07ER41457.

References

1. S. P. Merkuriev and L. D. Faddeev, *Quantum scattering theory for systems of few bodies* (Nauka Publishes, Moscow, 1985) (*in Russian*).
2. R. I. Jibuti, Elem. Part. Atom. Nucl. **14**, 741 (1983).
3. R. I. Jibuti and N. B. Krupennikova, *Hyperspherical harmonics method in quantum mechanics of few bodies* (Metsniereba, Tbilisi, 1984) (*in Russian*).
4. M. V. Zhukov, B. V. Danilin, D. V. Fedorov, J. M. Bang, I. J. Thompson, and J. S. Vaagen, Phys. Rep. **231**, 151 (1993).

5. A. M. Shirokov, in *Proc. Int. Conf. Perspectives of Nuclear Physics in the Late Nineties, Hanoi, Vietnam, 1994* (Ed. Nguyen Dinh Dang, Da Hsuan Feng, Nguyen Van Giai, Nguyen Dinh Tu), 295 (Singapore, World Scientific, 1995); nucl-th/9407019 (1994).
6. J. M. Bang, B. V. Danilin, V. D. Efros, J. S. Vaagen, M. V. Zhukov, and I. J. Thompson, Phys. Rep. **264**, 27 (1996).
7. E. Nielsen, D. V. Fedorov, A. S. Jensen, and E. Garrido, Phys. Rep. **347**, 373 (2001).
8. Yu. F. Smirnov and A. M. Shirokov, Preprint ITF-88-47R (Kiev, 1988); A. M. Shirokov, Yu. F. Smirnov, and S. A. Zaytsev, in *Modern Problems in Quantum Theory* (Ed. V. I. Savrin and O. A. Khrustalev), (Moscow, 1998), 184; Teoret. Mat. Fiz. **117**, 227 (1998) [Theor. Math. Phys. **117**, 1291 (1998)].
9. T. Ya. Mikhelashvili, Yu. F. Smirnov, and A. M. Shirokov, Yad. Fiz. **48**, 969 (1988) [Sov. J. Nucl. Phys. **48**, 617 (1988)]; J. Phys. **G 16**, 1241 (1990).
10. Yu. A. Lurie, Yu. F. Smirnov, and A. M. Shirokov, Izv. Ros. Akad. Nauk, Ser. Fiz. **57**, 193 (1993) [Bull. Rus. Acad. Sci., Phys. Ser. **57**, 943 (1993)].
11. A. M. Shirokov, Yu. A. Lurie, and Yu. F. Smirnov, *Halo'94 NORDITA/NorFa Study Weekend on Halo Nuclei, Copenhagen, 1994* (Ed. J. M. Bang and J. S. Vaagen).
12. Yu. F. Smirnov, Yu. A. Lurie, and A. M. Shirokov, Revista Mex. Fis. **40**, Supl. 1, 63 (1994).
13. Yu. A. Lurie and A. M. Shirokov, Izv. Ros. Akad. Nauk, Ser. Fiz. **61**, 2121 (1997) [Bull. Rus. Acad. Sci., Phys. Ser. **61**, 1665 (1997)].
14. Yu. A. Lurie and A. M. Shirokov, Ann. Phys. (NY) **312**, 284 (2004).
15. D. E. Lanskoy, Yu. A. Lurie, and A. M. Shirokov, Z. Phys. **A 357**, 95 (1997).
16. V. Vasilevsky, A. V. Nesterov, F. Arickx, J. Broeckhove, Phys. Rev. **C 63**, 034606 (2001); F. Arickx, J. Broeckhove, A. Nesterov, V. Vasilevsky, W. Vanroose, The Modified J-Matrix Approach for Cluster Description of Light Nuclei, Part IV, Chapter 3, this volume.
17. V. Vasilevsky, A. V. Nesterov, F. Arickx, J. Broeckhove, Phys. Rev. **C 63**, 034607 (2001).
18. L. Johansen, A. S. Jensen, and P. G. Hansen, Phys. Lett. **B 244**, 356 (1990).
19. H. Sagawa and M. Houma, Phys. Lett. **B 251**, 17 (1990); H. Sagawa, N. Takigawa, and N. van Giai, Nucl. Phys. **A 543**, 575 (1992).
20. M. V. Zhukov, B. V. Danilin, D. V. Fedorov, J. S. Vaagen, F. A. Gareev, and J. Bang, Phys. Lett. **B 265**, 19 (1991).
21. B. V. Danilin, M. V. Zhukov, A. A. Korsheninnikov, and L. V. Chulkov, Yad. Fiz. **53**, 71 (1991) [Sov. J. Nucl. Phys. **53**, 45 (1991)].
22. B. V. Danilin, M. V. Zhukov, J. S. Vaagen, I J Thompson, and J. M. Bang, Surrey University Preprint CNP92/17 (1992).
23. B. V. Danilin, M. V. Zhukov, J. S. Vaagen, and J. M. Bang, Phys. Lett. **B 302**, 129 (1993).
24. L. S. Ferreira, E. Maglione, J. M. Bang, I. J. Thompson, B. V. Danilin, M. V. Zhukov, and J. S. Vaagen, Phys. Lett. **B 316**, 23 (1993).
25. S. Funada, H. Kameyama, and Y. Sakuragi, Nucl. Phys. **A 575**, 93 (1994).
26. V. I. Kukulin, V. N. Pomerantsev, Kh. D. Razikov, V. T. Voronchev, and G. G. Ryzhikh, Nucl. Phys. **A 586**, 151 (1995).
27. J. S. Al-Khalili, J. A. Tostevin, and I. J. Thompson, Phys. Rev. **C 54**, 1843 (1996).
28. J. S. Al-Khalili and J. A. Tostevin, Phys. Rev. Lett. **76**, 3903 (1996).
29. K. Katō, S. Aoyama, S. Mukai, and I. Ikeda, Nucl. Phys. **A 588**, 29c (1995); S. Aoyama, S. Mukai, K. Katō, and I. Ikeda, Progr. Theor. Phys. **93**, 99 (1995).
30. E. Hiyama, M. Kamimura, Nucl. Phys. **A 588**, 35c (1995).
31. E. Garrido, D. V. Fedorov, and A. S. Jensen, Nucl. Phys. **A 617**, 153 (1997).
32. B. V. Danilin, T. Rogde, S. N. Ershov, H. Heiberg-Andersen, J. S. Vaagen, I. J. Thompson, and M. V. Zhukov, Phys. Rev. **C 55**, R577 (1997).
33. T. Aumann, L. V. Chulkov, V. N. Pribora, and M. H. Smedberg, Nucl. Phys. **A 640**, 24 (1998).
34. M. Hesse, D. Baye and J.-M. Sparenberg, Phys. Lett. **B 455**, 1 (1999).
35. I. J. Thompson, B. V. Danilin, V. D. Efros, J. S. Vaagen, J. M. Bang, and M. V. Zhukov, Phys. Rev. **C 61**, 024318 (2000).

36. N. Vinh Mau, Nucl. Phys. **A 722**, C104 (2003).
37. E. M. Tursunov, D. Baye, and P. Descouvemont, Phys. Rev. **C 73**, 014303 (2006); P. Descouvemont, E. M. Tursunov, and D. Baye, Nucl. Phys. **A 765**, 370 (2006); D. Baye, E. M. Tursunov, and P. Descouvemont, Phys. Rev. **C 74**, 064302 (2006).
38. I. Tanihata et al., RIKEN Preprint AF-NP-60 (1987).
39. I. Tanihata, T. Kobayashi, O. Yamakawa, S. Shimoura, K. Ekuni, K. Sugimoto, N. Takahashi, T. Shimoda, and H. Sato, Phys. Lett. **B 206**, 592 (1988).
40. L. V. Chulkov, B. V. Danilin, V. D. Efros, A. A. Korsheninnikov, and M. V. Zhukov, Europhys. Lett. **8**, 245 (1989).
41. I. Tanihata, D. Hirata, T. Kobayashi, S. Shimoura, K. Sugimoto, and H. Toki, Phys. Lett. **B 289**, 261 (1992).
42. D. Sackett et al., Phys. Rev. **C 48**, 118 (1993); K. Ieki et al., Phys. Rev. Lett. **70**, 730 (1993); A. Galonsky et al., Michigan State University Preprint MSUCL-899 (1993).
43. W. Benenson, Nucl. Phys. **A 588**, 11c (1995).
44. M. Zinser et al., Nucl. Phys. **A 619**, 151 (1997).
45. D. Aleksandrov et al., Nucl. Phys. **A 633**, 234 (1998); T. Aumann et al., Phys. Rev. **C 59**, 1252 (1999); D. Aleksandrov et al., Nucl. Phys. **A 669**, 51 (2000).
46. T. Nakamura et al., Nucl. Phys. **A 687**, 97 (2001); T. Nakamura, Nucl. Phys. **A 788**, 243 (2007).
47. H. Simon et al., Nucl. Phys. **A 734**, 323 (2004); L.V. Chulkov et al., Nucl. Phys. **A 759**, 23 (2005).
48. N. P. Harrington et al., Phys. Rev. **C 75**, 044311 (2007).
49. I. P. Okhrimenko, Few-body Syst. **2**, 169 (1987).
50. A. M. Shirokov, Yu. F. Smirnov, and L. Ya. Stotland, in *Proc. XIIth Europ. Conf. on Few-Body Phys., Uzhgorod, USSR, 1990* (Ed. V. I. Lengyel and M. I. Haysak), 173 (Uzhgorod, 1990).
51. Yu. F. Smirnov, L. Ya. Stotland, and A. M. Shirokov, Izv. Akad. Nauk SSSR, Ser. Fiz. **54**, No 5, 897 (1990) [Bull. Acad. Sci. USSR, Phys. Ser. **54**, No 5, 81 (1990)].
52. J. T. Broad and W. P. Reinhardt, Phys. Rev. **A 14**, 2159 (1976); [J. Phys. **B 9**, 1491 (1976)].
53. S. A. Zaytsev, Teoret. Mat. Fiz. **115**, 263 (1998) [Theor. Math. Phys. **115**, 575 (1998); **121**, 424 (1999)] [Theor. Math. Phys. **121**, 1617 (1999)].
54. A. M. Shirokov, A. I. Mazur, S. A. Zaytsev, J. P. Vary, and T. A. Weber, Phys. Rev. **C 70**, 044005 (2004); "Nucleon-Nucleon Interaction in the J-Matrix Inverse Scattering Approach and Few-Nucleon Systems", Part IV, Chapter 2, this volume.
55. A. M. Shirokov, J. P. Vary, A. I. Mazur, S. A. Zaytsev, and T. A. Weber, Phys. Lett. **B 621**, 96 (2005).
56. A. M. Shirokov, J. P. Vary, A. I. Mazur, and T. A. Weber, Phys. Lett. **B 644**, 33 (2007).
57. Yu. F. Smirnov and K. V. Shitikova, Fiz. Elem. Chast. At. Yad. **8**, 847 (1977) [Sov. J. Part. Nucl. **8**, 344 (1977)].
58. V. D. Efros, W. Leidemann, G. Orlandini, Phys. Lett. **B 338**, 130 (1994); S. Bacca, M. A. Marchisio, N. Barnea, W. Leidemann, G. Orlandini, Phys. Rev. Lett. **89**, 052502 (2002); N. Barnea, W. Leidemann, G. Orlandini, V. D. Efros, E. L. Tomusiak, Few Body Syst. **39**, 1 (2006); N. Barnea, W. Leidemann, G. Orlandini, Phys.Rev. **C 74**, 034003 (2006).
59. H. A. Yamani and L. Fishman, J. Math. Phys. **16**, 410 (1975).
60. J. M. Bang, A. I. Mazur, A. M. Shirokov, Yu. F. Smirnov, and S. A. Zaytsev, Ann. Phys. (NY) **280**, 299 (2000).
61. M. Abramowitz and I. A. Stegun (eds.), *Handbook on Mathematical Functions* (Dover, New York, 1972).
62. F. Ajzenberg-Selove, Nucl. Phys. **A 490**, 1 (1988).
63. A. H. Wapstra, G. Audi, and R. Hoekstra, At. Data Nucl. Data Tables **39**, 281 (1988).
64. P. Navrátil, J. P. Vary, and B. R. Barrett, Phys. Rev. Lett. **84**, 5728 (2000); Phys. Rev. **C 62**, 054311 (2000).
65. G. E. Brown and A. D. Jackson, *The Nucleon-Nucleon Interaction* (North-Holland, Amsterdam, 1976).

66. T. Kaneko, M. LeMere, and L. C. Tang, University of Minnesota Preprint TPI-MINN-91/32-T (1991).
67. J. Bang and C. Gignoux, Nucl. Phys. **A 313**, 119 (1979).
68. D. Gogny, P. Pires, and R. de Tourreil, Phys. Lett. **B 32**, 591 (1970).
69. R. V. Reid, Jr., Ann. Phys. (NY) **50**, 411 (1968).
70. O. F. Nemets, V. G. Neudatchin, A. E. Rudchik, Yu. F. Smirnov, and Yu. M. Tchuvil'sky, *Nucleon associations in atomic nuclei and multi-nucleon transfer reactions* (Naukova Dumka Publishers, Kiev, 1988) (*in Russian*).
71. M. Moshinsky and Yu. F. Smirnov, *The harmonic oscillator in modern physics* (Harwood Academic Publishers, Amsterdam, 1996).
72. R. G. Newton, *Scattering theory of waves and particles, 2nd. ed.* (Springer-Verlag, New York, 1982).
73. A. M. Shirokov and V. N. Sidorenko, Yad. Fiz. **63**, 2085 (2000) [Phys. At. Nucl. **63** 1993 (2000)].
74. F. Cooper, A. Khare, and U. Sukhatme, Phys. Rep. **251**, 267 (1995).
75. E. Garrido, D. V. Fedorov, and A. S. Jensen, Nucl. Phys. **A 650**, 247 (1999).
76. F. Coester, S. Cohen, B. Day, and C. W. Vincent, Phys. Rev. **C 1**, 769 (1970); M. I. Haftel and F. Tabakin, Phys. Rev. **C 3**, 921 (1971).
77. H. C. Pradhan, P. U. Sauer, and J. P. Vary, Phys. Rev. **C 6**, 407 (1972).
78. A. M. Shirokov, J. P. Vary, T. A. Weber, and A. I. Mazur, Izv. Ros. Akad. Nauk, Ser. Fiz. **65**, 1572 (2001) [Bul. Rus. Acad. Sci, Phys. Ser. **65**, v. 65, 1709 (2001)].
79. W. N. Polyzou, Phys. Rev. **C 58**, 91 (1998).

Nucleon–Nucleon Interaction in the J-Matrix Inverse Scattering Approach and Few-Nucleon Systems

A.M. Shirokov, A.I. Mazur, S.A. Zaytsev, J.P. Vary and T.A. Weber

Abstract The nucleon–nucleon interaction is constructed by means of the J-matrix version of inverse scattering theory. Ambiguities of the interaction are eliminated by postulating tridiagonal and quasi-tridiagonal forms of the potential matrix in the oscillator basis in uncoupled and coupled waves, respectively. The obtained interaction is very accurate in reproducing the NN scattering data and deuteron properties. The interaction is used in the no-core shell model calculations of ^{3}H and ^{4}He nuclei. The resulting binding energies of ^{3}H and ^{4}He are very close to experimental values.

1 Introduction

Nucleon–nucleon potentials conventionally referred to as 'realistic', are derived from the meson exchange theory. Modern realistic NN potentials like Bonn [1], Argonne [2], Nijmegen [3], etc., are carefully fitted to the existing experimental data on NN scattering and deuteron properties. Unfortunately, none of the known NN interactions provides a completely satisfactory description of the trinucleon and other light nuclei. To overcome this deficiency, meson exchange [4] or phenomenological [5] three-nucleon (NNN) forces are usually introduced. Impressive progress has been achieved recently in the description of the trinucleon and ^{4}He binding energies with realistic NN and NNN forces [6]. However, the NNN force parameters in such studies are sometimes fitted to the trinucleon binding and some of them may not be consistent with the parameters of the two-body interaction. In one very detailed study, when the NNN interaction parameters were chosen

A.M. Shirokov
Skobeltsyn Institute of Nuclear Physics, Moscow State University, Moscow 119992, Russia
e-mail: shirokov@nucl-th.sinp.msu.ru

Reprinted with permission from: Nucleon-nucleon interaction in the J-matrix inverse scattering approach and few-nucleon systems, Phys. Rev. C Vol. 70, 044005, 24 pgs, (2004) by A.M. Shirokov et al. Copyright (2004) by the American Physical Society. **URL**: http://link.aps.org/abstract/PRC/v70/e044005, with revisions, as well as a discussion and new references added.

A.D. Alhaidari et al. (eds.), *The J-Matrix Method*, 219–268.

consistently with the two-body parameters, the three-nucleon force contribution to the triton binding energy was shown to be negligible [7]. Rather than construct NNN forces, we will develop NN forces in which we exploit the off-shell freedom to improve the description of light nuclei. We defer the development of consistent NNN forces to a future effort.

Impressive progress using effective field theory has recently been reported (see review in Ref. [8]). The versions that provide the most accurate fit to the nucleon-nucleon properties [9] use a momentum-space cutoff and are still quite strong at short distances. The match with the nuclear many body model space cutoff is unclear; additional renormalization is required for typical model spaces that are feasible. We aim in this chapter to have high quality descriptions of the phase shifts with softer potentials whose cutoff is well-matched to the anticipated application in many-body systems.

Various microscopic models have been designed for the studies of few-body systems. It was demonstrated in Ref. [10] that all modern realistic microscopic models provide approximately the same results for the ^{4}He ground state. The no-core shell model [11, 12], which we adopt here, is one of these models. This model can be used not only for the few-body nuclear applications but also, with modern computer facilities, for microscopic studies of heavier nuclei with the number of nucleons A up to $A \sim 12$ [12]. The no-core shell model is based on a wave function expansion in a many-body oscillator function series with the aim to describe bound states and narrow resonances treated as bound states.

The oscillator basis matrices of the modern realistic NN potentials are very large and cannot be directly used without a severe truncation in the many-body no-core shell model calculations. As a result, the convergence of the calculations appears to be slow. This deficiency is conventionally addressed by constructing the so-called *effective NN interaction* (see, e.g., [11]). Ideally the effective NN interaction should reproduce in the finite model space the results of the infinite model space calculation. In a realistic application, the construction of the effective NN interaction is a complicated problem involving various approximations. In the present work, we do not adopt the effective interaction approach. Rather, we retain the bare interaction and carry out large space calculations sufficient to obtain converged ground state energies.

In this contribution, we construct the NN interaction by means of the J-matrix version of inverse scattering theory [13–15]. The matrix of the NN potential in the oscillator basis is obtained for each partial wave independently. Therefore, in our approach we derive the NN interaction as a set of potential matrices for different partial waves. We reproduce the experimental NN scattering data and deuteron properties with small potential matrices. Our NN interaction can be imagined as an effective interaction since its matrix can be directly used in the no-core shell model calculations of light nuclei. However, our NN interaction reproduces the energy spectrum and other observables in a many-body system as well as deuteron properties and NN scattering data. From this point of view, our NN interaction can be treated as a realistic one as well. Our interaction is not related to meson exchange theory, however we shall see that we obtain the deuteron and scattering wave functions that are very close to the ones obtained with realistic meson exchange potentials.

The potential derived by the J-matrix inverse scattering approach is ambiguous. The ambiguity originates from the phase-equivalent transformation suggested in Ref. [16] (see also [17, 18] and references therein). The ambiguity is eliminated in the present approach by a phenomenological ansatz that the potential matrix in the uncoupled partial waves is tridiagonal. Therefore our potentials are *Inverse Scattering Tridiagonal Potentials* (ISTP). The non-central nature of the NN interaction is manifested in the coupling of some partial waves, and the tridiagonal potential ansatz should be extended to allow for the coupling of these partial waves. We postulate phenomenologically the simplest generalization of the tridiagonal form of the potential matrix in this case; however, we refer to our potentials as ISTP in the cases of both uncoupled and coupled partial waves (though, strictly speaking, it is not correct in the later case). It is just the tridiagonal ansatz that brings us to the scattering wave functions which are very close to the ones provided by the meson exchange realistic NN potentials. However, in the case of the coupled sd waves we perform a phase equivalent potential transformation to improve the description of the deuteron properties.

The ansatz of a tridiagonal form represents a very economical version of an inverse scattering potential in the relative harmonic oscillator basis since it has the minimum number of off-diagonal two-body potential matrix elements for a given basis size. More complicated forms are easily imagined and may be obtained either by a unitary transformation [16–18] from the tridiagonal form or from direct inversion techniques that might be developed for each proposed form.

The suggested ISTP are used in the no-core shell model calculations of ^{3}H and ^{4}He. We shall see that the predicted ^{3}H and ^{4}He binding energies are very close to the experimental values. We do not use NNN interactions, yet our predictions of the ^{3}H and ^{4}He bindings are approximately of the same accuracy as the predictions based on the best realistic meson exchange two-nucleon plus three-nucleon forces.

Here we would like to mention some recent chapters where other approaches to the problem of constructing high-quality effective interaction were utilized. The authors of Refs. [19,20] added phenomenological non-local terms to a cut-off Yukawa tail of the realistic NN potentials. The obtained interaction reproduces the ^{3}H binding energy. The additional non-local terms do not reduce the rank of the potential energy matrix in the oscillator basis of the underlying realistic NN interaction. Therefore the use of this interaction in the shell model studies requires the construction of the shell model effective interaction.

A very interesting approach is the construction of the low momentum NN potential V_{low-k} from the realistic NN interactions (see the review in Ref. [21]). The use of V_{low-k} in the shell model applications still requires the construction of the shell model effective interaction but this problem is simplified. The effective interaction obtained from V_{low-k} was used successfully in various shell model applications (see, e.g., [22]). It is unclear whether this interaction provides the correct binding of three-body and four body nuclear systems. Contrary to V_{low-k}, our ISTP is designed for the direct use in shell model applications for light nuclei.

The chapter is organized as follows. In the next Section we present the single channel J-matrix inverse scattering approach, derive ISTP in the uncoupled partial waves and discuss their properties. The derivation and discussion of the ISTP properties in the coupled partial waves can be found in the Section 3. The results of the ^{3}H and ^{4}He calculations are presented in the Section 4. A short summary of the results can be found in Section 5.

2 Single Channel J-Matrix Inverse Scattering Approach and ISTP in Uncoupled NN Partial Waves

The J-matrix formalism in the quantum scattering theory was initially proposed in atomic physics [23]. Within the J-matrix formalism, the continuum spectrum wave function is expanded in an infinite series of L^2 functions. This approach was shown to be one of the most efficient and precise methods in calculations of photoionization [24–26] and electron scattering by atoms [27]. In nuclear physics the same approach has been developed independently [28, 29] as the method of the harmonic oscillator representation of scattering theory. This method has been successfully used in various nuclear applications allowing for the two-body continuum, e.g. nucleus–nucleus scattering has been studied in the algebraic version of RGM based on the J-matrix formalism (see the review chapters [30, 31]); the effect of Λ and neutron decay channels in hypernuclei production reactions has been investigated in Refs. [32, 33], etc. The approach was extended to the case of true few-body scattering in Ref. [34] and utilized in the studies of the monopole excitations of the ^{12}C nucleus in the 3α cluster model in Ref. [35]. It was also used in the studies of double-Λ hypernuclei in Ref. [36] and of weakly bound nuclei in the three-body cluster model in Refs. [16–18].

The J-matrix version of the inverse scattering theory was suggested in Refs. [13–15]. The discussion of the general formalism below follows the ideas of Refs. [13–15], however, some formulas are presented here in a manner that should be more convenient for the current application. The tridiagonalization of the interaction obtained by the inverse scattering methods have not previously been discussed in the literature, hence the corresponding theory and results are new.

The oscillator-basis J-matrix formalism is discussed in detail elsewhere (see, e.g., [23, 37]). We present here only some relations needed for understanding the inverse scattering J-matrix approach.

The Schrödinger equation in the partial wave with orbital angular momentum l reads

$$H^l\,\Psi_{lm}(E,\boldsymbol{r}) = E\,\Psi_{lm}(E,\boldsymbol{r}). \tag{1}$$

The wave function is given by

$$\Psi_{lm}(E,\boldsymbol{r}) = \frac{1}{r}\,u_l(E,r)\,Y_{lm}(\hat{\boldsymbol{r}}), \tag{2}$$

where $Y_{lm}(\hat{\boldsymbol{r}})$ is the spherical function. Within the J-matrix formalism, the radial wave function $u_l(E,r)$ is expanded in an oscillator function series

$$u_l(E,r) = \sum_{n=0}^{\infty} a_{nl}(E)\, R_{nl}(r), \tag{3}$$

where

$$R_{nl}(r) = (-1)^n \sqrt{\frac{2n!}{r_0\,\Gamma(n+l+3/2)}} \left(\frac{r}{r_0}\right)^{l+1} \exp\!\left(-\frac{r^2}{2r_0^2}\right) L_n^{l+\frac{1}{2}}\!\left(\frac{r^2}{r_0^2}\right), \tag{4}$$

$L_n^{\alpha}(x)$ is the associated Laguerre polynomial, the oscillator radius $r_0 = \sqrt{\hbar/m\omega}$, and m is the reduced mass. All energies are given in the units of the oscillator basis parameter $\hbar\omega$.

The wave function in the oscillator representation $a_{nl}(E)$ is a solution of the infinite set of algebraic equations

$$\sum_{n'=0}^{\infty} (H^l_{nn'} - \delta_{nn'} E)\, a_{n'l}(E) = 0, \tag{5}$$

where the Hamiltonian matrix elements $H^l_{nn'} = T^l_{nn'} + V^l_{nn'}$, the kinetic energy matrix elements

$$T^l_{n,\,n-1} = -\frac{1}{2}\sqrt{n(n+l+1/2)}, \tag{6a}$$

$$T^l_{n,\,n} = \frac{1}{2}\,(2n+l+3/2), \tag{6b}$$

$$T^l_{n,\,n+1} = -\frac{1}{2}\sqrt{(n+1)(n+l+3/2)}, \tag{6c}$$

and the potential energy V^l within the J-matrix formalism is approximated by the truncated matrix with elements

$$\widetilde{V}^l_{nn'} = \begin{cases} V^l_{nn'} & \text{if } n \text{ and } n' \leq N; \\ 0 & \text{if } n \text{ or } n' > N. \end{cases} \tag{7}$$

In the inverse scattering J-matrix approach, the potential energy is constructed in the form of the finite matrix of the type (7); therefore the J-matrix solutions with such an interaction are exact.

In the *external part of the model space* spanned by the functions (4) with $n \geq N$, Eq. (5) takes the form of a three-term recurrence relation

$$T^l_{n,\,n-1}\, a_{n-1,\,l}(E) + (T^l_{nn} - E)\, a_{nl}(E) + T^l_{n,\,n+1}\, a_{n+1,\,l}(E) = 0. \tag{8}$$

Any solution of Eq. (8) is a superposition of the fundamental regular $S_{nl}(E)$ and irregular $C_{nl}(E)$ solutions [23, 37],

$$a_{nl}(E) = \cos\delta(E)\, S_{nl}(E) + \sin\delta(E)\, C_{nl}(E), \tag{9}$$

where

$$S_{nl}(E) = \sqrt{\frac{\pi\, r_0\, n!}{\Gamma(n+l+3/2)}}\, q^{l+1} \exp\left(-\frac{q^2}{2}\right) L_n^{l+\frac{1}{2}}(q^2), \tag{10}$$

$$C_{nl}(E) = (-1)^l \sqrt{\frac{\pi\, r_0\, n!}{\Gamma(n+l+3/2)}}\, \frac{q^{-l}}{\Gamma(-l+1/2)} \exp\left(-\frac{q^2}{2}\right) \times \Phi(-n-l-1/2,\, -l+1/2;\, q^2), \tag{11}$$

$\Phi(a,\, b;\, z)$ is a confluent hypergeometric function [38], $q = \sqrt{2E}$, and $\delta(E)$ is the scattering phase shift.

The wave function in the oscillator representation $a_{nl}(E)$ in the *internal part of the model space* spanned by the functions (4) with $n \leq N$, can be expressed through the external solution $a_{N+1,l}(E)$:

$$a_{nl}(E) = \mathcal{G}_{nN}\, T^l_{N,\,N+1}\, a_{N+1,l}(E). \tag{12}$$

The matrix elements,

$$\mathcal{G}_{nn'} = -\sum_{\lambda'=0}^{N} \frac{\langle n|\lambda'\rangle\langle\lambda'|n'\rangle}{E_{\lambda'} - E}, \tag{13}$$

are expressed through the eigenvalues E_λ and eigenvectors $\langle n|\lambda\rangle$ of the truncated Hamiltonian matrix, i. e. E_λ and $\langle n|\lambda\rangle$ are obtained by solving the algebraic problem

$$\sum_{n'=0}^{N} H^l_{nn'} \langle n'|\lambda\rangle = E_\lambda \langle n|\lambda\rangle, \qquad n \leq N. \tag{14}$$

The matrix element $\mathcal{G}_{NN}$ is of primary importance in the calculation of the phase shift $\delta(E)$:

$$\tan\delta(E) = -\frac{S_{Nl}(E) - \mathcal{G}_{NN}\, T^l_{N,\,N+1}\, S_{N+1,\,l}(E)}{C_{Nl}(E) - \mathcal{G}_{NN}\, T^l_{N,\,N+1}\, C_{N+1,\,l}(E)}. \tag{15}$$

In the direct J-matrix approach, we first solve Eq. (14) and next calculate the phase shift $\delta(E)$ by means of Eq. (15). In the inverse scattering J-matrix approach, the phase shift $\delta(E)$ is taken to be known at any energy E and, instead of solving (14), we extract the eigenvalues E_λ and the eigenvectors $\langle n|\lambda\rangle$ from this information.

First we assign some value to N, the rank of the desired potential matrix [see Eq. (7)]. Generally, with a finite rank potential matrix it is possible to reproduce the phase shift $\delta(E)$ only in a finite energy interval; larger N supports a larger energy interval. However, from the point of view of many-body applications, it is desirable to have N as small as possible.

The components $a_{nl}(E)$ of the wave function in the oscillator representation, should be finite at arbitrary energy E. This is seen from Eqs. (12) and (13) to be possible at the energies $E = E_\lambda$, $\lambda = 0, 1, \ldots, N$ only if

$$a_{N+1,l}(E_\lambda) = 0. \tag{16}$$

Knowing the phase shift, we can calculate $a_{N+1,l}(E)$ at any energy E using Eq. (9). Therefore we can solve numerically the transcendental equation (16) and find the eigenvalues E_λ, $\lambda = 0, 1, \ldots, N$.

Due to Eq. (16),

$$a_{N+1,l}(E) \underset{E\to E_\lambda}{\longrightarrow} \alpha_l^\lambda (E - E_\lambda), \tag{17}$$

where

$$\alpha_l^\lambda = \left. \frac{d\, a_{N+1,l}(E)}{d\, E} \right|_{E=E_\lambda}. \tag{18}$$

Now it is easy to derive from Eqs. (12) and (13) the following equation:

$$a_{Nl}(E_\lambda) = |\langle N|\lambda\rangle|^2 \alpha_l^\lambda \, T^l_{N,\,N+1}, \tag{19}$$

or, equivalently,

$$|\langle N|\lambda\rangle|^2 = \frac{a_{Nl}(E_\lambda)}{\alpha_l^\lambda \, T^l_{N,\,N+1}}. \tag{20}$$

Within the J-matrix formalism, both $a_{Nl}(E)$ and $a_{N+1,l}(E)$ fit Eq. (9) and can be calculated using this equation at any energy E. Hence, one can also calculate α_l^λ by means of Eq. (18). Therefore the components $\langle N|\lambda\rangle$ can be obtained from Eq. (20) (the sign of the components $\langle N|\lambda\rangle$ is of no importance).

Equations (16) and (20) provide the general solution of the J-matrix inverse scattering problem: solving these equations we obtain the sets of E_λ and $\langle N|\lambda\rangle$, and these quantities completely determine the phase shifts $\delta(E)$. However $\langle N|\lambda\rangle$ are supposed to be the components of the eigenvectors $\langle n|\lambda\rangle$ of the truncated Hamiltonian matrix [see Eq. (14)] that should fit the completeness relation

$$\sum_{\lambda=0}^{N} \langle n|\lambda\rangle\langle\lambda|n'\rangle = \delta_{nn'}, \tag{21}$$

hence we should have

$$\sum_{\lambda=0}^{N} \langle N|\lambda\rangle\langle\lambda|N\rangle = 1. \tag{22}$$

Generally the set of $\langle N|\lambda\rangle$ obtained by means of Eq. (20) violates the completeness relation (22). Therefore this set of $\langle N|\lambda\rangle$ ideally describing the phase shifts, cannot be treated as the set of last components of the normalized eigenvectors $\langle n|\lambda\rangle$ of any truncated Hermitian Hamiltonian matrix; in other words, the set of $\langle N|\lambda\rangle$ violating Eq. (22) cannot be used to construct a Hermitian Hamiltonian matrix.

To overcome this difficulty, we fit Eq. (22) by changing the value of the component $\langle N|\lambda = N\rangle$ corresponding to the highest eigenvalue $E_{\lambda=N}$. This modification spoils the description of the phase shifts $\delta(E)$ at energies E different from E_λ, $\lambda = 0, 1, \dots, N$. We restore the phase shift description in the energy interval $[0, E_{\lambda=N-1}]$ by variation of $E_{\lambda=N}$. From the above consideration it is clear that larger N values make it possible to reproduce phase shifts in larger energy intervals $[0, E_{\lambda=N-1}]$.

There is an ambiguity in determining the potential matrix describing the given phase shifts $\delta(E)$: any of the phase equivalent transformations discussed in Refs. [16–18] [see also Eqs. (66)–(68) below] that do not change the truncated Hamiltonian eigenvalues E_λ and respective eigenvector components $\langle N|\lambda\rangle$, results in a potential matrix that brings us to the same phase shifts $\delta(E)$ at any energy E. Additional model assumptions are needed to resolve this ambiguity. As was already mentioned, we assume the tridiagonal form of the potential matrix. We now discuss the construction of the tridiagonal potential matrix supposing N and the sets of E_λ and $\langle N|\lambda\rangle$ to be known.

If the potential matrix is tridiagonal, the Eq. (14) can be rewritten as

$$H^l_{00}\langle 0|\lambda\rangle + H^l_{01}\langle 1|\lambda\rangle = E_\lambda\langle 0|\lambda\rangle, \tag{23a}$$

$$H^l_{n,n-1}\langle n-1|\lambda\rangle + H^l_{nn}\langle n|\lambda\rangle + H^l_{n,n+1}\langle n+1|\lambda\rangle = E_\lambda\langle n|\lambda\rangle$$
$$(n = 1, 2, \dots, N-1), \tag{23b}$$

$$H^l_{N,N-1}\langle N-1|\lambda\rangle + H^l_{NN}\langle N|\lambda\rangle = E_\lambda\langle N|\lambda\rangle. \tag{23c}$$

The unknown quantities in Eq. (23c) are the component $\langle N-1|\lambda\rangle$ and the Hamiltonian matrix elements $H^l_{N,N-1}$ and H^l_{NN}. We multiply Eq. (23c) by $\langle\lambda|N\rangle$, sum the result over λ, and use the completeness relation (21) to obtain the formula for the calculation of H^l_{NN}:

$$H^l_{NN} = \sum_{\lambda=0}^{N} E_\lambda \langle N|\lambda\rangle^2. \tag{24}$$

The Hermitian conjugate of Eq. (23c) reads:

$$\langle\lambda|N-1\rangle H^l_{N,N-1} + \langle\lambda|N\rangle H^l_{NN} = \langle\lambda|N\rangle E_\lambda. \tag{23c-c}$$

We multiply Eq. (23c) by Eq. (23c-c), sum the result over λ, and use the completeness relation (21) to obtain the following expression for the calculation of $H^l_{N,N-1}$:

$$H^l_{N,N-1} = -\sqrt{\sum_{\lambda=0}^{N} E^2_\lambda \langle N|\lambda\rangle^2 - \left(H^l_{NN}\right)^2}. \tag{25}$$

Generally, the sign in the right-hand-side of Eq. (25) is arbitrary. Here we use an additional assumption that the off-diagonal Hamiltonian matrix elements $H^l_{n,n\pm1}$ are dominated by the kinetic energy so that the sign of these matrix elements is the same as the kinetic energy matrix elements $T^l_{n\,n\pm1}$ [see Eqs. (6)]. This assumption brings us to the minus sign in the right-hand-side of Eq. (25).

Now Eq. (23c) can be used to calculate the last unknown quantity,

$$\langle N-1|\lambda\rangle = \frac{1}{H^l_{N,N-1}}\left(E_\lambda\langle N|\lambda\rangle - H^l_{NN}\langle N|\lambda\rangle\right). \tag{26}$$

We now turn to Eq. (23b) with $n = N-1$. This equation contains one more term than Eq. (23c), however this term does not include unknown quantities. We perform with Eq. (23b) exactly the same manipulations to obtain expressions for $H^l_{N-1,N-1}$, $H^l_{N-2,N-1}$ and $\langle N-2|\lambda\rangle$. Setting $n = N-2$ in Eq. (23b), we obtain the expressions for $H^l_{N-2,N-2}$, $H^l_{N-3,N-2}$ and $\langle N-3|\lambda\rangle$, etc. Equation (23a) is needed only to calculate the last matrix element H^l_{00}. As a result, we obtain the following generalization of Eq. (24) valid at $n = N, N-1, \ldots, 0$:

$$H^l_{nn} = \sum_{\lambda=0}^{N} E_\lambda \langle n|\lambda\rangle^2. \tag{27}$$

The equations

$$H^l_{n,n-1} = -\sqrt{\sum_{\lambda=0}^{N} E^2_\lambda \langle n|\lambda\rangle^2 - \left(H^l_{nn}\right)^2 - \left(H^l_{n,n+1}\right)^2} \tag{28}$$

and

$$\langle n-1|\lambda\rangle = \frac{1}{H^l_{n,n-1}}\left(E_\lambda\langle n|\lambda\rangle - H^l_{n,n}\langle n|\lambda\rangle + H^l_{n,n+1}\langle n+1|\lambda\rangle\right) \tag{29}$$

are valid at $n = N-1, N-2, \ldots, 1$. Equations (25)–(29) make it possible to calculate all unknown quantities. After calculating the Hamiltonian matrix elements $H^l_{nn'}$, we derive the ISTP matrix elements by the obvious equations

$$V^l_{nn} = H^l_{nn} - T^l_{nn}, \tag{30a}$$

$$V^l_{n,n\pm1} = H^l_{n,n\pm1} - T^l_{n,n\pm1}. \tag{30b}$$

The above theory is used to construct the NN ISTP matrix elements in uncoupled partial waves. We use as input the np scattering phase shifts reconstructed from the experimental data by the Nijmegen group [3]. The oscillator basis parameter $\hbar\omega = 40\,\mathrm{MeV}$. Usually in the shell model calculations, the complete $\varkappa\hbar\omega$ model space is used, i. e. all many-body oscillator basis states (configurations) with $\sum_i \varkappa_i \leq \varkappa$ where the single-particle state oscillator quanta $\varkappa_i = 2n_i + l_i$, are included in the calculation. Thus, to be applicable to all p-shell nuclei in accessible model spaces, we suggest the $8\hbar\omega$ and $7\hbar\omega$ ISTP, i.e. the rank of the ISTP matrix N is chosen so that $2N + l = 8$ in the partial waves with even orbital angular momentum l and $2N + l = 7$ in the partial waves with odd orbital angular momentum l.

The non-zero matrix elements of the obtained ISTP in uncoupled partial waves are presented in Tables 1–8 (in $\hbar\omega = 40\,\mathrm{MeV}$ units).

In Figs. 1–16 we present the results of the phase shift and scattering wave function calculations with our ISTP in the uncoupled partial waves. The phase shifts are seen to be better reproduced by ISTP up to the laboratory energy $E_{\mathrm{lab}} = 350\,\mathrm{MeV}$

Table 1 Non-zero matrix elements in $\hbar\omega$ units of the $8\hbar\omega$ ISTP matrix in the $^1s_0$ partial wave

n	V^l_{nn}	$V^l_{n,n+1} = V^l_{n+1,n}$
0	−0.370692591051	0.134054681241
1	−0.159916088622	0.016474369170
2	0.139593205593	−0.133446192137
3	0.266824207307	−0.078690196129
4	0.041490933216	

Table 2 Non-zero matrix elements in $\hbar\omega$ units of the $7\hbar\omega$ ISTP matrix in the $^1p_1$ partial wave

n	V^l_{nn}	$V^l_{n,n+1} = V^l_{n+1,n}$
0	0.106199364772	−0.094411509693
1	0.321832027399	−0.198614230564
2	0.382278903019	−0.125293001922
3	0.088186662748	

Table 3 Non-zero matrix elements in $\hbar\omega$ units of the $8\hbar\omega$ ISTP matrix in the $^1d_2$ partial wave

n	V^l_{nn}	$V^l_{n,n+1} = V^l_{n+1,n}$
0	−0.041824646289	0.038312478836
1	−0.112960462645	0.068735184648
2	−0.127611509816	0.040422120683
3	−0.025546698405	

Table 4 Non-zero matrix elements in $\hbar\omega$ units of the $7\hbar\omega$ ISTP matrix in the $^1f_3$ partial wave

n	V^l_{nn}	$V^l_{n,n+1} = V^l_{n+1,n}$
0	0.042387100374	−0.027905560992
1	0.074740011106	−0.028153835497
2	0.025116180890	

Table 5 Non-zero matrix elements in $\hbar\omega$ units of the $7\hbar\omega$ ISTP matrix in the $^3p_0$ partial wave

n	V^l_{nn}	$V^l_{n,n+1} = V^l_{n+1,n}$
0	−0.136747520574	0.015115026047
1	0.087868702261	−0.105904971180
2	0.236248878650	−0.080401020753
3	0.049099156034	

Table 6 Non-zero matrix elements in $\hbar\omega$ units of the $7\hbar\omega$ ISTP matrix in the $^3p_1$ partial wave

n	V^l_{nn}	$V^l_{n,n+1} = V^l_{n+1,n}$
0	0.088933281276	−0.092880110751
1	0.338999430587	−0.211115182274
2	0.361586494817	−0.098285652220
3	0.051672685711	

Table 7 Non-zero matrix elements in $\hbar\omega$ units of the $8\hbar\omega$ ISTP matrix in the $^3d_2$ partial wave

n	V^l_{nn}	$V^l_{n,n+1} = V^l_{n+1,n}$
0	−0.200240578055	0.119332193872
1	−0.288987898733	0.146304772643
2	−0.255222029014	0.079227780212
3	−0.054213944378	

Table 8 Non-zero matrix elements in $\hbar\omega$ units of the $7\hbar\omega$ ISTP matrix in the $^3f_3$ partial wave

n	V^l_{nn}	$V^l_{n,n+1} = V^l_{n+1,n}$
0	0.026292148118	−0.013940970302
1	0.034636722707	−0.012592178851
2	0.011196241352	

than by one of the best realistic meson exchange potentials Nijmegen-II. Some discrepancies are seen only at large energies. These discrepancies can be eliminated by using larger N values. This is illustrated in phase shifts of odd partial waves presented in Figs. 3, 7, 9, 11, 15. These are the results of the phase shift calculations with the $9\hbar\omega$ ISTP in addition to the $7\hbar\omega$ ISTP phase shifts. It is interesting that the differences between the $7\hbar\omega$ ISTP and $9\hbar\omega$ ISTP wave functions in odd partial waves are too small to be seen in Figs. 4, 8, 10, 12, 16 even at large energies. We note also that the use of $7\hbar\omega$ ISTP instead of $9\hbar\omega$ ISTP in the ^{3}H and ^{4}He calculations, result in negligible differences of the binding energies, wave functions, etc. The ISTP np scattering wave functions at different energies are very close to the Nijmegen-II wave functions both in odd and even partial waves. In other words, these ISTP wave functions can be regarded as realistic.

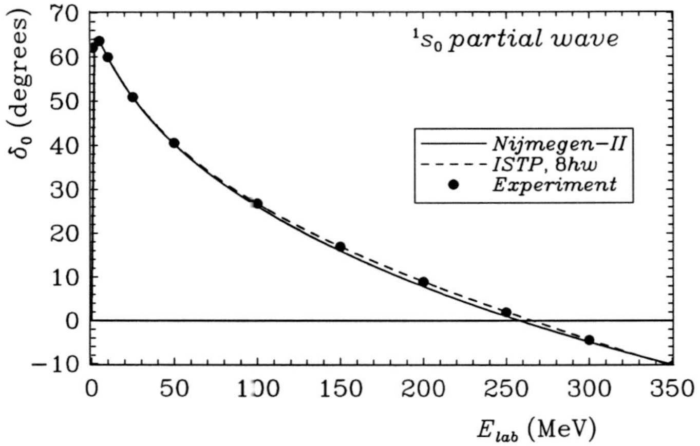

Fig. 1 $^1s_0$ *np* scattering phase shifts. *Filled circles* — experimental data of Ref. [3]; *solid line* — realistic meson exchange Nijmegen-II potential [3] phase shifts; *dashed line* — ISTP phase shifts

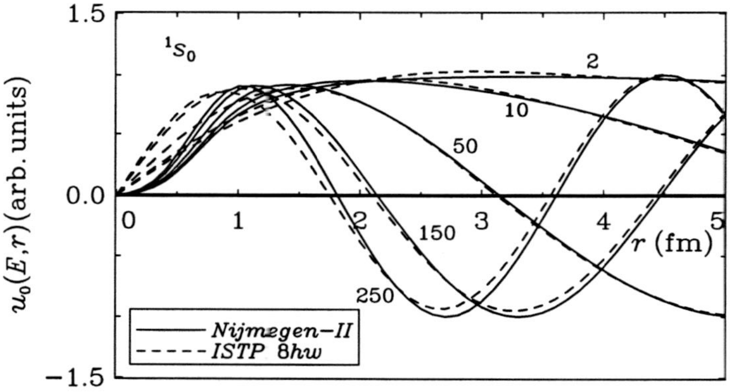

Fig. 2 $^1s_0$ *np* scattering wave functions at the laboratory energies $E_{\text{lab}} = 2, 10, 50, 150, 250$ MeV. *Solid line* — realistic meson exchange Nijmegen-II potential [3] wave functions; *dashed line* — ISTP wave functions

3 Two-Channel J-Matrix Inverse Scattering Approach and ISTP in Coupled NN Partial Waves

In the case of the nucleon–nucleon scattering, the spins of two nucleons can couple to the total spin $S = 0$ (singlet spin state) or to the total spin $S = 1$ (triplet spin state). In the case of the singlet spin state, we have only uncoupled partial waves in the nucleon–nucleon scattering. In the case of the triplet spin state, the total angular momentum $j = l + 1$ can be obtained by the coupling of the total spin $S = 1$ with the orbital angular momentum l. On the other hand, the higher triplet-spin partial wave of the same parity with the orbital angular momentum $l' = l + 2$, can

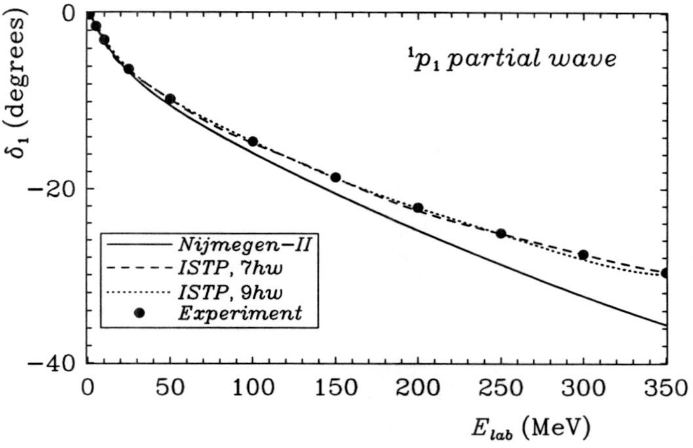

Fig. 3 $^1p_1$ *np* scattering phase shifts. *Filled circles* — experimental data of Ref. [3]; *solid line* — realistic meson exchange Nijmegen-II potential [3] phase shifts; *dashed line* — $7\hbar\omega$ ISTP phase shifts; *dotted line* — $9\hbar\omega$ ISTP phase shifts

have the same total angular momentum $j = l + 1 = l' - 1$. Such partial waves are coupled due to the non-central nature of the NN interaction. The *sd* coupled partial waves (the coupling of the $^3s_1$ and $^3d_1$ partial waves) and *pf* coupled partial waves (the coupling of the $^3p_2$ and $^3f_2$ partial waves) are of special interest for applications. The case of the *sd* coupled partial waves is of primary importance due to the existence of the only *np* bound state (the deuteron). The coupled equations describing the NN system in the coupled partial waves, are of the same structure with the coupled equations describing the two-channel system. In other words, the

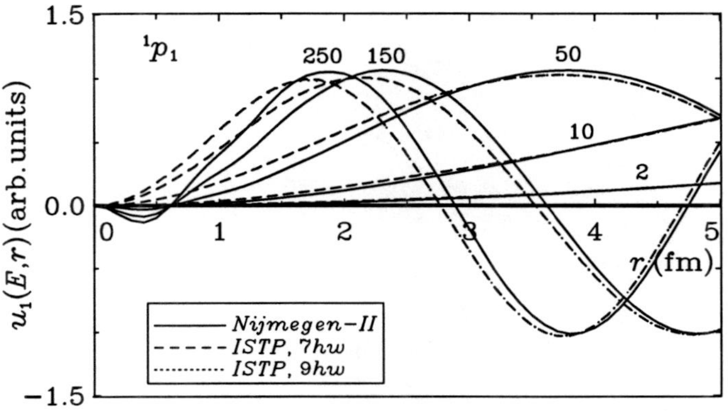

Fig. 4 $^1p_1$ *np* scattering wave functions at the laboratory energies $E_{\rm lab} = 2, 10, 50, 150, 250$ MeV. *Solid line* — realistic meson exchange Nijmegen-II potential [3] wave functions; *dashed line* — $7\hbar\omega$ ISTP wave functions; *dotted line* — $9\hbar\omega$ ISTP wave functions

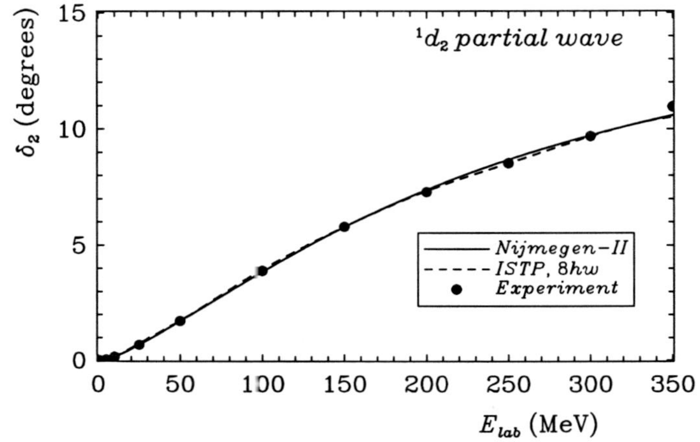

Fig. 5 $^1d_2$ *np* scattering phase shifts. See Fig. 1 for details

description of the coupled waves in the NN scattering is formally equivalent with the description of the two-channel scattering.

The wave function in the coupled waves case is

$$\Psi = \sum_{\Gamma} \frac{1}{r} u_{\Gamma}(E,r) \, |\Gamma\rangle, \tag{31}$$

where $|\Gamma\rangle$ is the spin-angle wave function which includes the spin variables of two nucleons coupled to the total spin $S = 1$, the spherical function $Y_{l_\Gamma m}(\hat{\boldsymbol{r}})$, and the coupling of the channel orbital momentum l_Γ with the total spin S into the total angular momentum j; $u_\Gamma(E,r)$ is the radial wave function in the given formal

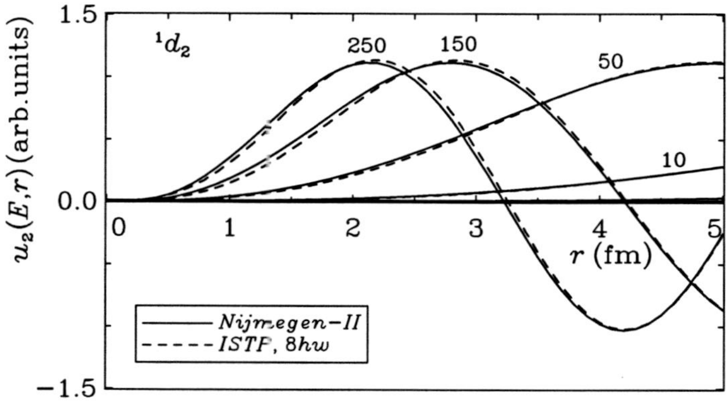

Fig. 6 $^1d_2$ *np* scattering wave functions at the laboratory energies $E_{\text{lab}} = 2, 10, 50, 150, 250$ MeV. See Fig. 2 for details

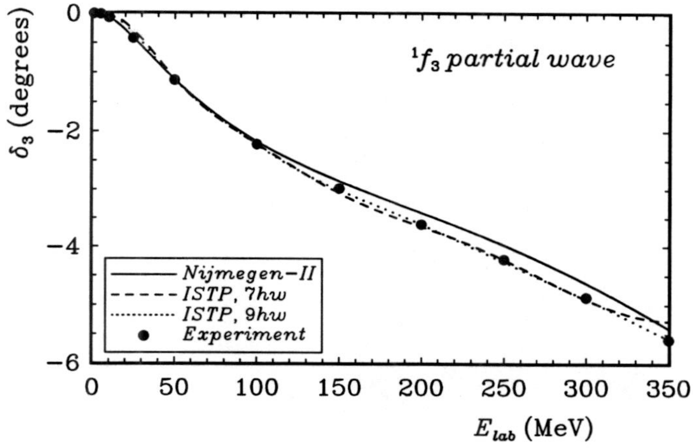

Fig. 7 $^1f_3$ *np* scattering phase shifts. See Fig. 3 for details

channel $\Gamma = \{l_\Gamma, j\}$. Generally there are two independent solutions for each radial wave function $u_\Gamma(E,r)$. To distinguish these solutions it is convenient to employ the K-matrix formalism associated with the standing wave asymptotics of the wave function:

$$u_{\Gamma(\Gamma_i)}(E,r) \underset{r\to\infty}{\longrightarrow} \frac{qr}{r_0}\left(\delta_{\Gamma\Gamma_i}\, j_{l_\Gamma}\!\left(\frac{qr}{r_0}\right) - K_{\Gamma\Gamma_i}(E)\, n_{l_\Gamma}\!\left(\frac{qr}{r_0}\right)\right). \tag{32}$$

Here the index Γ_i distinguishes independent radial functions $u_{\Gamma(\Gamma_i)}(E,r)$ in the channel Γ, $K_{\Gamma\Gamma_i}(E)$ is the K-matrix, and $j_l(x)$ and $n_l(x)$ are spherical Bessel and Neumann functions. The advantage of the K-matrix formalism is that the radial functions $u_{\Gamma(\Gamma_i)}(E,r)$ defined according to their standing wave asymptotics (32)

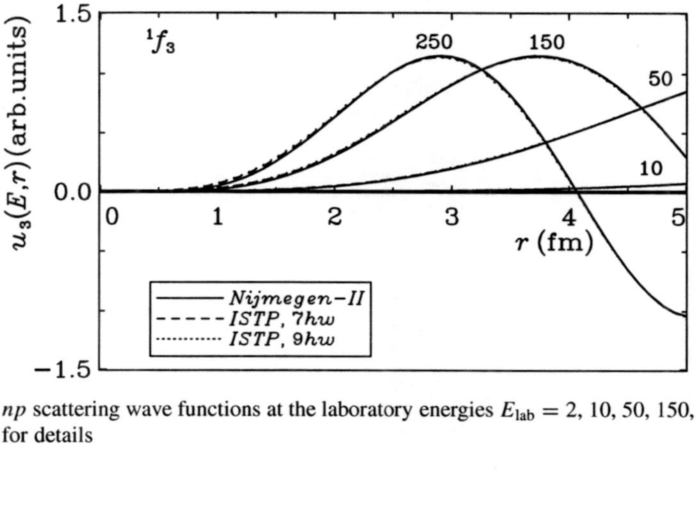

Fig. 8 $^1f_3$ *np* scattering wave functions at the laboratory energies $E_{\text{lab}} = 2, 10, 50, 150, 250$ MeV. See Fig. 4 for details

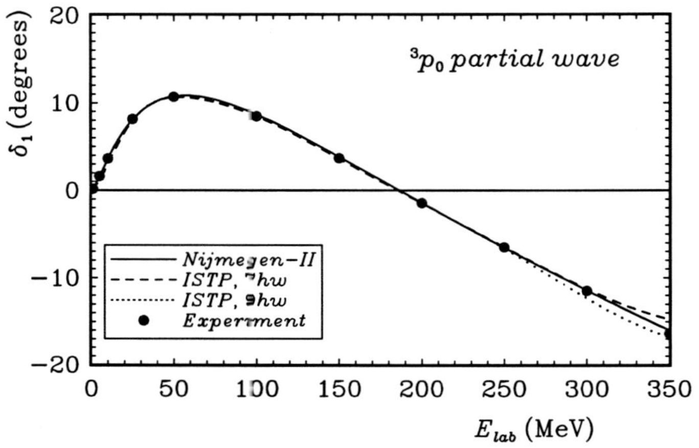

Fig. 9 $^3p_0$ *np* scattering phase shifts. See Fig. 3 for details

are real contrary to the more conventional S-matrix formalism with complex radial wave functions which are asymptotically a superposition of ingoing and outgoing spherical waves. The K-matrix $K_{\Gamma\Gamma_i}(E)$, of course, can be expressed through the S-matrix. However it is not the S-matrix but the so-called phase shifts δ_Γ and δ_{Γ_i} in each of the coupled partial waves Γ and Γ_i and the mixing parameter ε that are usually published as functions of the energy E in the experimental and theoretical investigations. The S-matrix can be parametrized in terms of δ_Γ, δ_{Γ_i} and ε. However for the present application it is more convenient to express the K-matrix elements directly through δ_Γ, δ_{Γ_i} and ε (see Refs. [39,40]):

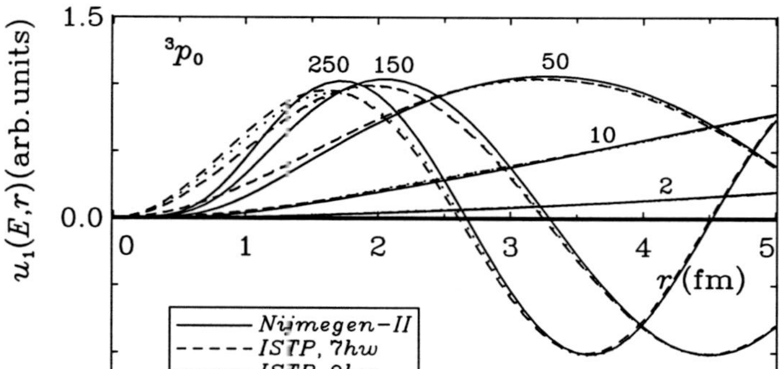

Fig. 10 $^3p_0$ *np* scattering wave functions at the laboratory energies $E_{\mathrm{lab}} = 2, 10, 50, 150, 250$ MeV. See Fig. 4 for details

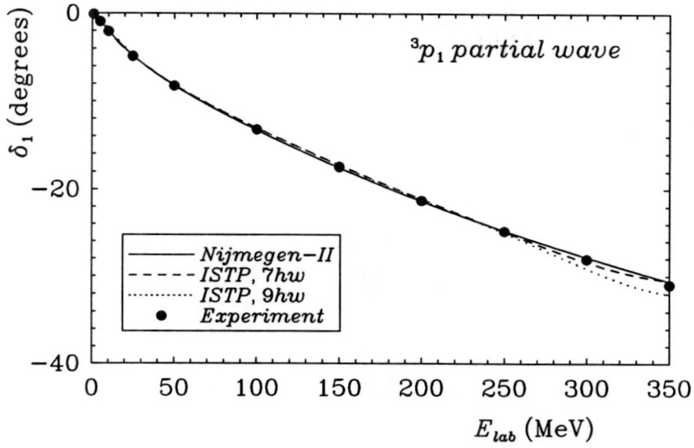

Fig. 11 $^3p_1$ np scattering phase shifts. See Fig. 3 for details

$$K_{ss}(E) = \frac{\tan\delta_s + \tan^2\varepsilon\cdot\tan\delta_d}{1-\tan^2\varepsilon\cdot\tan\delta_s\cdot\tan\delta_d}, \tag{33a}$$

$$K_{dd}(E) = \frac{\tan\delta_d + \tan^2\varepsilon\cdot\tan\delta_s}{1-\tan^2\varepsilon\cdot\tan\delta_s\cdot\tan\delta_d}, \tag{33b}$$

$$K_{sd}(E) = K_{ds}(E) = \frac{\tan\varepsilon}{\cos\delta_s\cdot\cos\delta_d\cdot(1-\tan^2\varepsilon\cdot\tan\delta_s\cdot\tan\delta_d)}. \tag{33c}$$

To be specific, we have specified the case of the coupled sd waves where the channel indexes Γ and Γ_i take the values s or d. In the case of the coupled pf waves, one

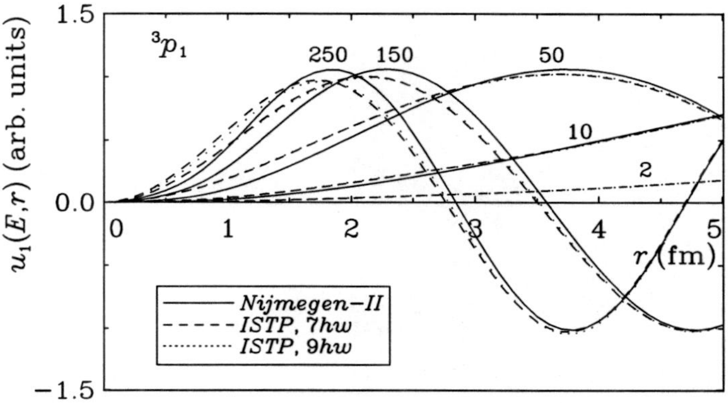

Fig. 12 $^3p_1$ np scattering wave functions at the laboratory energies $E_{\rm lab}$ = 2, 10, 50, 150, 250 MeV. See Fig. 4 for details

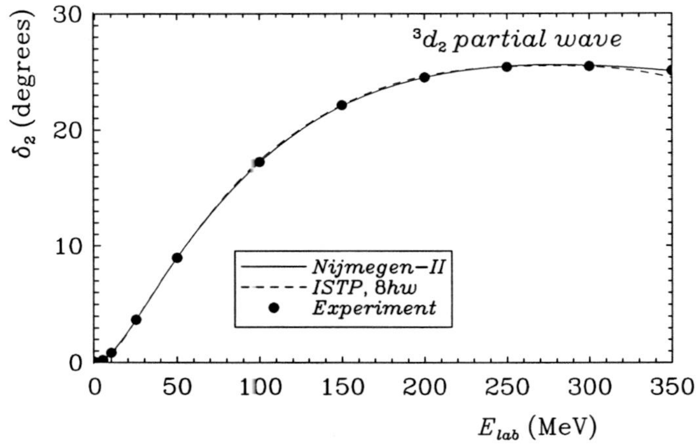

Fig. 13 $^3d_2$ *np* scattering phase shifts. See Fig. 1 for details

substitutes the indexes s and d by the indexes p and f in the above expressions and in other formulas in this section.

Within the inverse scattering J-matrix approach, the potential in the coupled partial waves is fitted with the form:

$$V = \sum_{\Gamma,\,\Gamma'} \sum_{n=0}^{N_\Gamma} \sum_{n'=0}^{N_{\Gamma'}} |n\Gamma\rangle\, V_{nn'}^{\Gamma\Gamma'}\, \langle n'\Gamma'|. \tag{34}$$

Here $V_{nn'}^{\Gamma\Gamma'} \equiv \langle n\Gamma|V|n'\Gamma'\rangle$ is the potential energy matrix element in the oscillator basis

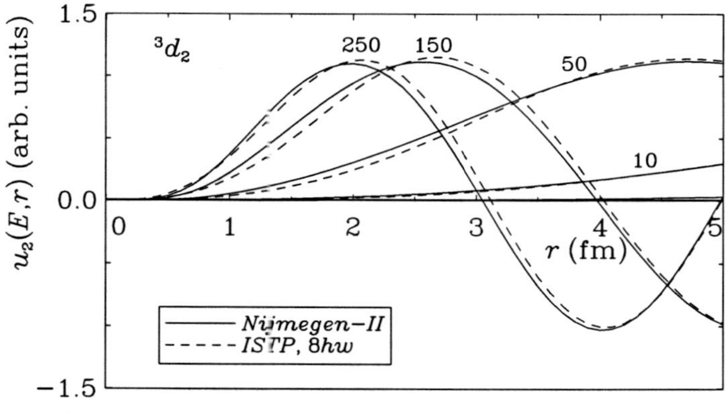

Fig. 14 $^3d_2$ *np* scattering wave functions at the laboratory energies $E_{\text{lab}} = 2, 10, 50, 150, 250$ MeV. See Fig. 2 for details

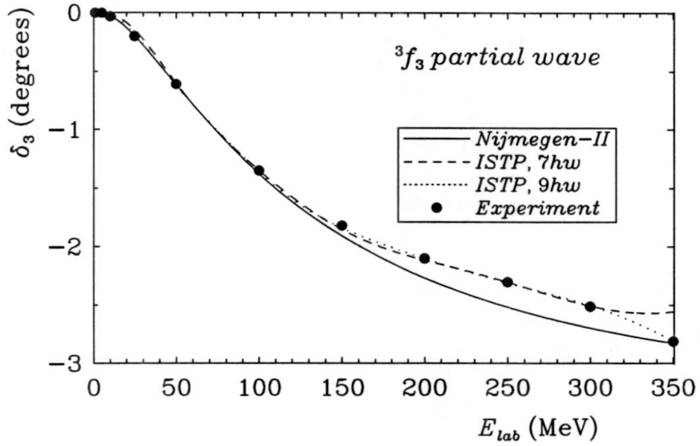

Fig. 15 $^3f_3$ *np* scattering phase shifts. See Fig. 3 for details

$$|n\Gamma\rangle = R_{nl_\Gamma}(r)\,|\Gamma\rangle, \tag{35}$$

where the radial oscillator function $R_{nl_\Gamma}(r)$ is given by Eq. (4) and $|\Gamma\rangle$ is the spin-angle function. Different truncation boundaries N_Γ can be used in different partial waves Γ.

The multi-channel J-matrix formalism is well known (see, e.g., [23, 37]) and we will not discuss it here in detail. The formalism provides exact solutions for the continuum spectrum wave functions in the case when the finite-rank potential V of the type (34) is employed. In the case of the discrete spectrum states, the exact solutions are obtained by the calculation of the corresponding S-matrix poles as is discussed in Refs. [17, 18, 34]. In particular, the deuteron ground state energy E_d

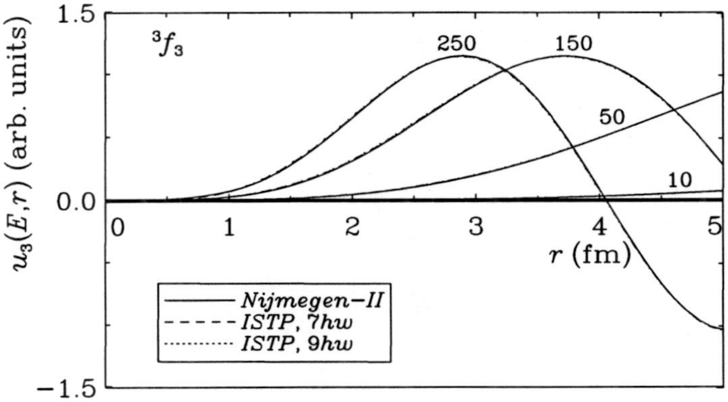

Fig. 16 $^3f_3$ *np* scattering wave functions at the laboratory energies $E_{\rm lab} = 2, 10, 50, 150, 250$ MeV. See Fig. 4 for details

should be associated with the S-matrix pole and its wave function is calculated by means of the J-matrix formalism applied to the negative energy $E = E_d$.

Within the J-matrix formalism, the radial wave function $u_{\Gamma(\Gamma_i)}(E, r)$ is expanded in the oscillator function series

$$u_{\Gamma(\Gamma_i)}(E, r) = \sum_{n=0}^{\infty} a_{n\,\Gamma(\Gamma_i)}(E)\, R_{nl_\Gamma}(r). \tag{36}$$

In the external part of the model space spanned by the functions (35) with $n \geq N_\Gamma$, the oscillator representation wave function $a_{n\,\Gamma(\Gamma_i)}(E)$ fits the three-term recurrence relation (8). Its solutions corresponding to the asymptotics (32) are

$$a_{n\Gamma(\Gamma_i)}(E) = \delta_{\Gamma\Gamma_i}\, S_{nl_\Gamma}(E) + K_{\Gamma\Gamma_i}(E)\, C_{nl_\Gamma}(E). \tag{37}$$

Equation (37) can be used for the calculation of $a_{n\Gamma(\Gamma_i)}(E)$ with $n \geq N_\Gamma$ if the coupled wave phase shifts δ_Γ and δ_{Γ_i} and the mixing parameter ε are known.

The oscillator representation wave function $a_{n\,\Gamma(\Gamma_i)}(E)$ in the internal part of the model space spanned by the functions (35) with $n \leq N_\Gamma$, can be expressed through the external oscillator representation wave functions $a_{N_\Gamma+1,\Gamma(\Gamma_i)}(E)$ as

$$a_{n\,\Gamma(\Gamma_i)}(E) = \sum_{\Gamma} \mathcal{G}^{\Gamma\Gamma'}_{nN_{\Gamma'}}\, T^{l_{\Gamma'}}_{N_{\Gamma'},\,N_{\Gamma'}+1}\, a_{N_{\Gamma'}+1,\Gamma'(\Gamma_i)}(E). \tag{38}$$

The matrix elements,

$$\mathcal{G}^{\Gamma\Gamma'}_{nn'} = -\sum_{\lambda'=0}^{N} \frac{\langle n\Gamma|\lambda'\rangle\, \langle\lambda'|n'\Gamma'\rangle}{E_{\lambda'} - E}, \tag{39}$$

where $N = N_\Gamma + N_{\Gamma'} + 1$, are expressed within the direct J-matrix formalism through the eigenvalues E_λ and eigenvectors $\langle n\Gamma|\lambda\rangle$ of the truncated Hamiltonian matrix, i. e. E_λ and $\langle n\Gamma|\lambda\rangle$ are obtained by solving the algebraic problem

$$\sum_{\Gamma'} \sum_{n'=0}^{N_{\Gamma'}} H^{\Gamma\Gamma'}_{nn'}\, \langle n'\Gamma'|\lambda\rangle = E_\lambda\, \langle n\Gamma|\lambda\rangle, \qquad n \leq N_\Gamma. \tag{40}$$

Here $H^{\Gamma\Gamma'}_{nn'} \equiv \langle n\Gamma|H|n'\Gamma'\rangle$ are the Hamiltonian matrix elements.

Within the inverse J-matrix approach, we start with assigning some values to the potential truncation boundaries N_Γ [see Eq. (34)] in each of the partial waves Γ. As a next step, we calculate the sets of eigenvalues E_λ and respective eigenvector components $\langle N_\Gamma\Gamma|\lambda\rangle$. This can be done using the set of the J-matrix matching conditions which are obtained from Eq. (38) supposing $n = N_\Gamma$. In more detail, these matching conditions are (to be specific, we again take the case of the coupled sd waves so the channel indexes Γ and Γ_i take the values s or d)

$$a_{N_s s(s)}(E) = \sum_{\Gamma'=s,d} \mathcal{G}_{s\Gamma'}\, T^{\Gamma'}_{N_{\Gamma'},N_{\Gamma'}+1}\, a_{N_{\Gamma'}+1,\Gamma'(s)}(E), \tag{41a}$$

$$a_{N_d d(s)}(E) = \sum_{\Gamma'=s,d} \mathcal{G}_{d\Gamma'}\, T^{\Gamma'}_{N_{\Gamma'},N_{\Gamma'}+1}\, a_{N_{\Gamma'}+1,\Gamma'(s)}(E), \tag{41b}$$

$$a_{N_s s(d)}(E) = \sum_{\Gamma'=s,d} \mathcal{G}_{s\Gamma'}\, T^{\Gamma'}_{N_{\Gamma'},N_{\Gamma'}+1}\, a_{N_{\Gamma'}+1,\Gamma'(d)}(E), \tag{41c}$$

and

$$a_{N_d d(d)}(E) = \sum_{\Gamma'=s,d} \mathcal{G}_{d\Gamma'}\, T^{\Gamma'}_{N_{\Gamma'},N_{\Gamma'}+1}\, a_{N_{\Gamma'}+1,\Gamma'(d)}(E), \tag{41d}$$

where we introduced the shortened notation

$$\mathcal{G}_{\Gamma\Gamma'} \equiv \mathscr{G}^{\Gamma\Gamma'}_{N_\Gamma N_{\Gamma'}} = -\sum_{\lambda'=0}^{N} \frac{\langle N_\Gamma \Gamma|\lambda'\rangle\, \langle\lambda'|N_{\Gamma'}\Gamma'\rangle}{E_{\lambda'} - E}. \tag{42}$$

To calculate $a_{N_\Gamma \Gamma(\Gamma_i)}(E)$ and $a_{N_\Gamma+1,\,\Gamma(\Gamma_i)}(E)$ entering Eqs. (41), we can use Eq. (37) with the K-matrix elements expressed through the experimental data by Eqs. (33). Therefore $\mathcal{G}_{ss}$, $\mathcal{G}_{sd}$, $\mathcal{G}_{ds}$ and $\mathcal{G}_{dd}$ are the only unknown quantities in Eqs. (41) and they can be obtained as the solutions of the algebraic problem (41) at any positive energy E.

These solutions may be expressed as

$$\mathcal{G}_{ss} = \frac{\Delta_{ss}(E)}{T^s_{N_s,N_s+1}\,\Delta(E)}, \tag{43a}$$

$$\mathcal{G}_{dd} = \frac{\Delta_{dd}(E)}{T^d_{N_d,N_d+1}\,\Delta(E)}, \tag{43b}$$

and

$$\mathcal{G}_{sd} = \mathcal{G}_{ds} = -\frac{r_0\sqrt{2E}\,K_{sd}}{2\,T^s_{N_s,N_s+1}\,T^d_{N_d,N_d+1}\,\Delta(E)}, \tag{43c}$$

where

$$\begin{aligned}\Delta_{ss}(E) = \Big(S_{N_s s}(E) + K_{ss}(E)\,C_{N_s s}(E)\Big)\Big(S_{N_d+1,d}(E) + K_{dd}(E)\,C_{N_d+1,d}(E)\Big)\\ -K^2_{sd}(E)\,C_{N_s s}(E)\,C_{N_d+1,d}(E),\end{aligned} \tag{44a}$$

$$\begin{aligned}\Delta_{dd}(E) = \Big(S_{N_s+1,s}(E) + K_{ss}(E)\,C_{N_s+1,s}(E)\Big)\Big(S_{N_d d}(E) + K_{dd}(E)\,C_{N_d d}(E)\Big)\\ -K^2_{sd}(E)\,C_{N_s+1,s}(E)\,C_{N_d d}(E),\end{aligned} \tag{44b}$$

and

$$\Delta(E) = \Big(S_{N_s+1,s}(E) + K_{ss}(E)\,C_{N_s+1,s}(E)\Big)\Big(S_{N_d+1,d}(E) + K_{dd}(E)\,C_{N_d+1,d}(E)\Big) - K^2_{sd}(E)\,C_{N_s+1,s}(E)\,C_{N_d+1,d}(E). \tag{44c}$$

To derive Eq. (43c), we used the following expression for the Casoratian determinant [34,37]:

$$\mathscr{K}^l_n(C,S) \equiv C_{n+1,l}(E)\,S_{nl}(E) - S_{n+1,l}(E)\,C_{nl}(E) = \frac{r_0\sqrt{2E}}{2\,T^l_{n,n+1}}. \tag{45}$$

It is obvious from Eqs. (42) and (43) that the eigenvalues E_λ can be found by solving the following equation:

$$\Delta(E_\lambda) = 0. \tag{46}$$

The eigenvector components $\langle N_\Gamma \Gamma|\lambda\rangle$ can be obtained from Eqs. (43a) and (43b) in the limit $E \to E_\lambda$ in the same manner as Eq. (20) in the single-channel case:

$$|\langle N_s s|\lambda\rangle|^2 = \frac{\Delta_{ss}(E_\lambda)}{T^s_{N_s,N_s+1}\,\Delta^\lambda} \tag{47}$$

and

$$|\langle N_d d|\lambda\rangle|^2 = \frac{\Delta_{dd}(E_\lambda)}{T^d_{N_d,N_d+1}\,\Delta^\lambda}, \tag{48}$$

where

$$\Delta^\lambda = \left.\frac{d\Delta(E)}{dE}\right|_{E=E_\lambda}. \tag{49}$$

Equations (47) and (48) make it possible to calculate the absolute values of $\langle N_s s|\lambda\rangle$ and $\langle N_d d|\lambda\rangle$ only. However the relative sign of these eigenvector components is important. This relative sign can be established using the relation

$$\frac{\langle N_s s|\lambda\rangle\,T^s_{N_s,N_s+1}}{\langle N_d d|\lambda\rangle\,T^d_{N_d,N_d+1}} = -\frac{a_{N_d+1,d(s)}(E_\lambda)}{a_{N_s+1,s(s)}(E_\lambda)} = -\frac{a_{N_d+1,d(d)}(E_\lambda)}{a_{N_s+1,s(d)}(E_\lambda)} \tag{50}$$

that can be easily obtained from Eqs. (41).

Using Eqs. (46)–(50) we obtain all eigenvalues $E_\lambda > 0$ and corresponding eigenvector components $\langle N_\Gamma \Gamma|\lambda\rangle$. For example, in the case of the coupled pf waves when the NN system does not have a bound state, all eigenvalues E_λ are positive and by means of Eqs. (46)–(50) we obtain a complete set of eigenvalues $E_\lambda = 0, 1, \ldots, N$ and the complete set of the eigenvector's last components $\langle N_\Gamma \Gamma|\lambda\rangle$ providing the best description of the 'experimental' (obtained by means of phase shift analysis) phase shifts $\delta_1(E)$ and $\delta_3(E)$ and mixing parameter ε. However, as in the case of the uncoupled waves, we should take care of fitting the completeness relation for the eigenvectors $\langle n\Gamma|\lambda\rangle$ that in the coupled wave case takes the form

$$\sum_{\lambda=0}^{N} \langle n\Gamma|\lambda\rangle\langle\lambda|n'\Gamma'\rangle = \delta_{nn'}\delta_{\Gamma\Gamma'}. \tag{51}$$

Due to Eq. (51), in the two-channel case, we should perform variation of the components $\langle N_\Gamma \Gamma|\lambda\rangle$ associated with the two largest eigenenergies $E_{\lambda=N}$ and $E_{\lambda=N-1}$ to fit three relations

$$\sum_{\lambda=0}^{N} \langle N_{\Gamma_1}\Gamma_1|\lambda\rangle\langle\lambda|N_{\Gamma_1}\Gamma_1\rangle = 1, \tag{52a}$$

$$\sum_{\lambda=0}^{N} \langle N_{\Gamma_1}\Gamma_1|\lambda\rangle\langle\lambda|N_{\Gamma_2}\Gamma_2\rangle = 0, \tag{52b}$$

and

$$\sum_{\lambda=0}^{N} \langle N_{\Gamma_2}\Gamma_2|\lambda\rangle\langle\lambda|N_{\Gamma_2}\Gamma_2\rangle = 1. \tag{52c}$$

This immediately spoils the description of the scattering data that can be restored by the additional variation of the eigenenergies $E_{\lambda=N}$ and $E_{\lambda=N-1}$. As a result, in the case of the coupled pf waves, we perform a standard fit to the data by minimizing χ^2 by the variation of $\langle N_p p|\lambda = N\rangle$, $\langle N_p p|\lambda = N-1\rangle$, $\langle N_f f|\lambda = N\rangle$, $\langle N_f f|\lambda = N-1\rangle$, $E_{\lambda=N}$ and $E_{\lambda=N-1}$. These six parameters should fit three relations (52), hence we face a simple problem of a three-parameter fit.

In the case of the coupled sd waves, the np system has a bound state (the deuteron) at the energy E_d ($E_d < 0$) and one of the eigenvalues E_λ is negative: $E_0 < 0$. We should extend the above theory to the case of a system with bound states. For the coupled sd waves case when the np system has only one bound state, we need three additional equations to calculate E_0 and the components $\langle N_s s|\lambda = 0\rangle$ and $\langle N_d d|\lambda = 0\rangle$.

The deuteron energy E_d should be associated with the S-matrix pole. As it was already noted, the technique of the S-matrix pole calculation within the J-matrix formalism is discussed together with some applications in Refs. [17, 18]. In the case of the finite-rank potentials of the type (34), one can obtain the exact value of the bound state energy E_d and the exact bound state wave function by the S-matrix pole calculation within the J-matrix formalism. To calculate the S-matrix, we use the standard outgoing-ingoing spherical wave asymptotics and the respective expression for the J-matrix oscillator space wave function in the external part of the model space discussed, e.g., in Refs. [17, 18, 34, 37] instead of the standing wave asymptotics (32) and respectively modified expression (37) for the J-matrix oscillator space wave function. Using the expressions for the multi-channel S-matrix within the J-matrix formalism presented in Refs. [17,18,34,37], it is easy to obtain the following expressions [14] for the two-channel S-matrix elements:

$$\begin{aligned} S_{ss} = \frac{1}{D(E)}\Big\{ & \Big(C^{(-)}_{N_s s}(E) - \mathcal{G}_{ss}\, T^s_{N_s,N_s+1}\, C^{(-)}_{N_s+1,s}(E)\Big) \\ & \times \Big(C^{(+)}_{N_d d}(E) - \mathcal{G}_{dd}\, T^d_{N_d,N_d+1}\, C^{(+)}_{N_d+1,d}(E)\Big) \\ & - \mathcal{G}^2_{sd}\, T^s_{N_s,N_s+1}\, T^d_{N_d,N_d+1}\, C^{(-)}_{N_s+1,s}(E)\, C^{(+)}_{N_d+1,d}(E)\Big\}, \end{aligned} \tag{53a}$$

$$\begin{aligned} S_{dd} = \frac{1}{D(E)}\Big\{ & \Big(C^{(+)}_{N_s s}(E) - \mathcal{G}_{ss}\, T^s_{N_s,N_s+1}\, C^{(+)}_{N_s+1,s}(E)\Big) \\ & \times \Big(C^{(-)}_{N_d d}(E) - \mathcal{G}_{dd}\, T^d_{N_d,N_d+1}\, C^{(-)}_{N_d+1,d}(E)\Big) \\ & - \mathcal{G}^2_{sd}\, T^s_{N_s,N_s+1}\, T^d_{N_d,N_d+1}\, C^{(+)}_{N_s+1,s}(E)\, C^{(-)}_{N_d+1,d}(E)\Big\}, \end{aligned} \tag{53b}$$

and

$$S_{sd} = S_{ds} = -\frac{i r_0 \sqrt{2E}\, \mathcal{G}_{sd}}{D(E)}, \tag{53c}$$

where

$$\begin{aligned} D(E) = & \Big(C^{(+)}_{N_s s}(E) - \mathcal{G}_{ss}\, T^s_{N_s,N_s+1}\, C^{(+)}_{N_s+1,s}(E)\Big) \\ & \times \Big(C^{(+)}_{N_d d}(E) - \mathcal{G}_{dd}\, T^d_{N_d,N_d+1}\, C^{(+)}_{N_d+1,d}(E)\Big) \\ & - \mathcal{G}^2_{sd}\, T^s_{N_s,N_s+1}\, T^d_{N_d,N_d+1}\, C^{(+)}_{N_s+1,s}(E)\, C^{(+)}_{N_d+1,d}(E) \end{aligned} \tag{54}$$

and

$$C^{(\pm)}_{nl}(E) = C_{nl}(E) \pm i\, S_{nl}(E). \tag{55}$$

We need to calculate $C^{(\pm)}_{nl}(E)$ at negative energy $E = E_d$ which can be done using Eqs. (55), (10) and (11) where imaginary values of $q = q_d = i\sqrt{2|E_d|}$ are employed. Extension of these expressions to the complex q plane is discussed in Ref. [34].

Since we associate the deuteron energy E_d with the S-matrix pole, from Eqs. (53) we have

$$D(E_d) = 0. \tag{56}$$

Assigning the experimental deuteron ground state energy to E_d in Eq. (56) and substituting $D(E_d)$ in this formula by its expression (54), we obtain one of the equations needed to calculate E_0, $\langle N_s s|\lambda = 0\rangle$ and $\langle N_d d|\lambda = 0\rangle$.

Two other equations utilize information about the asymptotic normalization constants of the deuteron bound state $\mathscr{A}_s$ and $\mathscr{A}_d$. If the S-matrix is treated as a function of the complex momentum q, then its residue can be expressed through $\mathscr{A}_s$ and $\mathscr{A}_d$ [41,42]:

$$i \operatorname*{Res}_{q=iq_d} S_{l_\Gamma l_{\Gamma'}} = r_0\, e^{i\frac{\pi}{2}(l_\Gamma + l_{\Gamma'})} \mathscr{A}_{l_\Gamma} \mathscr{A}_{l_{\Gamma'}}. \tag{57}$$

(the factor r_0 in the right-hand-side originates from the use of the dimensionless momentum q). $\mathscr{A}_s$ and $\eta = \dfrac{\mathscr{A}_d}{\mathscr{A}_s}$ are determined experimentally. Therefore it is useful to rewrite Eq. (57) as

$$i \lim_{q\to iq_d} (q - iq_d) S_{ss} = r_0 \mathscr{A}_s^2 \tag{58a}$$

and

$$i \lim_{q\to iq_d} (q - iq_d) S_{sd} = -r_0 \eta \mathscr{A}_s^2. \tag{58b}$$

Substituting S_{ss} and S_{sd} by its expressions (53)–(54), we obtain two additional equations for the calculation of E_0, $\langle N_s s|\lambda = 0\rangle$ and $\langle N_d d|\lambda = 0\rangle$.

Clearly, in the case of coupled sd waves, we should also fit the completeness relation (51). We employ the following method of calculation of the sets of the eigenvalues E_λ and the components $\langle N_s s|\lambda\rangle$ and $\langle N_d d|\lambda\rangle$. The E_λ values with $\lambda = 1, 2, \ldots, N-2$ are obtained by solving Eq. (46) while the respective eigenvector's last components $\langle N_s s|\lambda\rangle$ and $\langle N_d d|\lambda\rangle$ are calculated using Eqs. (47)–(50). Next we perform a χ^2 fit to the scattering data of the parameters E_0, $E_{\lambda=N-1}$, $E_{\lambda=N}$, $\langle N_s s|\lambda = 0\rangle$, $\langle N_s s|\lambda = N-1\rangle$, $\langle N_d d|\lambda = N\rangle$, $\langle N_d d|\lambda = 0\rangle$, $\langle N_d d|\lambda = N-1\rangle$, and $\langle N_d d|\lambda = N\rangle$. These nine parameters fit six relations (52a), (52b), (52c), (56), (58a) and (58b), i.e. we should perform a three-parameter fit as in the case of coupled pf waves.

Now we turn to the calculation of the remaining eigenvector components $\langle n\Gamma|\lambda\rangle$ with $n < N_\Gamma$ and the Hamiltonian matrix elements $H_{nn'}^{\Gamma\Gamma'}$ with $n \leq N_\Gamma$ and $n' \leq N_{\Gamma'}$ entering Eq. (40). The coupled waves Hamiltonian matrix obtained by the general J-matrix inverse scattering method is ambiguous; the ambiguity originates from the multi-channel generalization of the phase equivalent transformation mentioned in the single channel case. As in the single channel case, we eliminate the ambiguity by adopting a particular form of the potential energy matrix.

As in the case of uncoupled partial waves, we construct $8\hbar\omega$ ISTP in the coupled sd waves. Therefore $2N_\Gamma + l_\Gamma = 8$, or $2N_s + 0 = 8$ and $2N_d + 2 = 8$; hence $N_s = N_d + 1$. In the coupled pf waves, we construct $7\hbar\omega$ and $9\hbar\omega$ ISTP; clearly we again have $N_p = N_f + 1$. Thus the potential matrix $V_{nn'}^{\Gamma\Gamma'}$ has the following structure:

the submatrices $V^{\Gamma\Gamma}_{nn'}$ coupling the oscillator components of the same partial wave are quadratic [e.g., $(N_p+1)\times(N_p+1)$ submatrix $V^{pp}_{nn'}$ in the $^3p_2$ wave] while the submatrices $V^{\Gamma\Gamma'}_{nn'}$ with $\Gamma\neq\Gamma'$ coupling the oscillator components of different partial waves are $(N_\Gamma+1)\times N_\Gamma$ or $N_\Gamma\times(N_\Gamma+1)$ matrices [e.g., $(N_p+1)\times(N_p)$ submatrix $V^{pf}_{nn'}$ coupling the $^3p_2$ and $^3f_2$ waves]. Our assumptions are: we adopt (i) the tridiagonal form of the quadratic submatrices $V^{\Gamma\Gamma}_{nn'}$ and (ii) the simplest two-diagonal form of the non-quadratic submatrices $V^{\Gamma\Gamma'}_{nn'}$ with $\Gamma\neq\Gamma'$ coupling the oscillator components of different partial waves. The structure of the ISTP matrices in coupled partial waves is illustrated by Fig. 17.

Due to these assumptions, the algebraic problem (40) takes the following form:

$$H^{ss}_{00}\langle 0\,s|\lambda\rangle + H^{ss}_{01}\langle 1s|\lambda\rangle + H^{sd}_{00}\langle 0d|\lambda\rangle = E_\lambda\langle 0s|\lambda\rangle, \tag{59a}$$

$$H^{ds}_{00}\langle 0s|\lambda\rangle + H^{ds}_{01}\langle 1s|\lambda\rangle + H^{dd}_{00}\langle 0d|\lambda\rangle + H^{dd}_{01}\langle 1d|\lambda\rangle = E_\lambda\langle 0d|\lambda\rangle, \tag{59b}$$

$$H^{ss}_{n,n-1}\langle n-1,s|\lambda\rangle + H^{ss}_{nn}\langle ns|\lambda\rangle + H^{ss}_{n,n+1}\langle n+1,s|\lambda\rangle + H^{sd}_{n,n-1}\langle n-1,d|\lambda\rangle + H^{sd}_{nn}\langle nd|\lambda\rangle = E_\lambda\langle ns|\lambda\rangle \qquad (n=1,\,2,\,\dots,\,N_s-1), \tag{59c}$$

$$H^{ds}_{nn}\langle ns|\lambda\rangle + H^{ds}_{n,n+1}\langle n+1,s|\lambda\rangle + H^{dd}_{n,n-1}\langle n-1,d|\lambda\rangle + H^{dd}_{nn}\langle nd|\lambda\rangle + H^{dd}_{n,n+1}\langle n+1,d|\lambda\rangle = E_\lambda\langle nd|\lambda\rangle \qquad (n=1,\,2,\,\dots,\,N_d-1), \tag{59d}$$

$$H^{ss}_{N_s,N_s-1}\langle N_s-1,s|\lambda\rangle + H^{ss}_{N_sN_s}\langle N_ss|\lambda\rangle + H^{sd}_{N_sN_d}\langle N_dd|\lambda\rangle = E_\lambda\langle N_ss|\lambda\rangle, \tag{59e}$$

and

$$H^{ds}_{N_d,N_s-1}\langle N_s-1,s|\lambda\rangle + H^{ds}_{N_dN_s}\langle N_ss|\lambda\rangle + H^{dd}_{N_d,N_d-1}\langle N_d-1,d|\lambda\rangle + H^{dd}_{N_dN_d}\langle N_dd|\lambda\rangle = E_\lambda\langle N_dd|\lambda\rangle. \tag{59f}$$

Even though this set of equations is more complicated than the set (23) discussed in the uncoupled waves case, it can be solved in the same manner.

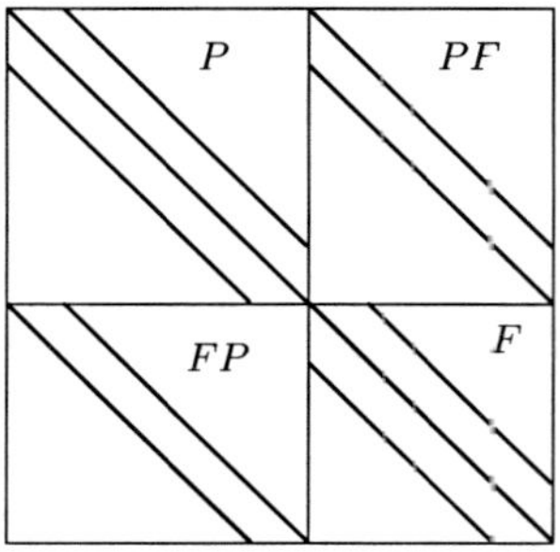

Fig. 17 Structure of the ISTP matrix in the coupled pf waves and of the Version 0 ISTP in the coupled sd waves. The location of non-zero matrix elements is schematically illustrated by solid lines

Multiplying Eqs. (59e)–(59f) by $\langle N_s s|\lambda\rangle$ and $\langle N_d d|\lambda\rangle$, summing the results over λ and using the completeness relation (51) we obtain

$$H^{ss}_{N_sN_s} = \sum_{\lambda=0}^{N} E_\lambda \langle N_s s|\lambda\rangle^2, \tag{60a}$$

$$H^{dd}_{N_dN_d} = \sum_{\lambda=0}^{N} E_\lambda \langle N_d d|\lambda\rangle^2, \tag{60b}$$

and

$$H^{sd}_{N_sN_d} = \sum_{\lambda=0}^{N} E_\lambda \langle N_s s|\lambda\rangle\langle\lambda|N_d\, d\rangle. \tag{60c}$$

Now we multiply each of the Eqs. (59e) and (59f) by its Hermitian conjugate and one of these equations by the Hermitian conjugate of the other, sum the results over λ and use (51) to obtain

$$H^{ss}_{N_s,N_s-1} = -\sqrt{\sum_{\lambda=0}^{N} E^2_\lambda \langle N_s s|\lambda\rangle^2 - \left(H^{ss}_{N_sN_s}\right)^2 - \left(H^{sd}_{N_sN_d}\right)^2}, \tag{61a}$$

$$H^{ds}_{N_d,N_s-1} = \frac{1}{H^{ss}_{N_s,N_s-1}}\left[\sum_{\lambda=0}^{N} E^2_\lambda \langle N_s s|\lambda\rangle\, \langle\lambda|N_d\, d\rangle - H^{sd}_{N_sN_d}\left(H^{ss}_{N_sN_s} + H^{dd}_{N_dN_d}\right)\right], \tag{61b}$$

and

$$H^{dd}_{N_d,N_d-1} = -\sqrt{\sum_{\lambda=0}^{N} E^2_\lambda \langle N_d\, d|\lambda\rangle^2 - \left(H^{dd}_{N_dN_d}\right)^2 - \left(H^{sd}_{N_sN_d}\right)^2 - \left(H^{ds}_{N_d,N_s-1}\right)^2}. \tag{61c}$$

As in the case of uncoupled waves, we take the off-diagonal matrix elements-$H^{ss}_{N_s,N_s\pm1}$ and $H^{dd}_{N_d,N_d\pm1}$ to be dominated by the respective kinetic energy matrix elements $T^{s}_{N_s,N_s\pm1}$ and $T^{d}_{N_d,N_d\pm1}$ and therefore choose the minus sign in the right-hand-sides of Eqs. (61a) and (61c).

By means of Eqs. (60) and (61) we obtain all matrix elements $H^{\Gamma\Gamma'}_{nn'}$ entering Eqs. (59e) and (59f). Using this information, the eigenvector components $\langle N_s - 1, s|\lambda\rangle$ and $\langle N_d - 1, d|\lambda\rangle$ can be extracted directly from Eqs. (59e) and (59f):

$$\langle N_s - 1, s|\lambda\rangle = \frac{1}{H^{ss}_{N_s,N_s-1}}\left(E_\lambda \langle N_s s|\lambda\rangle - H^{ss}_{N_sN_s}\langle N_s s|\lambda\rangle - H^{sd}_{N_sN_d}\langle N_d d|\lambda\rangle\right) \tag{62a}$$

and

$$\langle N_d - 1, d|\lambda\rangle = \frac{1}{H^{dd}_{N_d,N_d-1}}\Big(E_\lambda\,\langle N_d\,d|\lambda\rangle - H^{dd}_{N_dN_d}\langle N_d d|\lambda\rangle - H^{ds}_{N_dN_s}\langle N_s s|\lambda\rangle - H^{ds}_{N_d,N_s-1}\langle N_s - 1, s|\lambda\rangle\Big). \tag{62b}$$

Now we can perform the same manipulations with Eqs. (59a)–(59d). We take $n = N_s - 1, N_s - 2, \ldots, 1$ in Eq. (59c) and $n = N_d - 1, N_d - 2, \ldots, 1$ in Eq. (59d). Equations (59c) and (59d) are a bit more complicated than Eqs. (59e) and (59f), however the additional terms in Eqs. (59c) and (59d) include only the quantities calculated on the previous step. As a result, we obtain the following relations for the calculation of the matrix elements $H^{\Gamma\Gamma'}_{nn}$:

$$H^{ss}_{nn} = \sum_{\lambda=0}^{N} E_\lambda\,\langle ns|\lambda\rangle^2, \tag{63a}$$

$$H^{dd}_{nn} = \sum_{\lambda=0}^{N} E_\lambda\,\langle nd|\lambda\rangle^2, \tag{63b}$$

and

$$H^{sd}_{nn} = \sum_{\lambda=0}^{N} E_\lambda\,\langle ns|\lambda\rangle\langle\lambda|nd\rangle. \tag{63c}$$

Equation (63a) is valid for $n = N_s, N_s - 1, \ldots, 0$ while Eqs. (63b) and (63c) are valid for $n = N_d, N_d - 1, \ldots, 0$.

For the matrix elements $H^{\Gamma\Gamma'}_{n,n-1}$ we obtain

$$H^{dd}_{n,n-1} = -\sqrt{\sum_{\lambda=0}^{N} E^2_\lambda\,\langle nd|\lambda\rangle^2 - \left(H^{ds}_{nn}\right)^2 - \left(H^{ds}_{n,n+1}\right)^2 - \left(H^{dd}_{nn}\right)^2 - \left(H^{dd}_{n,n+1}\right)^2}, \tag{64a}$$

$$H^{sd}_{n,n-1} = \frac{1}{H^{dd}_{n,n-1}}\left[\sum_{\lambda=0}^{N} E^2_\lambda\,\langle ns|\lambda\rangle\langle\lambda|nd\rangle - H^{sd}_{nn}\left(H^{ss}_{nn} + H^{dd}_{nn}\right) - H^{ss}_{n,n+1}\,H^{ds}_{n,n+1}\right], \tag{64b}$$

and

$$H^{ss}_{n,n-1} = -\sqrt{\sum_{\lambda=0}^{N} E_\lambda^2 \langle ns|\lambda\rangle^2 - \left(H^{ss}_{nn}\right)^2 - \left(H^{ss}_{n,n+1}\right)^2 - \left(H^{sd}_{n,n-1}\right)^2 - \left(H^{sd}_{nn}\right)^2}. \tag{64c}$$

Equation (64a) is valid for $n = N_d - 1, N_d - 2, \ldots, 1$; Equation (64b) is valid for $n = N_d, N_d - 1, \ldots, 1$, and Eq. (64c) is valid for $n = N_s - 1, N_s - 2, \ldots, 1$.

The eigenvector components $\langle n-1, s|\lambda\rangle$ with $n = N_s - 1, N_s - 2, \ldots, 1$ and $\langle n-1, d|\lambda\rangle$ with $n = N_d - 1, N_d - 2, \ldots,\ 1$ can be calculated using the following expressions:

$$\langle n-1, s|\lambda\rangle = \frac{1}{H^{ss}_{n,n-1}}\Big(E_\lambda \langle ns|\lambda\rangle - H^{ss}_{nn}\langle ns|\lambda\rangle - H^{ss}_{n,n+1}\langle n+1, s|\lambda\rangle - H^{sd}_{n,n-1}\langle n-1, d|\lambda\rangle - H^{sd}_{nn}\langle nd|\lambda\rangle\Big) \tag{65a}$$

and

$$\langle n-1, d|\lambda\rangle = \frac{1}{H^{dd}_{n,n-1}}\Big(E_\lambda \langle nd|\lambda\rangle - H^{ds}_{nn}\langle ns|\lambda\rangle - H^{ds}_{n,n+1}\langle n+1, s|\lambda\rangle - H^{dd}_{nn}\langle nd|\lambda\rangle - H^{dd}_{n,n+1}\langle n+1, d|\lambda\rangle\Big). \tag{65b}$$

Having calculated the Hamiltonian matrix elements $H^{\Gamma\Gamma'}_{nn'}$, we obtain the potential energy matrix elements $V^{\Gamma\Gamma'}_{nn'}$ by subtracting the kinetic energy.

We recall here that we arbitrarily assigned the values s and d to the channel index Γ but the above theory can be applied to any pair of coupled partial waves. The only equations specific for the sd coupled partial waves case are Eqs. (56)–(58) that are needed to account for the experimental information about the bound state which is present in the np system in the sd coupled partial waves. In Eqs. (33), (41)–(50) and (59)–(65) one can substitute s and d by p and f, respectively, and use them for constructing the ISTP in the coupled pf waves.

We construct ISTP in the coupled NN partial waves using as input the np scattering phase shifts and mixing parameters reconstructed from the experimental data by the Nijmegen group [3]. We start the discussion from the ISTP in the coupled pf waves.

The non-zero potential energy matrix elements of the obtained $7\hbar\omega$ pf-ISTP are given in Table 9 (in $\hbar\omega = 40$ MeV units). The description of the phase shifts δ_p and δ_f and of the mixing parameter ε is shown in Figs. 18–20. The phenomenological data are seen to be well reproduced by the $7\hbar\omega$ ISTP up to the laboratory energy $E_{\text{lab}} \approx 270$ MeV; at higher energies there are discrepancies between the ISTP predictions and the experimental data that are most pronounced in the ${}^3p_2$ partial wave (note the very different scales in Figs. 18–20). These discrepancies are seen to be eliminated by constructing the $9\hbar\omega$ pf-ISTP.

Table 9 Non-zero matrix elements in $\hbar\omega$ units of the $7\hbar\omega$ ISTP matrix in the pf coupled partial waves

$V^{pp}_{nn'}$ matrix elements		
n	V^{pp}_{nn}	$V^{pp}_{n,n+1} = V^{pp}_{n+1,n}$
0	−0.083205863022	0.068281300876
1	−0.173337478975	0.097104660674
2	−0.163079253268	0.047370054433
3	−0.025144490505	
$V^{ff}_{nn'}$ matrix elements		
n	V^{ff}_{nn}	$V^{ff}_{n,n+1} = V^{ff}_{n+1,n}$
0	−0.018607311796	0.008146529481
1	−0.012301122585	0.002878668409
2	−0.002274165032	
$V^{pf}_{nn'}$ matrix elements		
n	$V^{pf}_{n,n-1} = V^{fp}_{n-1,n}$	$V^{pf}_{nn} = V^{fp}_{nn}$
0		0.031138374332
1	−0.027310965160	0.026548899815
2	−0.005320397951	−0.007039900978
3	0.009906839670	

Generally, for the coupled pf waves, we have four radial wave function components $u_{p(p)}(E,r)$, $u_{p(f)}(E,r)$, $u_{f(p)}(E,r)$ and $u_{f(f)}(E,r)$ defined according to their standing wave asymptotics (32). We present in Figs. 21–30 the plots of these components at the laboratory energies $E_{\text{lab}} = 2$, 10, 50, 150 and 250 MeV obtained with the $7\hbar\omega$ and $9\hbar\omega$ ISTP in comparison with the respective Nijmegen-II wave function components.

It is seen from the figures that the $9\hbar\omega$ ISTP and Nijmegen-II 'large' (diagonal) wave function components $u_{p(p)}(E,r)$ and $u_{f(f)}(E,r)$ are indistinguishable. The same $7\hbar\omega$ ISTP components differ a little from those of Nijmegen-II at high energies. At the same time, the 'small' (non-diagonal) ISTP wave function

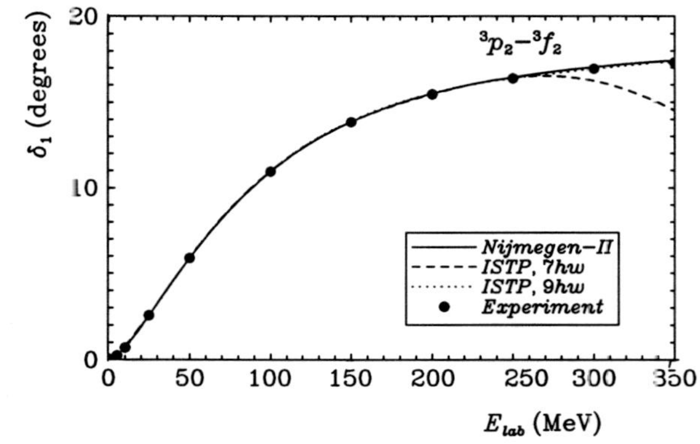

Fig. 18 $^3p_2$ np scattering phase shifts δ_p (coupled pf waves). *Filled circles* — experimental data of Ref. [3]; *solid line* — realistic meson exchange Nijmegen-II potential [3] phase shifts; *dashed line* — $7\hbar\omega$ ISTP phase shifts; *dotted line* — $9\hbar\omega$ ISTP phase shifts

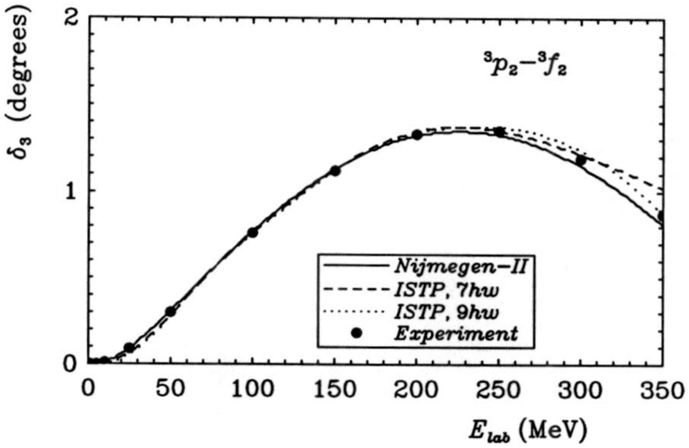

Fig. 19 $^3f_2$ *np* scattering phase shifts δ_f (coupled *pf* waves). See Fig. 18 for details

components $u_{p(f)}(E,r)$ and $u_{f(p)}(E,r)$ differ essentially at small distances from the Nijmegen-II ones. It is a clear indication of a very different nature of the ISTP tensor interaction.

Now we apply the inverse scattering J-matrix approach to the coupled *sd* partial waves and obtain the $8\,\hbar\omega$ ISTP hereafter refered to as Version 0 ISTP. The description of the phenomenological data by this potential (and other ISTP versions discussed later) is shown in Figs. 31–33. The *np* *s* wave and *d* wave phase shifts δ_s and δ_d are excellently reproduced up to the laboratory energy of 350 MeV. There is a small discrepancy between the experimental and the Version 0 ISTP mixing parameter ε at the laboratory energy of $E_{\text{lab}} \approx 25$ MeV. However, the overall Version 0 ISTP description of experimental scattering data (including the mixing parameter ε) over the full energy interval $E_{\text{lab}} = 0 \div 350$ MeV is seen from Figs. 31–33 to be competitive with the Nijmegen-II, one of the best realistic meson exchange potentials.

The Version 0 ISTP is constructed by fitting the experimental scattering data, the deuteron ground state energy E_d, the *s* wave asymptotic normalization constant

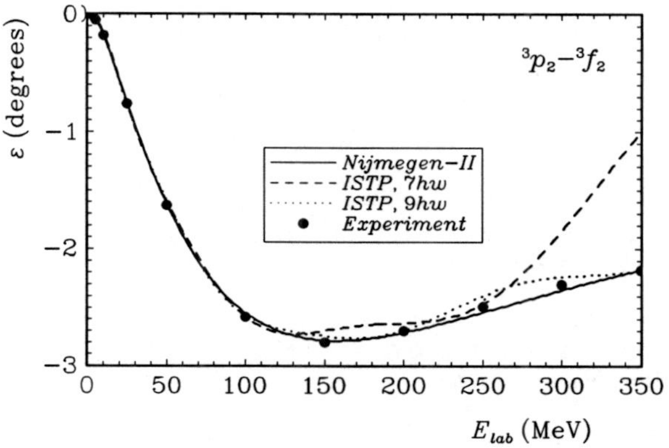

Fig. 20 *np* scattering mixing parameter ε in the coupled *pf* waves. See Fig. 18 for details

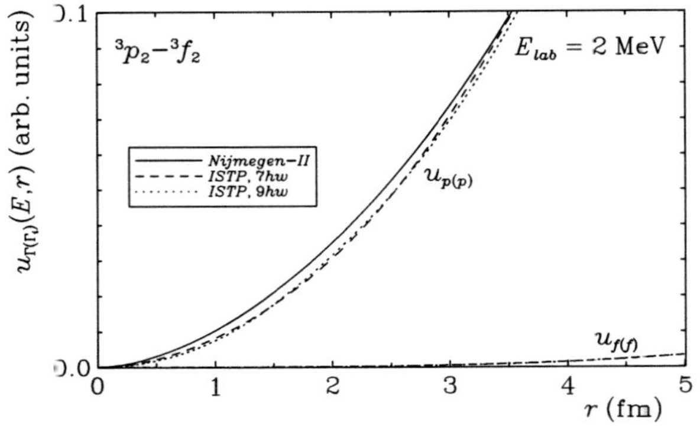

Fig. 21 Large components $u_{p(p)}(E,r)$ and $u_{f(f)}(E,r)$ of the coupled pf waves np scattering wave function at the laboratory energy $E_{\rm lab} = 2\,{\rm MeV}$. See Fig. 18 for details

$\mathscr{A}_s$ and $\eta = \dfrac{\mathscr{A}_d}{\mathscr{A}_s}$. However, there are other important deuteron observables known experimentally such as the deuteron rms radius $\langle r^2\rangle^{-1/2}$ and the probability of the d state. Various deuteron properties obtained with the Version 0 ISTP (and other ISTP versions discussed later) are compared in Table 10 with the predictions obtained with Nijmegen-II potential and with recent compilations of the experimental data [43, 44]. It is seen from the table that the Version 0 ISTP overestimates the deuteron rms radius and underestimates the d state probability.

The deuteron wave functions can be calculated by utilizing the J-matrix formalism at the negative energy E_d as is discussed in Ref. [17, 18]. The plots of the deuteron wave functions are presented in Fig. 34. It is seen that the Version 0 ISTP s wave component is very close to that of Nijmegen-II. The Version 0 ISTP d wave component coincides with that of Nijmegen-II at large distances since both potentials provide the same $\mathscr{A}_d$ value; however at the distances less than 5 fm the Version 0 ISTP d wave component is suppressed. We note also that the Version 0 ISTP scattering wave functions (not shown in the figures below) are significantly different from those of Nijmegen-II at short distances.

Our conclusion is that the Version 0 ISTP does not seem to be a realistic NN potential.

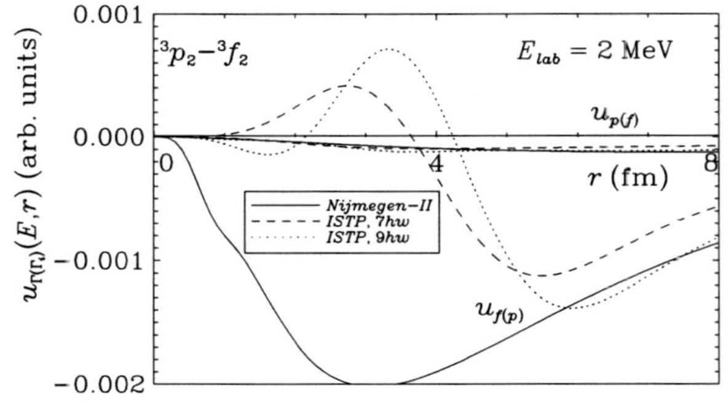

Fig. 22 Small components $u_{p(f)}(E,r)$ and $u_{f(p)}(E,r)$ of the coupled pf waves np scattering wave function at the laboratory energy $E_{\rm lab} = 2\,{\rm MeV}$. See Fig. 18 for details

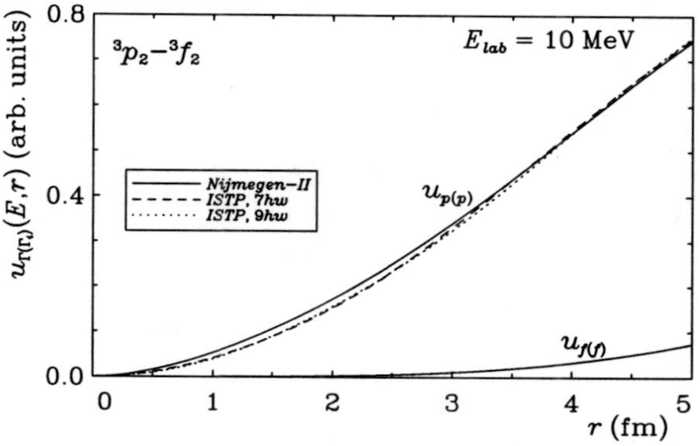

Fig. 23 Large components $u_{p(p)}(E, r)$ and $u_{f(f)}(E, r)$ of the coupled pf waves np scattering wave function at the laboratory energy $E_{\text{lab}} = 10\,\text{MeV}$. See Fig. 18 for details

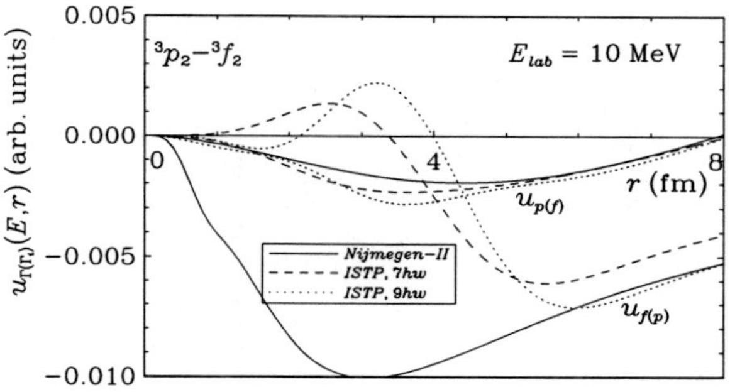

Fig. 24 Small components $u_{p(f)}(E, r)$ and $u_{f(p)}(E, r)$ of the coupled pf waves np scattering wave function at the laboratory energy $E_{\text{lab}} = 10\,\text{MeV}$. See Fig. 18 for details

To improve the description of the deuteron properties, it appears natural to apply to our Version 0 ISTP a phase equivalent transformation that leaves unchanged the scattering observables δ_s, δ_d, ε, the deuteron ground state energy E_d and the deuteron asymptotic normalization constants $\mathscr{A}_s$ and $\mathscr{A}_d$. The phase equivalent transformation discussed in Refs. [16–18] is very convenient for our purposes since it is defined in the oscillator basis. This transformation gives rise to an ambiguity of

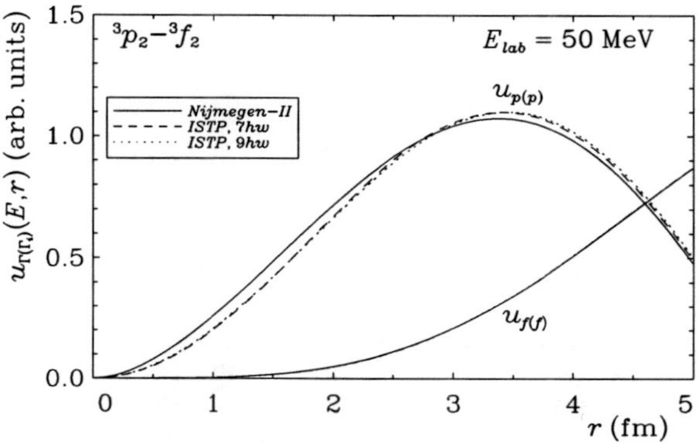

Fig. 25 Large components $u_{p(p)}(E, r)$ and $u_{f(f)}(E, r)$ of the coupled pf waves np scattering wave function at the laboratory energy $E_{\text{lab}} = 50\,\text{MeV}$. See Fig. 18 for details

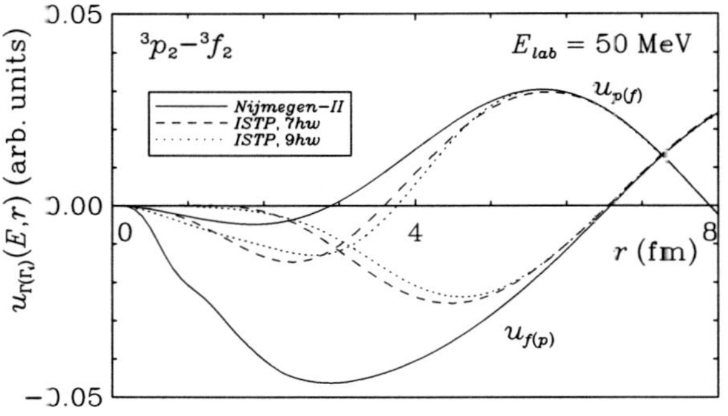

Fig. 26 Small components $u_{p(f)}(E,r)$ and $u_{f(p)}(E,r)$ of the coupled pf waves np scattering wave function at the laboratory energy $E_{\text{lab}} = 50\,\text{MeV}$. See Fig. 18 for details

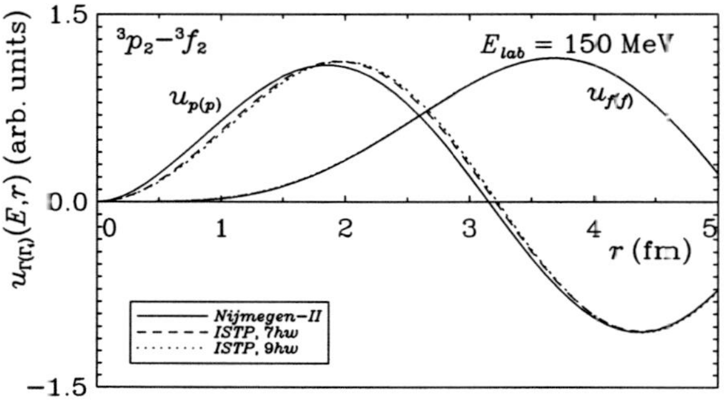

Fig. 27 Large components $u_{p(p)}(E,r)$ and $u_{f(f)}(E,r)$ of the coupled pf waves np scattering wave function at the laboratory energy $E_{\text{lab}} = 150\,\text{MeV}$. See Fig. 18 for details

the potential fit within the inverse scattering J-matrix approach, which have been mentioned several times already. We now need to discuss this in more detail.

This phase equivalent transformation is based on the unitary transformation

$$U = \sum_{\Gamma=s,d} \sum_{\Gamma'=s,d} \sum_{n=0}^{\infty} \sum_{n'=0}^{\infty} |n\Gamma\rangle \, U_{nn'}^{\Gamma\Gamma'} \, \langle n'\Gamma'|, \tag{66a}$$

where the unitary matrix $[U]$ with matrix elements $U_{nn'}^{\Gamma\Gamma'}$ should be of the form [16–18]

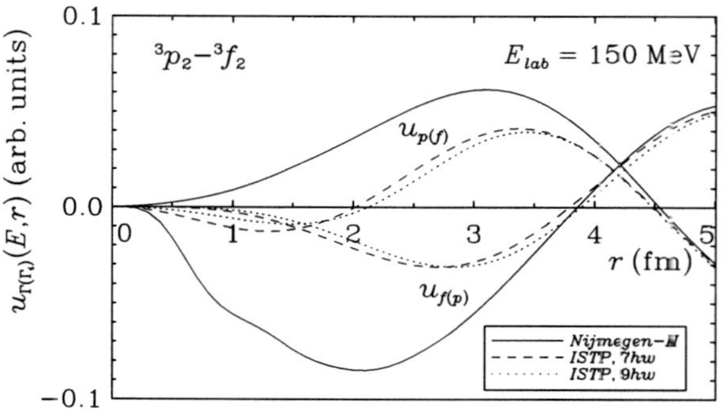

Fig. 28 Small components $u_{p(f)}(E,r)$ and $u_{f(p)}(E,r)$ of the coupled pf waves np scattering wave function at the laboratory energy $E_{\text{lab}} = 150\,\text{MeV}$. See Fig. 18 for details

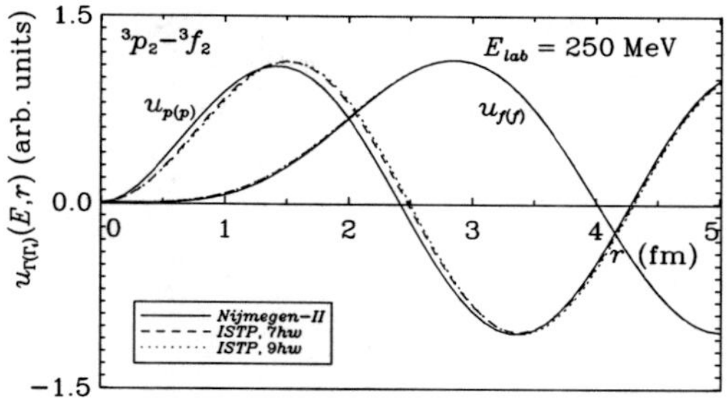

Fig. 29 Large components $u_{p(p)}(E,r)$ and $u_{f(f)}(E,r)$ of the coupled pf waves np scattering wave function at the laboratory energy $E_{\text{lab}} = 250\,\text{MeV}$. See Fig. 18 for details

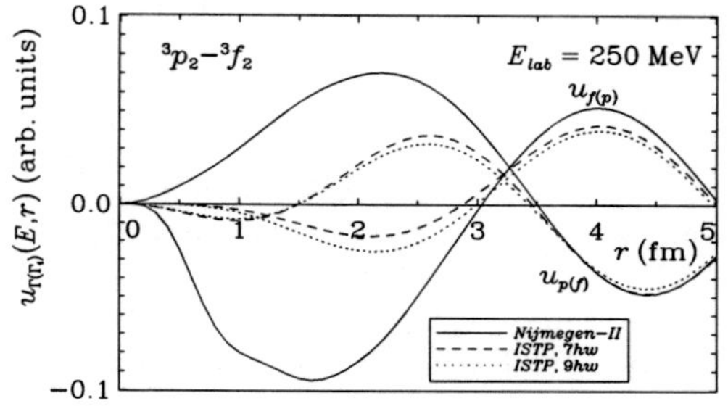

Fig. 30 Small components $u_{p(f)}(E,r)$ and $u_{f(p)}(E,r)$ of the coupled pf waves np scattering wave function at the laboratory energy $E_{\text{lab}} = 250\,\text{MeV}$. See Fig. 18 for details

$$[U] = [U_0] \oplus [I] = \left[\begin{array}{c|c} [U_0] & 0 \\ \hline 0 & [I] \end{array}\right] \tag{66b}$$

and $[I]$ is the infinite unit matrix. The unitary transformation (66) is applied to the infinite Hamiltonian matrix $[H]$ in the oscillator basis $\{|n\Gamma\rangle\}$:

$$[\widetilde{H}] = [U]\,[H]\,\big[U^{+}\big]. \tag{67}$$

The transformed Hamiltonian $\widetilde{H}$ is defined through its (infinite) matrix $[\widetilde{H}]$ with matrix elements $\widetilde{H}^{\Gamma\Gamma'}_{nn'} \equiv \langle n\Gamma|\widetilde{H}|n'\Gamma'\rangle$. That is, the matrix $[\widetilde{H}]$ is obtained by means of the unitary transformation (67) *in the original basis* $\{|n\Gamma\rangle\}$ *and not in*

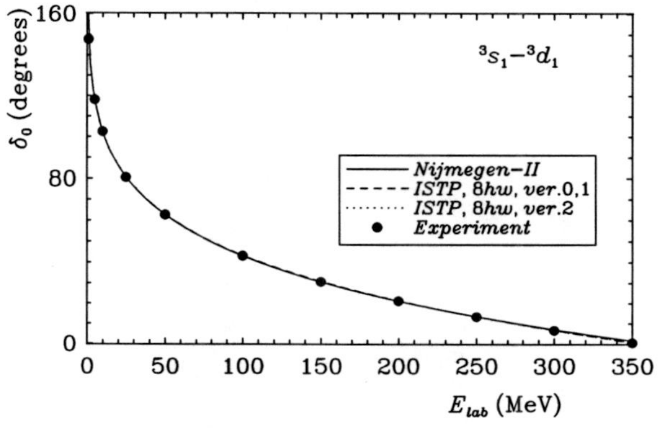

Fig. 31 $^3s_1$ np scattering phase shifts δ_s (coupled sd waves). *Filled circles* — experimental data of Ref. [3]; *solid line* — realistic meson exchange Nijmegen-II potential [3] phase shifts; *dashed line* — Version 0 and Version 1 ISTP phase shifts; *dotted line* — Version 2 ISTP phase shifts

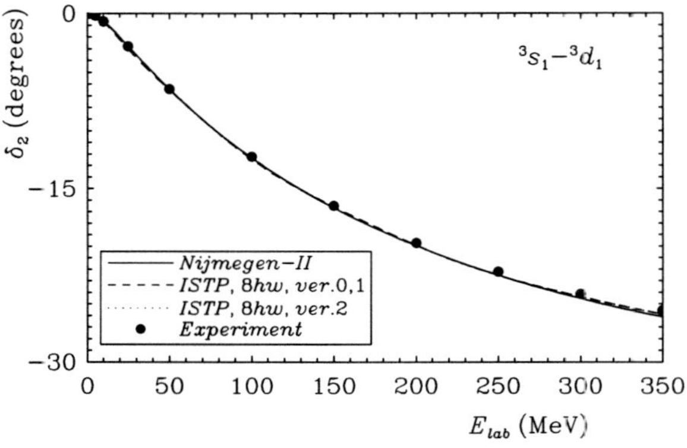

Fig. 32 $^3d_1$ *np* scattering phase shifts δ_d (coupled *sd* waves). See Fig. 31 for details

the transformed basis $\{|\widetilde{n\Gamma}\rangle\} \equiv U\{|n\Gamma\rangle\}$. Clearly the spectra of the Hamiltonians H and $\widetilde{H}$ are identical. If the submatrix $[U_0]$ is small enough, the unitary transformation (67) leaves unchanged the last components $\langle N_\Gamma\Gamma|\lambda\rangle$ of the eigenvectors $\langle n\Gamma|\lambda\rangle$ obtained by solving the algebraic problem (40) and hence it leaves unchanged the functions $\mathcal{G}_{\Gamma\Gamma'}$ that completely determine the K-matrix, the S-matrix, the phase shifts δ_s and δ_d, the mixing parameter ε, the asymptotic normalization constants $\mathscr{A}_s$ and $\mathscr{A}_d$, etc.

The potential $\widetilde{V}$ entering the Hamiltonian $\widetilde{H}$, phase equivalent to the initial potential V entering the Hamiltonian H, can be expressed as

$$\widetilde{V} = V + \Delta V, \tag{68a}$$

where

$$\Delta V = \widetilde{H} - H. \tag{68b}$$

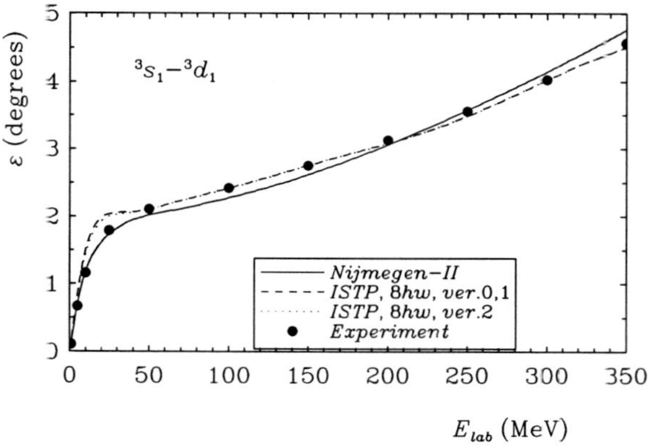

Fig. 33 *np* scattering mixing parameter ε in the coupled *sd* waves. See Fig. 31 for details

Table 10 Deuteron property predictions obtained with various $8\hbar\omega$ ISTP versions and with Nijmegen-II potential in comparison with recent compilations [43, 44]

Potential	E_d, MeV	d state probability, %	rms radius, fm	$\mathscr{A}_s$, fm$^{-1/2}$	$\eta = \frac{\mathscr{A}_d}{\mathscr{A}_s}$
Version 0	−2.224575	0.4271	1.9877	0.8845	0.0252
Version 1	−2.224575	5.620	1.9997	0.8845	0.0252
Version 2	−2.224575	5.696	1.968	0.8629	0.0252
Nijmegen-II	−2.224575	5.635	1.968	0.8845	0.0252
Compilation [43]	−2.224575(9)	5.67(11)	1.9676(10)	0.8845(8)	0.0253(2)
Compilation [44]	−2.224589	—	{1.9635; 1.9560; 1.950}	0.8781	0.0272

We should improve the tensor component of the NN interaction to increase the d state probability in the deuteron and reduce the rms radius. Therefore the only non-trivial submatrix $[U_0]$ of the matrix (66b) should couple the oscillator components $|ns\rangle$ and $|n'd\rangle$ of different partial waves. We take the simplest form of the submatrix $[U_0]$: a 2×2 matrix coupling the $|0s\rangle$ and $|0d\rangle$ basis functions. In other words, the non-trivial matrix elements $U_{nn'}^{\Gamma\Gamma'}$ constitute a 2×2 rotation matrix with a single continuous parameter ϑ:

$$[U_0] = \begin{bmatrix} U_{00}^{ss} & U_{00}^{sd} \\ U_{00}^{ds} & U_{00}^{dd} \end{bmatrix} = \begin{bmatrix} \cos\vartheta & +\sin\vartheta \\ -\sin\vartheta & \cos\vartheta \end{bmatrix}, \tag{69a}$$

while all the remaining matrix elements

$$U_{nn'}^{\Gamma\Gamma'} = \delta_{nn'}\,\delta_{\Gamma\Gamma'} \qquad \text{for } n > 0 \text{ or } n' > 0. \tag{69b}$$

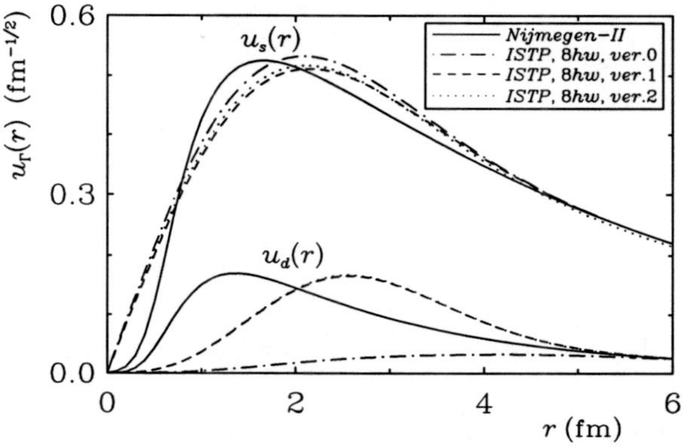

Fig. 34 Radial deuteron wave functions. *Solid line* — realistic meson exchange Nijmegen-II potential [3] wave functions; *dot-dash line* — Version 0 ISTP wave functions; *dashed line* — Version 1 ISTP wave functions; *dotted line* — Version 2 ISTP wave functions

Varying the parameter ϑ of the transformation (67)–(69), we obtain a family of phase equivalent potentials and examine which of them provides the better description of the deuteron properties and np scattering wave functions. The best result seems to be the potential obtained with $\vartheta = -14°$. This potential is hereafter referred to as Version 1 ISTP.

As a result of the transformation (67)–(69), the potential energy matrix acquires two additional non-zero matrix elements $V_{01}^{sd} = V_{10}^{ds}$. These additional matrix elements are schematically illustrated by filled circles in Fig. 35. The non-zero matrix elements of the Version 1 ISTP are given in Table 11 (in $\hbar\omega = 40\,\text{MeV}$ units).

The deuteron properties obtained with the Version 1 ISTP are presented in Table 10. The d state probability is improved by the phase equivalent transformation. However the phase equivalent transformation produces an increase of the deuteron rms radius; so this observable becomes even worse than that given by the Version 0 ISTP. We found it impossible to obtain an exact description of all deuteron properties by means of the phase equivalent transformation (67) with the simplest matrix (69).

The deuteron wave functions provided by the Version 1 ISTP are shown in Fig. 34. The Version 1 ISTP s wave component is seen to be very close to that of the Nijmegen-II. The maximum of the Version 1 ISTP d wave component is seen to be shifted to larger distances as compared with that of the Nijmegen-II. Of course, the shape of the d wave component of the wave function cannot be determined experimentally. Hence the shape of the Version 1 ISTP deuteron wave functions look realistic though these wave functions result in the slightly overestimated deuteron rms radius.

The Version 1 ISTP np scattering wave function components at the laboratory energies $E_{\text{lab}} = 2, 10, 50, 150$ and $250\,\text{MeV}$ are shown in Figs. 36–45 in comparison with those of Nijmegen-II potential. As in the case of the coupled pf partial waves, the large components $u_{s(s)}(E,r)$ and $u_{d(d)}(E,r)$ differ very little from the Nijmegen-II ones but the small components are essentially different at short distances due to the difference of the tensor interaction of these two potential models.

Generally we conclude that the Version 1 ISTP is very close to the realistic interaction. The most important discrepancy of this interaction is that it overestimates the deuteron rms radius by approximately 1.5%.

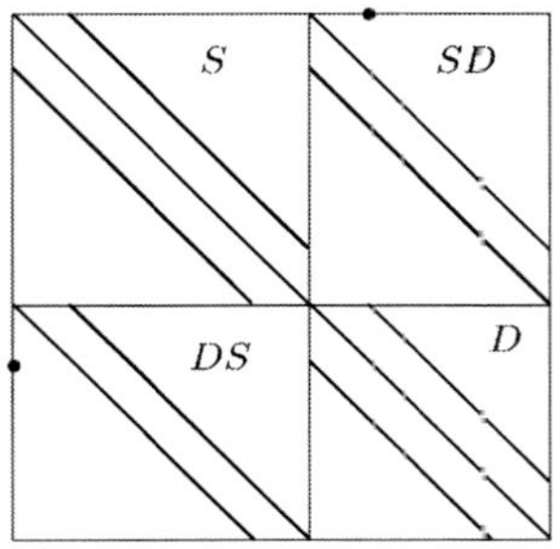

Fig. 35 Structure of the Version 1 and Version 2 ISTP matrix. The location of non-zero matrix elements is schematically illustrated by solid lines and filled circles

Table 11 Non-zero matrix elements in $\hbar\omega$ units of the Version 1 ISTP matrix in the sd coupled waves

$V^{ss}_{nn'}$ matrix elements

n	V^{ss}_{nn}	$V^{ss}_{n,n+1} = V^{ss}_{n+1,n}$
0	−0.457670450906	0.211126251530
1	−0.278324060593	0.078168834003
2	−0.011531530086	−0.053467071879
3	0.151447629416	−0.055928268627
4	0.036322781738	

$V^{dd}_{nn'}$ matrix elements

n	V^{dd}_{nn}	$V^{dd}_{n,n+1} = V^{dd}_{n+1,n}$
0	0.008456639592	−0.083373543646
1	0.322043907371	−0.178838809860
2	0.308493158866	−0.093044099373
3	0.061181660346	

$V^{sd}_{nn'} = V^{ds}_{n'n}$ matrix elements

n	$V^{sd}_{n,n-1} = V^{ds}_{n-1,n}$	$V^{sd}_{nn} = V^{ds}_{nn}$	$V^{sd}_{n,n+1} = V^{ds}_{n+1,n}$
0		−0.482407689587	0.254012350019
1	−0.068997529558	−0.061366928740	
2	0.067744180124	−0.080685245987	
3	0.049138732449	−0.020412912639	
4	−0.001715094993		

We attempted the phase equivalent transformation (67) with a more complicated matrix $[U]$ than Eq. (69). However, we did not manage to obtain a completely satisfactory interaction. It is possible to obtain the potential providing the required values of the deuteron rms radius and of the d state probability by increasing the dimension of the submatrix $[U_0]$ and introducing additional transformation parameters, but our attempts yielded unrealistic scattering wave functions.

To improve the sd-ISTP we suggest a slight change to the s wave asymptotic normalization constant $\mathscr{A}_s$ that is used as an input in our inverse scattering approach. The $\mathscr{A}_s$ value cannot be measured in a direct experiment. As was mentioned in

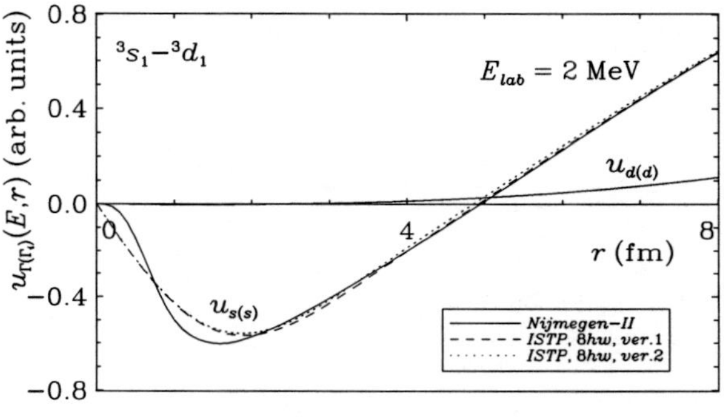

Fig. 36 Large components $u_{s(s)}(E, r)$ and $u_{d(d)}(E, r)$ of the coupled sd waves np scattering wave function at the laboratory energy $E_{\text{lab}} = 2\,\text{MeV}$. See Fig. 34 for details

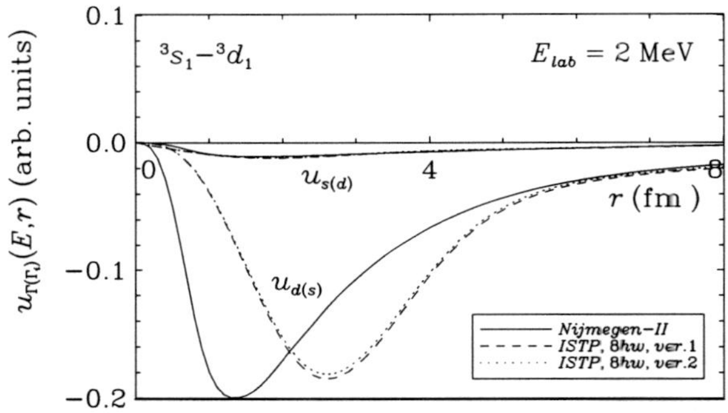

Fig. 37 Small components $u_{s(d)}(E,r)$ and $u_{d(s)}(E,r)$ of the coupled sd waves np scattering wave function at the laboratory energy $E_{\rm lab} = 2\,{\rm MeV}$. See Fig. 34 for details

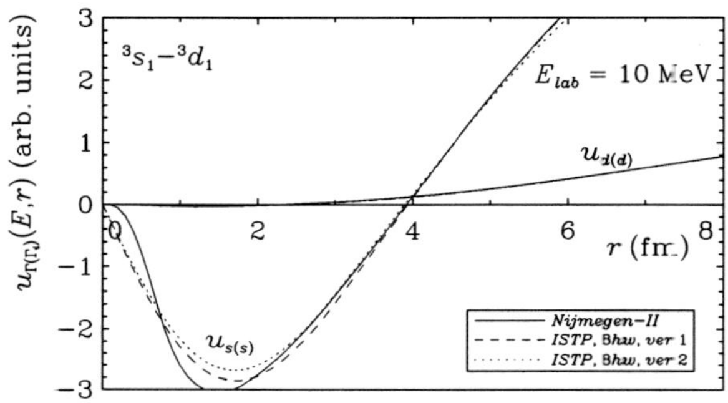

Fig. 38 Large components $u_{s(s)}(E,r)$ and $u_{d(d)}(E,r)$ of the coupled sd waves np scattering wave function at the laboratory energy $E_{\rm lab} = 10\,{\rm MeV}$. See Fig. 34 for details

Ref. [44], the $\mathscr{A}_s$ values discussed in the literature vary within a broad range from $0.7592\,{\rm fm}^{-1/2}$ to $0.9863\,{\rm fm}^{-1/2}$. Therefore, the modified value $\mathscr{A}_s = 0.8629\,{\rm fm}^{-1/2}$ that we use for the construction of the improved sd-ISTP, seems to be reasonable. We do not change the remaining inputs in our inverse scattering approach including $\eta = \dfrac{\mathscr{A}_d}{\mathscr{A}_s}$ (and hence we modify $\mathscr{A}_d$ together with $\mathscr{A}_s$) to obtain the ISTP of the type shown in Fig. 17 and apply to it the phase equivalent transformation (67) with the parameter $\vartheta = -14^\circ$ of the matrix (69). This potential is referred to as Version 2 ISTP. This potential has the structure schematically depicted in Fig. 35 and its matrix elements are listed in Table 12.

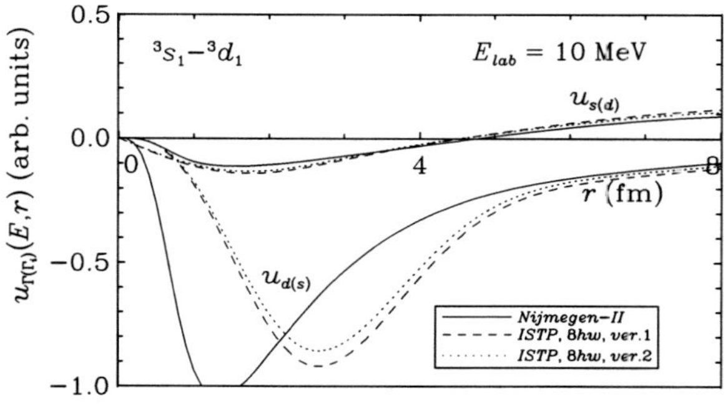

Fig. 39 Small components $u_{s(d)}(E,r)$ and $u_{d(s)}(E,r)$ of the coupled sd waves np scattering wave function at the laboratory energy $E_{\rm lab} = 10\,{\rm MeV}$. See Fig. 34 for details

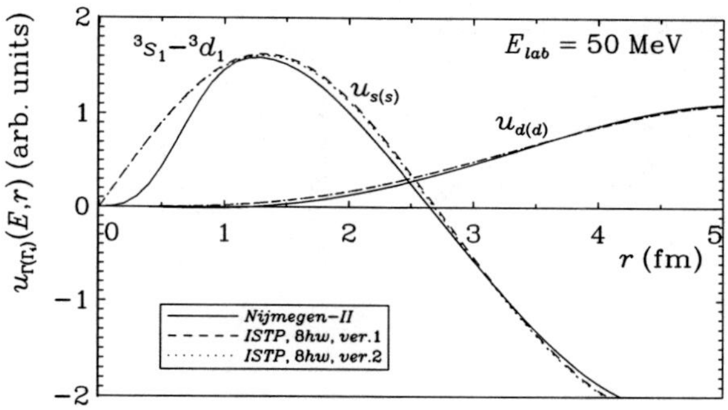

Fig. 40 Large components $u_{s(s)}(E,r)$ and $u_{d(d)}(E,r)$ of the coupled sd waves np scattering wave function at the laboratory energy $E_{\rm lab} = 50\,{\rm MeV}$. See Fig. 34 for details

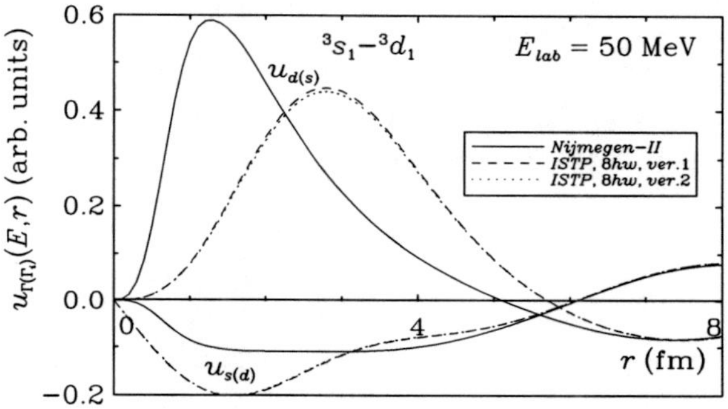

Fig. 41 Small components $u_{s(d)}(E,r)$ and $u_{d(s)}(E,r)$ of the coupled sd waves np scattering wave function at the laboratory energy $E_{\rm lab} = 50\,{\rm MeV}$. See Fig. 34 for details

The deuteron properties are seen from Table 10 to be well described by the Version 2 ISTP. The Version 2 ISTP scattering wave functions are very close to those of the Version 1 ISTP (see Figs. 36–45). Its deuteron wave functions are very close to those of Version 1 ISTP (see Fig. 34) and differ from those of Nijmegen-II in the position of the d wave component maximum.

We suppose that the Version 2 ISTP can be treated as a realistic interaction in the coupled sd partial waves.

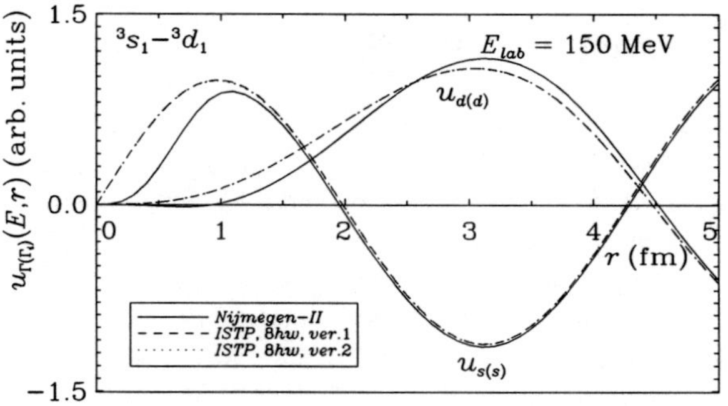

Fig. 42 Large components $u_{s(s)}(E,r)$ and $u_{d(d)}(E,r)$ of the coupled sd waves np scattering wave function at the laboratory energy $E_{\rm lab} = 150\,{\rm MeV}$. See Fig. 34 for details

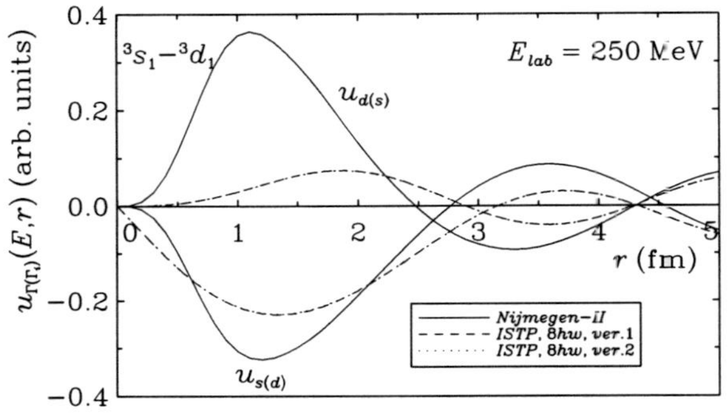

Fig. 43 Small components $u_{s(d)}(E,r)$ and $u_{d(s)}(E,r)$ of the coupled sd waves np scattering wave function at the laboratory energy $E_{\rm lab} = 150\,{\rm MeV}$. See Fig. 34 for details

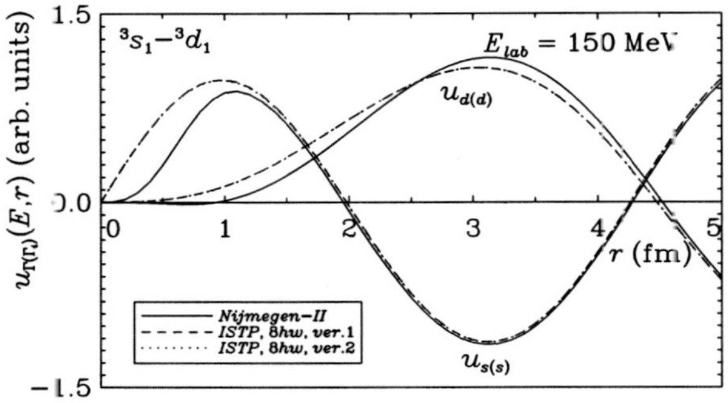

Fig. 44 Large components $u_{s(s)}(E,r)$ and $u_{d(d)}(E,r)$ of the coupled sd waves np scattering wave function at the laboratory energy $E_{\rm lab} = 250\,{\rm MeV}$. See Fig. 34 for details

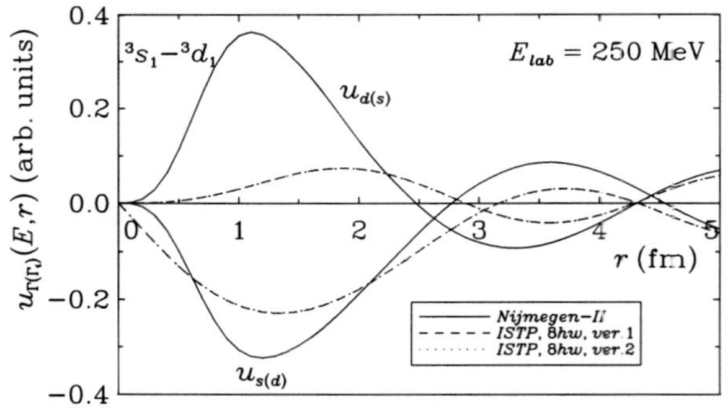

Fig. 45 Small components $u_{s(d)}(E,r)$ and $u_{d(s)}(E,r)$ of the coupled sd waves np scattering wave function at the laboratory energy $E_{\rm lab} = 250\,{\rm MeV}$. See Fig. 34 for details

4 Application of NN ISTP in ^{3}H and ^{4}He Calculations

We employ the obtained ISTP in the ^{3}H and ^{4}He calculations within the no-core shell model [11, 12] with $\hbar\omega = 40\,{\rm MeV}$. The same NN potentials are used to describe the neutron–neutron and neutron–proton interactions; in the proton–proton case these potentials are supplemented by the Coulomb interaction.

The calculations are performed in the complete $N\hbar\omega$ model spaces with $N \leq 14$. We use both $7\hbar\omega$-ISTP and $9\hbar\omega$-ISTP in odd partial waves. The ^{3}H and ^{4}He nuclei are slightly more bound in the case when we use the $7\hbar\omega$-ISTP in the odd waves. However, the differences are very small: less than 15 keV for ^{3}H and about 40 keV

Table 12 Non-zero matrix elements in $\hbar\omega$ units of the Version 2 ISTP matrix in the sd coupled waves

$V^{ss}_{nn'}$ matrix elements

n	V^{ss}_{nn}	$V^{ss}_{n,n+1} = V^{ss}_{n+1,n}$
0	−0.466063146350	0.216883948836
1	−0.276168029473	0.080907735691
2	−0.009473803659	−0.051881443108
3	0.152873734289	−0.055193589842
4	0.037547929880	

$V^{dd}_{nn'}$ matrix elements

n	V^{dd}_{nn}	$V^{dd}_{n,n+1} = V^{dd}_{n+1,n}$
0	0.008667454659	−0.083339374560
1	0.322126471805	−0.178808793641
2	0.308516673061	−0.093012604766
3	0.061200037193	

$V^{sd}_{nn'} = V^{ds}_{n'n}$ matrix elements

n	$V^{sd}_{n,n-1} = V^{ds}_{n-1,n}$	$V^{sd}_{nn} = V^{ds}_{nn}$	$V^{sd}_{n,n+1} = V^{ds}_{n+1,n}$
0		−0.483308500313	0.254003830709
1	−0.067221025404	−0.060476585693	
2	0.068044496963	−0.080187106458	
3	0.049400578816	−0.020205646231	
4	−0.001503998139		

for ^{4}He. The sequence of levels in the ^{4}He spectrum provided by the odd wave $7\hbar\omega$-ISTP and by the odd wave $9\hbar\omega$-ISTP is the same but the energies of excited ^{4}He states are shifted down in the case of the odd wave $7\hbar\omega$-ISTP by approximately 100 keV or less. Therefore the deviations of the $7\hbar\omega$-ISTP predictions from the experimental odd wave scattering data at high enough energies seem to produce a negligible effect in the ^{3}H and ^{4}He calculations. At the same time, $7\hbar\omega$-ISTP has a smaller matrix than $9\hbar\omega$-ISTP and hence is more convenient in applications. Below we present only the results obtained with the $7\hbar\omega$-ISTP in the odd partial waves.

We have presented various versions of ISTP in the coupled sd partial waves. The choice of ISTP in other partial waves is fixed. Using this fixed set of the non-sd-ISTP in combination with the Version M sd-ISTP, we have the set of potentials that is refered to as the Version M potential model in what follows.

The ^{3}H ground state energies E_t obtained with the Version 1 and the Version 2 potential models in $N\hbar\omega$ model spaces are presented in Fig. 46 as functions of $1/N$. It is seen that both potential models provide very similar E_t values. The convergence of the calculations with N appears adequate. The ground state energy E_t is seen from the figure to be nearly a linear function of $1/N$. Therefore it is natural to perform a linear extrapolation to the infinite $N\hbar\omega$ model space, i.e. to the point $1/N = 0$. The linear extrapolation using the two results at the highest N-values yields $E_t \approx -8.6$ MeV in the Version 1 potential model and in $E_t \approx -8.7$ MeV in the Version 2 potential model.

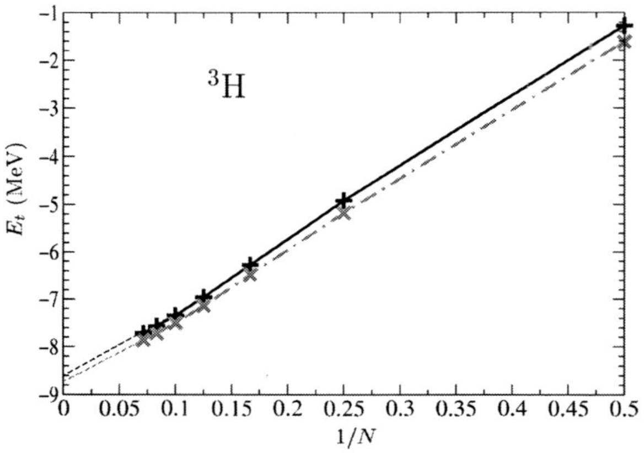

Fig. 46 ^{3}H ground state energy obtained in the $N\hbar\omega$ no-core shell model calculation vs $1/N$. + — Version 1 potential model; × — Version 2 potential model; *dashed line* — linear extrapolation to the infinite $N\hbar\omega$ model space based on the last two calculated points; *solid* and *dash-dot* lines are to guide the eye

In Fig. 47 we present the results of the ^{4}He ground state energy E_α calculations with the same potential models. In the ^{4}He case we also obtain very similar results with the Version 1 and the Version 2 potential models. It is interesting that the convergence of the ^{4}He ground state energy is better than that of ^{3}H. In this case the curves connecting the E_α values deviate from the straight lines. Nevertheless we also perform the linear extrapolations of $E_\alpha(1/N)$ to infinite N using the E_α values obtained in $12\hbar\omega$ and $14\hbar\omega$ calculations and obtain $E_\alpha \approx -26.6$ MeV in the Version 1 potential model and $E_\alpha \approx -27.0$ MeV in the Version 2 potential model.

The quality of the linear extrapolation of $E_{\text{g.s.}}(1/N)$ may be tested in the deuteron calculations. In the deuteron case, we know the exact result for the infinite $N\hbar\omega$ model space ground state energy $E_d = -2.244575$ MeV obtained by the S-matrix pole calculation with our potentials. The E_d results obtained in the $N\hbar\omega$ model spaces with $N \leq 14$ with the Version 1 and Version 2 sd-ISTP, are shown in Fig. 48. It is seen that $E_d(1/N)$ seems to be a linear function in the interval $4 \leq N \leq 14$. The linear extrapolation results in $E_d \approx -2.5$ MeV that differs from the exact energy.

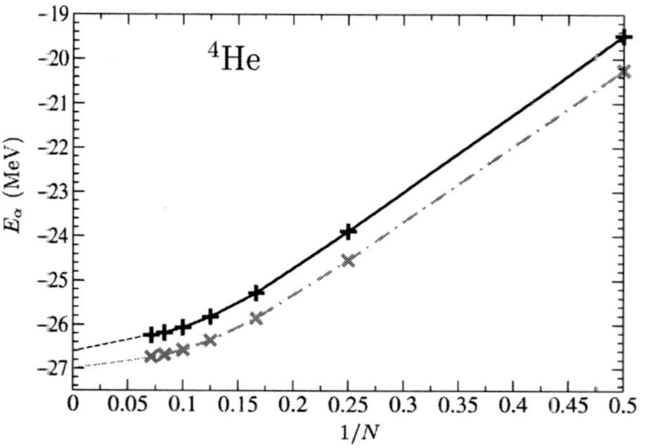

Fig. 47 ^{4}He ground state energy obtained in the $N\hbar\omega$ no-core shell model calculation vs $1/N$. See Fig. 46 for details

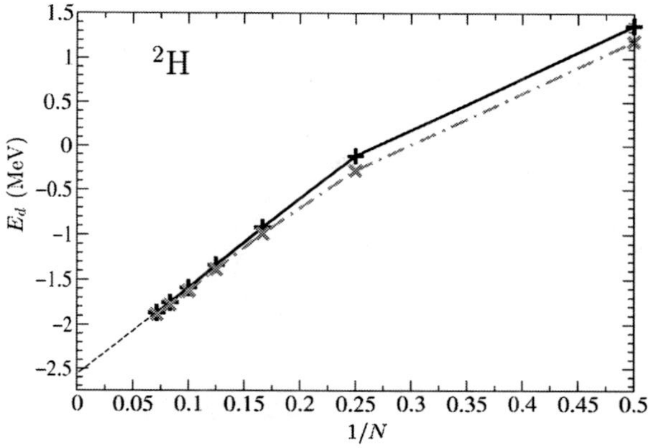

Fig. 48 Deuteron ground state energy obtained in the $N\hbar\omega$ no-core shell model calculation vs $1/N$. See Fig. 46 for details

Therefore the linear extrapolation results can be regarded only as a rough estimate of the binding energy. However in the ^{4}He case we achieved a reasonable convergence and by the linear extrapolation we increase the binding energy by approximately 0.3 MeV only. Therefore our estimate of the ^{4}He binding energy seems to be accurate enough.

The differences in convergence rates for the deuteron, ^{3}H and ^{4}He can be understood from the fact that $\hbar\omega = 40$ MeV is more optimal for the tighter bound ^{4}He than for the lesser bound systems.

Our results of the ^{3}H and ^{4}He ground state energy calculations are summarized in Table 13. We also present in the table the results obtained with the less realistic Version 0 potential model. Both ^{3}H and ^{4}He are essentially overbound in this potential model. With both Version 1 and Version 2 potential models we obtain a reasonable description of the ^{3}H and ^{4}He bindings. Our ^{4}He results are better than the ones obtained (see Ref. [6]) with any of the realistic meson exchange interactions without allowing for the three-body interactions. In the ^{3}H case, we have underbinding in the $14\hbar\omega$ model space and a small overbinding obtained by the linear extrapolation. Unfortunately, the difference between the $14\hbar\omega$ model space and the linear extrapolation results is rather large. Most probably the ^{3}H ground state energy curve in Fig. 46 will flatten out in larger model spaces. This will shift the extrapolated ground state energy upwards from our current result. Hence the expected ground state energy in the $N \to \infty$ limit lies between the $14\hbar\omega$ and the present linear extrapolation.

Table 13 ^{3}H and ^{4}He ground state energies (in MeV) obtained in $14\hbar\omega$ no-core shell model calculations and by the linear extrapolation to the infinite $N\hbar\omega$ model space

Potential model	^{3}H		^{4}He	
	$14\hbar\omega$	Extrapolation	$14\hbar\omega$	Extrapolation
Version 0	−9.091	−9.7	−33.223	−33.4
Version 1	−7.718	−8.6	−26.241	−26.6
Version 2	−7.860	−8.7	−26.734	−27.0
Nature	− 8.48		− 28.30	

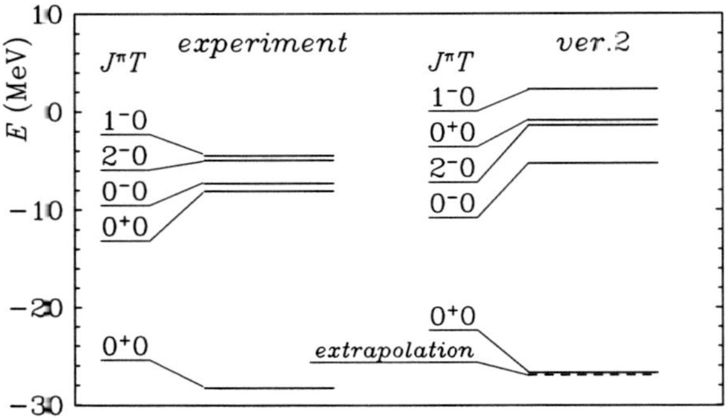

Fig. 49 ^{4}He spectrum obtained with Version 2 potential model in the no-core shell model in the $14\hbar\omega$ ($13\hbar\omega$) model space for even (odd) parity states. *Dashed line* shows the result of the linear extrapolation of the ground state energy to the infinite $N\hbar\omega$ model space. Experimental data are taken from [45]

In other words, our linear extrapolation and $14\hbar\omega$ results are expected to be the lower and upper boundaries for the exact results, respectively. An approximately 0.9 MeV difference between the $14\hbar\omega$ and the linear extrapolation ground state energies in the ^{3}H case indicates the 0.9 MeV uncertainty of our predictions. The ^{3}H ground state energy obtained in Faddeev calculations with CD-Bonn NN potential is -8.012 MeV (see Ref. [6]). All the remaining modern realistic meson exchange potentials predict the ^{3}H binding energy to be less than 7.4 MeV [6]. Therefore our ^{3}H binding energy predictions are not worse than those obtained with the realistic meson exchange potentials without allowing for the three-body forces while our ^{4}He binding energy predictions are better.

In Fig. 49 we present the spectrum of the lowest excited ^{4}He states of each J^π. The description of the excited states energies is reasonable though further from experiment than the ground state. On the other hand, we expect the excited states to be less converged and to drop more in larger model spaces. Of course, a full discussion of the states above breakup threshold must await proper extensions of the theory to the scattering domain.

5 Concluding Remarks

We obtained nucleon-nucleon ISTP potentials by means of the J-matrix version of the inverse scattering approach. The potentials accurately describe the scattering data. They are in the form of $8\hbar\omega$-truncated matrices in the oscillator basis with $\hbar\omega = 40$ MeV. The potential matrices are tridiagonal in the uncoupled partial waves. In the coupled partial waves, the potential matrices have two additional quasi-diagonals in each of the submatrices responsible for the channel coupling. The sd-ISTP of this type (Version 0) underestimates the deuteron d state probability and overestimates the deuteron rms radius. We designed two other sd-ISTP with two additional matrix elements providing the correct description of the d state probability, one of them (Version 1) overestimates the rms radius by approximately 1.5% while the other one (Version 2) provides the correct description of the deuteron rms radius. All other deuteron observables are reproduced by all sd-ISTP versions.

The ISTP potentials are used in the ^{3}H and ^{4}He no-core shell model calculations. Both Version 1 and Version 2 ISTP potential models provide very good predictions for the ^{3}H and ^{4}He binding energies and a reasonable ^{4}He spectrum. With the less realistic Version 0 potential model, we obtain overbound ^{3}H and ^{4}He nuclei. We note that there were other attempts to design the NN interaction providing the description of the triton binding energy together with the NN scattering data and the deuteron properties [19, 20]. Our interactions are much simplier and can be directly used in the shell model calculations of heavier nuclei.

Generally our approach is aimed at shell model applications in heavier nuclei. However our potentials are simple enough and can be used directly in other microscopic approaches, e.g. in Faddeev calculations. We hope that our interactions minimize the need for three-body forces. It is known [46] that the three-body force effect can be reproduced in a three-body system by the phase equivalent transformation of the two-body interaction. This phase equivalent transformation can also spoil the description of the deuteron observables, in particular, the deuteron rms radius can be arbitrary changed by phase equivalent transformations [47]. We expect that there exist transformations minimizing the need for three-body force effects, that do not significantly change the nucleon–nucleon interaction. That is, the deuteron properties, the deuteron and scattering wave functions of the transformed NN potential may remain very close to the ones developed here while achieving improved descriptions of other nuclei. In this context, it is worth noting that our approach does not assume either a particular operator structure to the interaction or locality.

From this point of view, the Version 2 ISTP accurately describing the deuteron properties and providing good predictions for the ^{3}H and ^{4}He bindings, can be regarded as such an interaction effectively accounting for effects that might otherwise be attributed to three-body forces. Clearly, additional efforts may provide superior NN interactions with less dependence on three-body forces for precision agreement with experiment.

Finally, we suggested a new approach to the construction of the high-quality NN interaction and examined the obtained ISTP NN interaction in three and four nucleon systems by means of the no-core shell model. The ^{3}H and ^{4}He binding energies are surprisingly well described. Obviously it will be very interesting to extend these studies on heavier nuclei, to investigate in detail not only their binding but the spectra of excited states as well. It is also important to investigate more carefully the ISTP description of the two-nucleon system since, for example, we have deferred the discussion of the deuteron quadrupole moment Q. We just mention here that the Version 2 ISTP prediction of $Q = 0.317$ fm^2 is not so far from the experimental value of $0.2875 \pm 20\,\text{fm}^2$ [48]. The phase equivalent transformations discussed above make it possible to improve the Q predictions and to examine the effect of such improvement in light nuclear systems. We plan to address this problem in future publications.

After the original publication of this work [49], the approach discussed in this chapter has been essentially improved and developed. The development was based on the use of phase equivalent transformations and extensive large-scale no-core shell model calculations. The J-matrix inverse scattering approach made it possible to construct a non-local NN potential providing an excellent description of the NN scattering data with χ^2/datum $= 1.03$ for the 1992 np data base (2514 data), and 1.05 for the 1999 np data base (3058 data) [50]. To compare, χ^2/datum $= 1.08$, 1.03 and 0.99 (the 1992 np data base) and 1.07, 1.02 and 0.99 (the 1999 np data base) [1] for respectively Argonne AV18, CD-Bonn potentials and the Nijmegen phase shift analysis utilized in the J-matrix inverse scattering approach to construct the NN interaction. The deuteron observables not involved in the inverse scattering formalism, provided additional information about the NN interaction. We fitted the deuteron rms radius and quadrupole moment to the experimental values using phase equivalent transformations. This resulted in the improved description of the ^{3}H and ^{4}He binding energies [51]. The binding energies and spectra of light nuclei provided additional information for further improvement of the NN interaction. To utilize this information, we employed phase equivalent transformations in various NN partial waves and performed a set of no-core shell model calculations of p-shell nuclei with the resulting NN interactions with the idea to fit the bindings and excitation energies in the spectra of p-shell nuclei. Following this route, we constructed J-matrix inverse scattering potentials JISP6 [51] and JISP16 [52] fitted to the properties of nuclei with mass numbers $A \leq 6$ and $A \leq 16$ respectively.

The nucleon–nucleon interaction JISP16 [52] appears to be competitive with the best realistic NN interactions nowadays. It is noteworthy that JISP16 is able to describe well not only the NN scattering data and deuteron properties but also the binding energies and spectra of light nuclei without introducing three-nucleon forces. The description by JISP16 of various observables in s and p shell nuclei (level energies, electromagnetic moments and transitions, etc.) also appears competitive with modern realistic interaction models involving three-nucleon forces [52, 53]. The JISP16 interaction has been introduced very recently [52], however it was already used by other authors in their own studies [54, 55].

Acknowledgments This work was supported in part by the State Program 'Russian Universities', by the Russian Foundation of Basic Research grants No 02-02-17316 and No 05-02-17429, by US DOE grants No DE-FG02-87ER40371 and No DE-FC02-07ER41457, and by US NSF grant No PHY-007-1027.

References

1. R. Machleidt, Phys. Rev. **C 63**, 024001 (2001).
2. R. B. Wiringa, V. G. J. Stoks, and R. Schiavilla, Phys. Rev. **C 51**, 38 (1995).
3. V. G. J. Stoks, R. A. M. Klomp, C. P. F. Terheggen, and J. J. de Swart, Phys. Rev. **C 49**, 2950 (1994).
4. S. A. Coon, M. D. Scadron, P. C. McNamee, B. R. Barrett, D. W. E. Blatt, and B. H. J. McKellar, Nucl. Phys. **A 317**, 242 (1979); J. L. Friar, D. Hüber, and U. van Kolck, Phys. Rev. **C 59**, 53 (1999); D. Hüber, J. L. Friar, A. Nogga, H. Witala, and U. van Kolck, Few Body Syst. **30**, 95 (2001).
5. B. S. Pudliner, V. R. Pandharipande, J. Carlson, S. C. Pieper, and R. B. Wiringa, Phys. Rev. **C 56**, 1720 (1997); R. B. Wiringa, Nucl. Phys. **A 631**, 70c (1998).
6. A. Nogga, H. Kamada, and W. Glöckle, Phys. Rev. Lett. **85**, 944 (2000).

7. A. Picklesimer, R. A. Rice, and R. Brandenburg, Phys. Rev. Lett. **68**, 1484 (1992); Phys. Rev. **C 45**, 547 (1992); Phys. Rev. **C 45**, 2045 (1992); Phys. Rev. **C 45**, 2624 (1992); Phys. Rev. **C 46**, 1178 (1992).
8. S. R. Beane, P. F. Bedaque, W. C. Haxton, D. R. Phillips, M. J. Savage, *From Hadrons to Nuclei: Crossing the Border*, in: M. Shifman (Ed.), *At the Frontier of Particle Physics*, Vol. 1, 133 (World Scientific, Singapore, 2001); nucl-th/0008064.
9. D. R. Entem and R. Machleidt, Phys. Lett. **B 524**, 93 (2002).
10. H. Kamada, A. Nogga, W. Glöckle, E. Hiyama, M. Kamimura, K. Varga, Y. Suzuki, M. Viviani, A. Kievsky, S. Rosati, J. Carlson, S. C. Pieper, R. B. Wiringa, P. Navrátil, B. R. Barrett, N. Barnea, W. Leidemann, and G. Orlandini, Phys. Rev. **C 64**, 044001 (2001).
11. D. C. Zheng, J. P. Vary, and B. R. Barrett, Phys. Rev. **C 50**, 2841 (1994); D. C. Zheng, B. R. Barrett, J. P. Vary, W. C. Haxton, and C. L. Song, Phys. Rev. **C 52**, 2488 (1995).
12. P. Navrátil, J. P. Vary, and B. R. Barrett, Phys. Rev. Lett. **84**, 5728 (2000); Phys. Rev. **C 62**, 054311 (2000).
13. S. A. Zaytsev, Teoret. Mat. Fiz. **115**, 263 (1998) [Theor. Math. Phys. **115**, 575 (1998)].
14. S. A. Zaytsev, *Inverse Scattering Approximate Method in Discrete Approach*, in *Proceedings of the XIV International Workshop on High Energy Physics and Quantum Field Theory, Moscow 1999* (Ed. B.B.Levchenko and V.I.Savrin), 666 (MSU-Press, Moscow, 2000); Teoret. Mat. Fiz. **121**, 424 (1999) [Theor. Math. Phys. **121**, 1617 (1999)].
15. S. A. Zaitsev and E. I. Kramar, J. Phys. **G 27**, 2037 (2001).
16. Yu. A. Lurie and A. M. Shirokov, Izv. Ros. Akad. Nauk, Ser. Fiz. **61**, 2121 (1997) [Bull. Rus. Acad. Sci., Phys. Ser. **61**, 1665 (1997)].
17. Yu. A. Lurie and A. M. Shirokov, *J-matrix Approach to Loosely-Bound Three-Body Nuclear Systems*, Part IV, Chapter 1, this volume.
18. Yu. A. Lurie and A. M. Shirokov, Ann. Phys. (NY) **312**, 284 (2004).
19. P. Doleschall and I. Borbély, Phys. Rev. **C 62**, 054004 (2000).
20. P. Doleschall, I. Borbély, Z. Papp, and W. Plessas, Phys. Rev. **C 67**, 064005 (2003).
21. S. K. Bogner, T. T. S. Kuo, and A. Schwenk, Phys. Rep. **386**, 1 (2003).
22. S. Bogner, T. T. S. Kuo, L. Coraggio, A. Covello, and N. Itaco, Phys. Rev. **C 65**, 051301 (2002).
23. H. A. Yamani and L. Fishman, J. Math. Phys., **16**, 410 (1975).
24. J. T. Broad and W. P. Reinhardt, Phys. Rev. **A 14**, 2159 (1976); J. Phys. **B 9**, 1491 (1976).
25. A. M. Shirokov, Yu. F. Smirnov, and L. Ya. Stotland, *H^- Ion and He Atom in J-Matrix Method: Continuum Spectra Wave Functions, S-matrix Poles, Bound States*, in *Proc. XIIth European Conference on Few-Body Physics, Uzhgorod, USSR, 1990* (Ed. V. I. Lengyel and M. I. Haysak), 173 (Uzhgorod, 1990).
26. Yu. F. Smirnov, L. Ya. Stotland, and A. M. Shirokov, Izv. Akad. Nauk SSSR, Ser. Fiz. **54**, No 5, 897 (1990) [Bull. Acad. Sci. USSR, Phys. Ser. **54**, No 5, 81 (1990)].
27. D. A. Konovalov and I. A. McCarthy, J. Phys. **B 27**, L407 (1994); J. Phys. **B 27**, L741 (1994).
28. G. F. Filippov and I. P. Okhrimenko, Yad. Fiz. **32**, 932 (1980) [Sov. J. Nucl. Phys. **32**, 480 (1980)]; G. F. Filippov, Yad. Fiz. **33**, 928 (1981) [Sov. J. Nucl. Phys. **33**, 488 (1981)].
29. Yu. F. Smirnov and Yu. I. Nechaev, Kinam **4**, 445 (1982); Yu. I. Nechaev and Yu. F. Smirnov, Yad. Fiz. **35**, 1385 (1982) [Sov. J. Nucl. Phys. **35**, 808 (1982)].
30. G. F. Filippov, V. S. Vasilevski, and L. L. Chopovski, Fiz. Elem. Chastits At. Yadra **15**, 1338 (1984); **16**, 349 (1985) [Sov. J. Part. Nucl. **15**, 600 (1984); **16**, 153 (1985)].
31. G. F. Filippov, Rivista Nuovo Cim. **12**, 1 (1989).
32. V. A. Knyr, A. I. Mazur, and Yu. F. Smirnov, Yad. Fiz. **52**, 754 (1990) [Sov. J. Nucl. Phys. **52**, 483 (1990)].
33. V. A. Knyr, A. I. Mazur, and Yu. F. Smirnov, Yad. Fiz. **54**, 1518 (1991) [Sov. J. Nucl. Phys. **54**, 927 (1991)].
34. Yu. F. Smirnov and A. M. Shirokov, Preprint ITF-88-47R (Kiev, 1988); A. M. Shirokov, Yu. F. Smirnov, and S. A. Zaytsev, in *Modern Problems in Quantum Theory* (Ed. V. I. Savrin and O. A. Khrustalev), (Moscow, 1998), 184; Teoret. Mat. Fiz. **117**, 227 (1998) [Theor. Math. Phys. **117**, 1291 (1998)].

35. T. Ya. Mikhelashvili, Yu. F. Smirnov, and A. M. Shirokov, Yad. Fiz. **48**, 969 (1988) [Sov. J. Nucl. Phys. **48**, 617 (1988)]; J. Phys. **G 16**, 1241 (1990).
36. D. E. Lanskoy, Yu. A. Lurie, and A. M. Shirokov, Z. Phys. **A 357**, 95 (1997).
37. J. M. Bang, A. I. Mazur, A. M. Shirokov, Yu. F. Smirnov, and S. A. Zaytsev, Ann. Phys. (NY), **280**, 299 (2000).
38. M. Abramowitz and I. A. Stegun (eds.), *Handbook on Mathematical Functions, Chapt. 13, pp. 504–535* (Dover, New York, 1972).
39. V. V. Babikov, *Method of Phase Functions in Quantum Mechanics* (Nauka Publishers, Moscow, 1976).
40. H. P. Stapp, T. I. Ypsilantis, and N. Metropolis, Phys. Rev. **105**, 302 (1957).
41. A. I. Baz, Ya. B. Zeldovitch, and A. M. Perelomov, *Scattering, Reactions and Decays in Nonrelativistic Quantum Mechanics* (Nauka Publishers, Moscow, 1971).
42. L. D. Blokhintsev, I. Borbely, and É. I. Dolinskii, Sov. J. Part. Nucl. 8, 485 (1977).
43. J. J. de Swart, C. P. F. Terheggen and V. G. J. Stoks, *Invited talk by J. J. de Swart at the 3rd Int. Symp. "Dubna Deuteron 95" Dubna, Russia, July 4–7, 1995*; nucl-th/9509032.
44. V. A. Babenko and N. M. Petrov, Yad. Fiz. **66**, 1359 (2003) [Phys. At. Nucl. **66**, 1319 (2003)].
45. LANL T-2 Nucl. Information Service, http://t2.lanl.gov/data/map.html.
46. W. N. Polyzou and W. Glöckle, Few-Body Syst. **9**, 97 (1990).
47. W. N. Polyzou, Phys. Rev. **C 58**, 91 (1998).
48. R. V. Reid, Jr. and M. L. Vaida, Phys. Rev. Lett. **29**, 494 (1972).
49. A. M. Shirokov, A. I. Mazur, S. A. Zaytsev, J. P. Vary, and T. A. Weber, Phys. Rev. **C 70**, 044005 (2004).
50. R. Machleidt, private communication (2005).
51. A. M. Shirokov, J. P. Vary, A. I. Mazur, S. A. Zaytsev, and T. A. Weber, Phys. Lett. **B 621**, 96 (2005); J. Phys. **G 31**, S1283 (2005).
52. A. M. Shirokov, J. P. Vary, A. I. Mazur, and T. A. Weber, Phys. Lett. **B 644**, 33 (2007).
53. J. P. Vary, O. V. Atramentov, B. R. Barrett, M. Hasan, A. C. Hayes, R. Lloyd, A. I. Mazur, P. Navrátil, A. G. Negoita, A. Nogga, W. E. Ormand, S. Popescu, B. Shehadeh, A. M. Shirokov, J. R. Spence, I. Stetcu, S. Stoica, T. A. Weber, and S. A. Zaytsev, Eur. Phys. J. **A 25**, Suppl. 1, 475 (2005); J. Phys. Conf. Ser. **20**, 71 (2005).
54. N. Barnea, W. Leidemann, and G. Orlandini, Phys. Rev. **C 74**, 034003 (2006).
55. T. Dytrych, K. D. Sviratcheva, C. Bahri, J. P. Draayer, and J. P. Vary, Phys. Rev. Lett. **98**, 162503 (2007).

The Modified J-Matrix Approach for Cluster Descriptions of Light Nuclei

F. Arickx, J. Broeckhove, A. Nesterov, V. Vasilevsky and W. Vanroose

Abstract We present a fully microscopic three-cluster nuclear model for light nuclei on the basis of a J-Matrix approach. We apply the Modified J-Matrix method on ^{6}He and ^{6}Be for both scattering and reaction problems, analyse the Modified J-Matrix calculation, and compare the results to experimental data.

1 Introduction

The J-Matrix method (JM) has proven very successful in microscopic nuclear calculations, particularly the Modified J-Matrix Method (MJM) [1–3] also called the Algebraic Model in some nuclear physics literature [4–9]. Both collective [10–12] and cluster descriptions [13–17] of light (p-shell) nuclei have been studied with the MJM, using an oscillator basis.

The Harmonic Oscillator basis has always been very popular in nuclear physics. For light nuclei, spherical oscillator states have often been used as a first approximation to the single-particle orbital wave functions in the popular nuclear shell-model. The nuclear many-body basis is then built up as a set of Slater determinants of single-particle oscillator orbital states to take the Pauli principle into account.

Many nuclear two-body potentials feature a superposition of Gaussian components. Matrix elements of two-body operators for Slater determinants reduce to a simple sum of two-body matrix elements, involving the single-particle oscillator states. A Gaussian form of the operator then easily leads to an analytical form for the matrix elements. This immediately shows the computational advantage of using an oscillator state (or even a superposition of oscillator states) for the single-particle wave functions in a fully microscopic many-body nuclear model.

One of the features of the oscillator basis is the Jacobi (tridiagonal) form of the matrix of the kinetic energy operator. This makes it a proper candidate for considering the JM approach to solve the Schrödinger equation expressed in matrix form.

F. Arickx
University of Antwerp, Group Computational Modeling and Programming, Antwerp, Belgium
e-mail: Frans.Arickx@ua.ac.be

A.D. Alhaidari et al. (eds.), *The* J-*Matrix Method*, 269–307.

The nuclear many-body system is very complex though, and requires huge superpositions of shell-model states to reproduce spectral properties over an important energy range. Also, the nuclear system exhibits several modes when the system is energetically excited, and possibly fragmented. There is an interplay between collective modes, such as monopole and quadrupole excitations, and cluster effects that are particularly pronounced when the nucleus disintegrates. One therefore often introduces very specific antisymmetrized forms for the wave function with a specific configuration of single-particle orbitals, that feature some specific collective behavior that one wants to study. In this way the Hilbert space is limited to a single (or a few coupled) nuclear model state(s) in which the collective coordinates are the only remaining dynamical coordinates. The dimensions of the Schrödinger equation are then strongly reduced.

Although the nuclear two-body interaction has a short range, the effective interaction as a function of the collective coordinates often displays a long range. If the collective behavior is described as a superposition of oscillator states, this is usually also reflected in the matrix elements of the nuclear potential, which display a slow decrease for increasing oscillator excitation. This clearly limits the applicability of the standard JM method, as too large energy matrices have to be considered in the internal, non-asymptotic, region. Indeed, the main computational cost for nuclear JM calculations often lies in the construction of the matrix elements of highly excited states. To account for this limitation the MJM approach was developed [1–3], introducing a semi-classical approximation for the matrix elements in the highly excited internal region. This leads to a modification of the standard three-term recursion relation for the far interaction and asymptotic region by including (semiclassical) potential contributions. A drastic reduction of the matching position for boundary condition is obtained, resulting in a remaining matrix equation of manageable dimensions.

The Coulomb contribution to the nuclear potential, which is known to have a long range, can be handled in the same way.

In the following sections we will elaborate on the cluster description of nuclei as an important collective description for the lightest nuclei, in which the internuclear distances then become the dynamical (collective) coordinates. We will link this model to the MJM approach to solve the Schrödinger equation, and present applications on 6-particle nuclear systems where three-cluster effects are important. We will compare the theoretical results to the experimental ones to test the validity of the approach.

2 JM Cluster Models for Light Nuclei

Two- and three-cluster configurations carry an important part of the low-energy physics in light nuclei, and directly relate to scattering and reaction experiments in which smaller fragments are used to study compound properties.

In this section we discuss the three-cluster description for light nuclei. Where appropriate we briefly present some two-cluster properties, as the three-cluster approach is essentially a generalization hereof.

The many-particle wave functions for a three-cluster system of A nucleons ($A = A_1 + A_2 + A_3$) can be written, using the anti-symmetrization operator $\mathcal{A}$, as follows

$$\Psi(\mathbf{q}_1, .., \mathbf{q}_{A-1}) = \mathcal{A}\left[\Psi_1(A_1)\ \Psi_2(A_2)\ \Psi_3(A_3)\ \Psi_R(R)\right] \tag{1}$$

where the centre of mass of the A-nucleon system has been eliminated by the use of Jacobi coordinates $\mathbf{q}_i$ so that only internal dynamics are described. The cluster wave functions $\Psi_i(A_i)$

$$\Psi_i(A_i) = \Psi_i\left(\mathbf{q}_1^{(i)}, .., \mathbf{q}_{A_i-1}^{(i)}\right) \quad (i = 1, 2, 3) \tag{2}$$

represent the internal structure of the i-th cluster, centered around its centre of mass $\mathbf{R}_i$. To limit the computational complexity of the problem, these cluster functions are fixed and they are Slater determinants of harmonic oscillator ($0s$)-states, corresponding to the ground state shell-model configuration of the cluster ($A_i \leq 4$ for all i). The $\Psi_R(R)$ wave function

$$\Psi_R(R) = \Psi_R\left(\mathbf{q}_1^{(R)}, \mathbf{q}_2^{(R)}\right) = \Psi_R(\mathbf{q}_1, \mathbf{q}_2) \tag{3}$$

represents the relative motion of the three clusters with respect to one another, and $\mathbf{q}_1$ and $\mathbf{q}_2$ represent Jacobi coordinates. In Fig. 1 we indicate an enumeration of possible Jacobi coordinates and their relation to the component clusters.

The state (3) is not limited to any particular type of orbital; on the contrary we will use a complete basis of harmonic oscillator states for the relative motion degrees of freedom. Thus the full A-particle state cannot be expressed as a single Slater determinant of single particle orbitals.

An important approximation, known as the "Folding" model, is obtained by breaking the Pauli principle between the individual clusters, but retaining a proper quantum-mechanical description of the clusters, which is described by the wave function

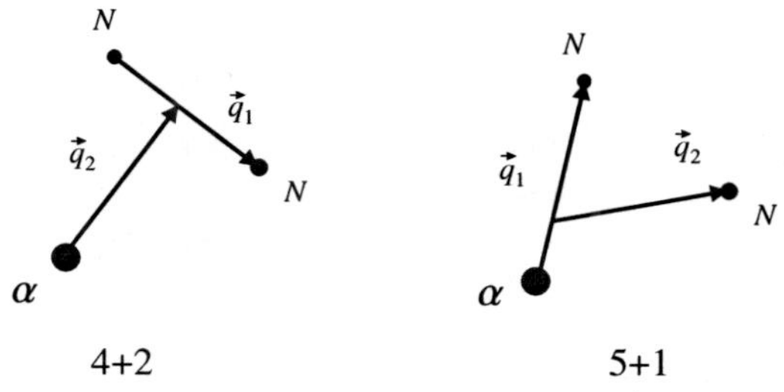

Fig. 1 Two configurations of Jacobi coordinates for the three-cluster system $\alpha + N + N$

$$\Psi_F(\mathbf{q}_1, .., \mathbf{q}_{A-1}) = \Psi_1(A_1)\ \Psi_2(A_2)\ \Psi_3(A_3)\ \Psi_R(R) \tag{4}$$

Because each cluster wave function is antisymmetric (they are Slater determinants) one is indeed neglecting the inter-cluster anti-symmetrization only. The Folding Model has the advantage of preserving the identities of the clusters and, if the intra-cluster structure is kept "frozen", it reduces the many-particle problem to that of the relative motion of the clusters.

For a two-cluster description the formulae simplify to:

$$\Psi(\mathbf{q}_1, .., \mathbf{q}_{A-1}) = \mathcal{A}\left[\Psi_1(A_1)\ \Psi_2(A_2)\ \Psi_R(R)\right] \tag{5}$$

with

$$\Psi_R(R) = \Psi_R\left(\mathbf{q}_0^{(R)}\right) = \Psi_R(\mathbf{q}_0) \tag{6}$$

The folding approximation will be the natural choice for calculating the asymptotic behavior of the cluster-system, i.e. the disintegration of the system in the two or three non-interacting individual clusters. This amounts to the situation that all clusters are a sufficient distance apart and inter-cluster antisymmetrization has a negligible effect.

Because the cluster states are fixed and built up of $(0s)$-orbitals, the problem of labeling the basis states with quantum numbers relates to the inter-cluster wave function only. This holds true whether one uses the full anti-symmetrization or the folding approximation. In a two-cluster case, the set of quantum numbers describing inter-cluster motion is unambiguously defined, and is obtained from the reduction of the symmetry group $U(3) \supset O(3)$ of the one-dimensional oscillator. This reduction provides the quantum numbers n for the radial excitation, and L, M for the angular momentum of the two-cluster system.

In a three-cluster case, several schemes can be used to classify the inter-cluster wave function in the oscillator representation. In [18,19] three distinct but equivalent schemes were considered. One of these used the quantum numbers provided by the Hyperspherical Harmonics (HH) method (see for instance [20–22]). This is the classification that we will adopt. Even within this particular scheme there are several ways to classify the basis states. We shall restrict ourselves to the so-called Zernike–Brinkman basis [23]. This corresponds to the following reduction of the unitary group $U(6)$, the symmetry group of the three-particle oscillator Hamiltonian,

$$U(6) \supset O(6) \supset O(3) \otimes O(3) \supset O(3) \tag{7}$$

This reduction provides the quantum numbers K, the hypermomentum, n, the hyper-radial excitation, l_1, the angular momentum connected with the first Jacobi vector, l_2, the angular momentum connected with the second Jacobi vector, and L and M the total angular momentum obtained from coupling the partial angular momenta l_1, l_2. Collectively these quantum numbers will be denoted by ν, i.e. $\nu = \{n,\ K,\ (l_1 l_2)LM\}$ in the remainder of the text.

There are a number of relations and constraints on these quantum numbers:

- the total angular momentum is the vector sum of the partial angular momenta $\mathbf{l}_1$ and $\mathbf{l}_2$, i.e. $\mathbf{L} = \mathbf{l}_1 + \mathbf{l}_2$ or $|l_1 - l_2| \leq L \leq l_1 + l_2$.
- by fixing the values of l_1 and l_2, we impose restrictions on the hypermomentum $K = l_1 + l_2,\ l_1 + l_2 + 2,\ l_1 + l_2 + 4, \ldots$ This condition implies that for certain values of hypermomentum K the sum of partial angular momenta $l_1 + l_2$ cannot exceed K.
- the partial angular momenta l_1 and l_2 define the parity of the three-cluster state by the relation $\pi = (-1)^{l_1+l_2}$.
- for the "normal" parity states $\pi = (-1)^L$ the minimal value of hypermomentum is $K_{\min} = L$, whereas $K_{\min} = L + 1$ for the so-called "abnormal" parity states $\pi = (-1)^{L+1}$.
- oscillator shells with N quanta are characterized by the constraint $N = 2n + K$.

Thus for a given hyperangular and rotational configuration the quantum number n ladders the oscillator shells of increasing oscillator energy.

Note that the HH basis is widely used to study nuclei with large proton or neutron excess within a three-body model [24–29].

2.1 Asymptotic Solutions in Coordinate Representation

The MJM method for solving the Schrödinger equation for quantum scattering systems is based on a matrix representation of the Schrödinger equation in terms of a square integrable basis. In a nuclear context the Harmonic Oscillator basis is appropriate. The boundary conditions are ultimately formulated in terms of the asymptotic behavior of the expansion coefficients of the wave function.

In the case of three-cluster calculations, one needs to determine the asymptotic behavior of the wave function. Consider an expansion of the relative wave function

$$\Psi_R(\mathbf{q}_1, \mathbf{q}_2) = \sum_{\nu} c_\nu \Psi_\nu(\mathbf{q}_1, \mathbf{q}_2) \tag{8}$$

with $\nu = \{n, K, (l_1 l_2)LM\}$ and $\{\Psi_\nu\}$ a complete basis of six-dimensional oscillator states. It covers all possible types of relative motion between the three clusters.

To obtain the asymptotic behavior of the three-cluster system the folding approximation is used. The assumption that antisymmetrization effects between clusters are absent in the asymptotic region is a natural one. It simplifies the many-body dynamics effectively to that of a three-particle system, as the cluster descriptions are frozen. The relative motion problem of the three clusters in the absence of a potential (i.e. considering the JM reference Hamiltonian only) can then be explicitly solved in the HH method (see for instance [20–22]). It involves the transformation of the Jacobi coordinates $\mathbf{q}_1$ and $\mathbf{q}_2$ to the hyperradius ρ and a set of hyperangles Ω. The hyperspherical coordinates ρ and Ω describe the geometry of the three-particle system in the same way that the spherical coordinates describe the geometry of two-particle systems. The inter-cluster wave function in coordinate representation is expanded in HH's $H_K^{\nu_0}(\Omega)$ where ν_0 has been chosen as a shorthand for $(l_1 l_2) LM$, and which are the generalization of the spherical harmonics $Y_{LM}(\theta, \varphi)$.

In the absence of the Coulomb interaction this leads to a set of equations for the hyperradial asymptotic solutions, with the kinetic energy operator as the JM reference Hamiltonian

$$\left\{-\frac{\hbar^2}{2m}\left[\frac{d^2}{d\rho^2}+\frac{5}{\rho}\frac{d}{d\rho}-\frac{K(K+4)}{\rho^2}\right]-E\right\}R_{K,\nu_0}(\rho)=0 \tag{9}$$

The solutions can be obtained analytically and are represented by a pair of Hänkel functions for the ingoing and outgoing solutions:

$$R_{K,\nu_0}^{(\pm)}(\rho)=\left\{\begin{array}{l} H_{K+2}^{(1)}(k\rho)/\rho^2 \\ H_{K+2}^{(2)}(k\rho)/\rho^2 \end{array}\right\} \tag{10}$$

where

$$k=\sqrt{\frac{2mE}{\hbar^2}}$$

One notices that these asymptotic solutions are independent of all quantum numbers ν_0, and are determined by the value of hypermomentum K only. In particular, different K-channels are uncoupled.

When charged clusters are considered the asymptotic reference Hamiltonian should consist of the kinetic energy and the Coulomb interaction:

$$\left\{-\frac{\hbar^2}{2m}\left[\frac{d^2}{d\rho^2}+\frac{5}{\rho}\frac{d}{d\rho}-\frac{\|\mathcal{K}\|}{\rho^2}\right]+\frac{\|Z_{eff}\|}{\rho}-E\right\}\|\mathcal{R}(\rho)\|=0 \tag{11}$$

The matrix $\|\mathcal{K}\|$ is diagonal with matrix elements $K(K+4)$, and $\|Z_{eff}\|$, the "effective charge", is off-diagonal in K and $(l_1 l_2)$. The solution matrix $\|\mathcal{R}(\rho)\|$ reflects the coupling of K-channels. A standard approximation for solving these equations is to decouple the K-channels, by assuming that the off-diagonal matrix-elements of $\|Z_{eff}\|$ are sufficiently small:

$$\left\{-\frac{\hbar^2}{2m}\left[\frac{d^2}{d\rho^2}+\frac{5}{\rho}\frac{d}{d\rho}-\frac{K(K+4)}{\rho^2}\right]+\frac{Z_{eff}}{\rho}-E\right\}R_{K,\nu_0}(\rho)=0 \tag{12}$$

The constants Z_{eff} depends on K and ν_0 and all parameters of the many-body system under consideration. We will restrict ourselves to this decoupling approximation, but it is to be understood that its validity has to be checked for any specific three-cluster system.

The asymptotic solutions then become

$$R_{K,\nu_0}^{(\pm)}(\rho) = \frac{1}{\sqrt{k}} W_{\pm i\eta, K+2}(\pm 2ik\rho)/\rho^{\frac{5}{2}} \tag{13}$$

where W is the Whittaker function and η is the well-known Sommerfeld parameter

$$\eta = \frac{m}{2\hbar^2}\frac{Z_{eff}}{k} \tag{14}$$

As η is a function of K, l_1 and l_2 through the parameter Z_{eff}, the asymptotic solutions will now be dependent on K and ν_0.

A two-cluster description leads to the well-known one-dimensional radial equation, with asymptotic solutions

$$R_L^{(\pm)}(\rho) = \frac{1}{\sqrt{k}} W_{\pm i\eta, L+\frac{1}{2}}(\pm 2ik\rho)/\rho$$

where ρ now is the inter-cluster radial distance and η is the Sommerfeld parameter with $Z_{eff} = Z_1 Z_2 e^2 \sqrt{\frac{A_1 A_2}{A_1 + A_2}}$

One can easily combine the two- and three cluster asymptotics in a single representation as

$$R_{\mathcal{L}}^{(\pm)}(\rho) = \frac{1}{\sqrt{k}} W_{\pm i\eta, \lambda}(\pm 2ik\rho)/\rho^{\frac{\sigma-1}{2}} \tag{15}$$

where the parameters $\mathcal{L}$, λ, σ and η, differ for the two- and three-cluster channels, and are summarized in Table 1.

2.2 Asymptotic Solutions in Oscillator Representation

The JM relies on an expansion in terms of oscillator functions, and the asymptotic behavior of the corresponding expansion coefficients c_ν. We will restrict ourselves to the three-cluster situation; the two-cluster results can be readily obtained from [2] and the results of the Section 2.1.

It was conjectured (see for instance [2]) that for very large values of the oscillator quantum number n the expansion coefficients for physically relevant wave-functions behave like

Table 1 Two- and three-cluster asymptotic solution parameters

	$\mathcal{L}$	σ	λ	η
two-cl. channel	L	3	$L+\frac{1}{2}$	$\frac{Z_1 Z_2 e^2}{k}\frac{m}{2\hbar^2}\sqrt{\frac{A_1 A_2}{A_1+A_2}}$
three-cl. channel	K	6	$K+2$	$\frac{Z_{eff}}{k}\frac{m}{2\hbar^2}$

$$c_n = \langle n|\psi\rangle \simeq \sqrt{2}\rho_n^2\psi(b\rho_n) \tag{16}$$

with b the oscillator parameter of the basis, $\rho_n = \sqrt{4n+2K+6}$ the classical turning point, and ψ the hyperradial wave function.

In the case of neutral clusters this leads after substitution of the hyperradial asymptotic solutions to the following expansion coefficients $c_n^{(\pm)}$

$$c_n^{(\pm)K} \simeq \sqrt{2}\left\{\begin{array}{l} H_{K+2}^{(1)}(kb\rho_n) \\ H_{K+2}^{(2)}(kb\rho_n) \end{array}\right\} \tag{17}$$

This result can be obtained in an alternative way [30, 31] by representing the Schrödinger equation, with the kinetic energy operator $\hat{T}$ as the reference Hamiltonian to describe the asymptotic situation, in a (hyperradial) oscillator representation

$$\sum_{m=0}^{\infty}\left\langle n,(K,\nu_0)\left|\hat{T}-E\right|m,(K,\nu_0)\right\rangle c_m^{K,\nu_0} = 0 \tag{18}$$

This matrix equation is of a three-diagonal form because of the properties of $\hat{T}$ and the oscillator basis. Solving for the expansion coefficients c_n^{K,ν_0} leads to a three-term recurrence relation

$$T_{n,n-1}^{K,\nu_0}c_{n-1}^{K,\nu_0} + \left(T_{n,n}^{K,\nu_0} - E\right)c_n^{K,\nu_0} + T_{n,n+1}^{K,\nu_0}c_{n+1}^{K,\nu_0} = 0 \tag{19}$$

where

$$T_{n,m}^{K,\nu_0} = \left\langle n,(K,\nu_0)\left|\hat{T}\right|m,(K,\nu_0)\right\rangle \tag{20}$$

The asymptotic solutions (i.e. for high n) of this recurrence relation are then precisely given by (17).

When the Coulomb interaction is present we again apply (16) to obtain

$$c_n^{(\pm)K} \simeq \sqrt{2}\left\{\begin{array}{l} W_{i\eta,\mu}(2ikb\rho_n)/\sqrt{\rho_n} \\ W_{-i\eta,\mu}(-2ikb\rho_n)/\sqrt{\rho_n} \end{array}\right\} \tag{21}$$

In this case the oscillator representation of the Schrödinger equation is no longer of a tridiagonal form, and cannot be solved analytically for the asymptotic solutions to corroborate this result.

It should be noted that the above elaborations are valid for relatively small values of momentum k to which correspond sufficiently large values of the discrete hyperradius ρ_n, which defines the asymptotic region. So for any value of k one can determine a minimum value for n to be safely in the asymptotic region.

As we consider an asymptotic decoupling in the (K,ν_0) quantum numbers, one will deal with asymptotic channels characterized by the (K,ν_0) values. So only in the internal (or interaction) region will states with different K and ν_0 be coupled by the short-range nuclear potential and the Coulomb potential. The three-cluster system can therefore be described by a coupled-channels approach, where the individual channels are characterized by a single K-value, and we will henceforth refer to these channels as "K -channels".

2.3 Multi-Channel JM Equations

In the current many-channel description of the JM formulation for three-cluster systems, the channels will be characterized by the a specific value of the set of quantum numbers K, ν_0, whereas the relative motion of clusters within the channel is connected to the oscillator index n. We will use K henceforth as an aggregate index for individual channels, and assume it represents all K, ν_0 quantum numbers.

The Schrödinger equation can be cast in a matrix equation of the form

$$\sum_{K',m} \left\langle n,\, K \left| \hat{H} - E \right| m,\, K' \right\rangle\, c_m^{K'} = 0 \tag{22}$$

As we will consider an S-matrix formulation of the scattering problem, the expansion coefficients are rewritten as

$$c_n^K = c_n^{(0)K} + \delta_{K_i K} c_n^{(-)K} - S_{K_i K} c_n^{(+)K} \tag{23}$$

where, for each channel K, the $c_n^{(0)K}$ are the so-called residual coefficients, the $c_n^{(\pm)K}$ are the incoming and outgoing asymptotic coefficients. The matrix element $S_{K_i K}$ describes the coupling between the current channel K and the entrance channel K_i.

As shown in [32–34], the $c_n^{(\pm)K}$ satisfy the following system of equations

$$\sum_{m=0}^{\infty} \left\langle n,\, K \left| \hat{H}_0 - E \right| m,\, K \right\rangle\, c_m^{(\pm)K} = \beta_0^{(\pm)K}\, \delta_{n,0} \tag{24}$$

$\hat{H}_0$ being the asymptotic reference Hamiltonian, consisting of the kinetic energy operator for uncharged clusters, and the kinetic energy operator plus Coulomb interaction for charged clusters. The right-hand side for the equation for $c_m^{(-)K}$ features $\beta_0^{(-)K}$ which is a regularization factor to account for the irregular behavior of the $c_0^{(-)K}$. This factor allows one to solve (24) for all values of n. The value of $\beta_0^{(-)K}$ can be obtained for both reference Hamiltonians (i.e. with or without Coulomb). We introduced $\beta_0^{(+)K} = 0$ to keep the equations in a homogeneous form.

The set of equations (24) for the asymptotic coefficients can then be solved numerically to different degrees of approximation depending on the requested precision. The $c_n^{(\pm)K}$ have the desired asymptotic behavior (cf. eqs (17) and (21)).

Substitution of (23) in the equations (22) then leads to the following system of dynamical equations for the many-channel system:

$$\sum_{K',m} \langle n, K \,|\hat{H} - E|\, m, K' \rangle c_m^{(0)K'} - \sum_{K'} S_{K_i K'} V_n^{(+)KK'} \tag{25}$$
$$= -\beta_0^{(-)K} \, \delta_{n,0}\delta_{K_i K} - V_n^{(-)KK_i}$$

where the coefficients $V_n^{(\pm)KK'}$, defined in [2], are given by

$$V_n^{(\pm)KK'} = \sum_{m=0}^{\infty} \langle n, K \,|\hat{V}|\, m, K' \rangle \; c_m^{(\pm)K'} \tag{26}$$

This system of equations should be solved for both the residual coefficients $c_n^{(0)K}$ and the S-matrix elements $S_{K'K}$.

To obtain an appropriate approximation to the exact solution of (25), we consider an internal region corresponding to $n < N$ and an asymptotic region with $n \geq N$. The choice of N is such that one can expect the residual expansion coefficients $\{c_n^{(0)K}\}$ to be sufficiently small in the asymptotic region. Under these assumptions (25) reduces to the following set of $N + 1$ equations ($n = 0..N$):

$$\sum_{K',m<N} \langle n, K \,|\hat{H} - E|\, m, K' \rangle c_m^{(0)K'} - \sum_{K'} S_{K_i K'} V_n^{(+)KK'} \tag{27}$$
$$= -\beta_0^{(-)K} \, \delta_{n,0}\delta_{K_i K} - V_n^{(-)KK_i}$$

The total number of equations for an entrance channel K_i amounts to $N_{ch} * (N+1)$, and solving the set of equations by traditional numerical linear algebra leads to $N_{ch} * N$ residual coefficients $\{c_n^{(0)K};\, K = K_{\min}..K_{\max};\, n = 0..N-1\}$ and N_{ch} S-matrix elements $\{S_{K_i K};\, K = K_{\min}..K_{\max}\}$. The set of equations has to be solved for all N_{ch} entrance channels labelled by K_i.

2.4 Matrix Elements and the Generating Function Method

We only present the general principles for calculating matrix elements in a three-cluster basis. We will do so by considering the Generating Function Method. The two main quantities of interest to solve the JM equations are: the overlap matrix, and the Hamiltonian matrix. The former is of importance because of the proper normalization of the basis states. The latter is decomposed into the kinetic energy operator, the matrix elements of which are obtained mainly by group-theoretical considerations, the potential energy operator, which in our case will be chosen to be a local, semi-realistic, two-body interaction based on a superposition of Gaussians, and the Coulomb interaction.

The basic principle of generating functions is well-known from mathematical physics. A generating function or generator state depends on a parameter, referred to as the generating coordinate, in such a way that an expansion with respect to that parameter yields basis states as expansion terms. Let us explain the principle using a familiar example: the single-particle translated Gaussian wave functions, which are appropriate for two-cluster descriptions

$$\phi(\mathbf{r}\,|\mathbf{R}) = \exp\left\{-\frac{1}{2}\mathbf{r}^2 + \sqrt{2}\,\mathbf{R}\cdot\mathbf{r} - \frac{1}{2}\mathbf{R}^2\right\} \tag{28}$$

with the translation parameter $\mathbf{R}$ acting as generator coordinate. The choice of parametrization of the generator coordinate influences the quantum numbers of the individual basis states that are generated. In a Cartesian parametrization $\mathbf{R} = (R_x, R_y, R_z)$ one generates the familiar Cartesian $\phi_{n_x}(R_x)\phi_{n_y}(R_y)\phi_{n_z}(R_z)$ oscillator states. With a radial parametrization $\mathbf{R} = R\check{\mathbf{R}}$ (where the inverted hat stands for a unit vector) the expansion yields

$$\phi(\mathbf{r}\,|\mathbf{R}) = \sum_{n,l,m} \mathcal{N}_{nl} R^{2n+l} Y_{lm}(\check{\mathbf{R}})\phi_{nlm}(\mathbf{r}) \tag{29}$$

An underlying mathematical connection exists between such expansions, group representation theory and coherent state analysis [35, 36]. We exploit the generating function principle to facilitate the computation of matrix elements. The matrix element of any operator between generating states is a function of the generating coordinates on the left and right

$$X(\mathbf{R}, \mathbf{R}') = \left\langle \phi\,(\mathbf{r}\,|\mathbf{R})\,\middle|\hat{\mathbf{X}}\middle|\,\phi\,(\mathbf{r}\,|\mathbf{R}')\right\rangle \tag{30}$$

Expansion of this function will yield the matrix elements between the basis states. They can be identified in the expansion by the appropriate dependence on the generator coordinates

$$X(\mathbf{R}, \mathbf{R}') = \sum_{nlm}\sum_{n'l'm'} \mathcal{N}_{nl}\mathcal{N}_{n'l'} R^{2n+l} R'^{(2n'+l')} Y^*_{lm}(\check{\mathbf{R}}) Y_{l'm'}(\check{\mathbf{R}}') \left\langle \phi_{nlm}(\mathbf{r})\,\middle|\hat{\mathbf{X}}\middle|\,\phi_{n'l'm'}(\mathbf{r})\right\rangle \tag{31}$$

For the three-cluster basis we consider the following generating function for inter-cluster basis functions (in what follows we shall use small $\mathbf{q}$ for the Jacobi vectors and capital $\mathbf{Q}$ for the corresponding generating coordinates)

$$\begin{aligned} &\Psi\,(\mathbf{q}_1, \mathbf{q}_2\,|\mathbf{Q}_1, \mathbf{Q}_2) \\ &= \exp\left\{-\frac{1}{2}\left(\mathbf{q}_1^2 + \mathbf{q}_2^2\right) + \sqrt{2}\,(\mathbf{Q}_1\cdot\mathbf{q}_1 + \mathbf{Q}_2\cdot\mathbf{q}_2) - \frac{1}{2}\left(\mathbf{Q}_1^2 + \mathbf{Q}_2^2\right)\right\} \end{aligned} \tag{32}$$

The choice of parametrization is linked to the basis states one intends to generate. Associated with our choice of basis (Zernike–Brinkman [23]), we introduce hyperspherical coordinates. The hyperradius and hyperangles, both for spatial coordinates and for generating parameters, are defined by:

$$\rho = \sqrt{\mathbf{q}_1^2 + \mathbf{q}_2^2},\ q_1 = \rho\cos\theta,\ q_2 = \rho\sin\theta;$$
$$R = \sqrt{\mathbf{Q}_1^2 + \mathbf{Q}_2^2},\ \ Q_1 = R\cos\Theta,\ Q_2 = R\sin\Theta, \tag{33}$$

Using these, one expands the generating function (32) in HH functions:

$$\Psi\left(\mathbf{q}_1, \mathbf{q}_2 \,|\mathbf{Q}_1, \mathbf{Q}_2\right) = \sum_{\nu} \Psi_\nu\left(\rho, \theta, \check{\mathbf{q}}_1, \check{\mathbf{q}}_2\right)\ \Xi_\nu^*\left(R, \Theta, \check{\mathbf{Q}}_1, \check{\mathbf{Q}}_2\right) \tag{34}$$

where the full set of quantum numbers ν (introduced previously) is involved in the summation. The oscillator basis functions are

$$\Psi_\nu\left(\rho, \theta, \check{\mathbf{q}}_1, \check{\mathbf{q}}_2\right) = \mathcal{N}_{n,K}\ \rho^K \exp\{-\rho^2/2\}\ L_n^{K+2}(\rho^2)\ H_K^{(l_1 l_2)LM}\left(\theta, \check{\mathbf{q}}_1, \check{\mathbf{q}}_2\right) \tag{35}$$

and the generator coordinate functions are

$$\Xi_\nu\left(R, \Theta, \check{\mathbf{Q}}_1, \check{\mathbf{Q}}_2\right) = \mathcal{N}_{n,K}\ R^{K+2n}\ H_K^{(l_1 l_2)LM}\left(\Theta, \check{\mathbf{Q}}_1, \check{\mathbf{Q}}_2\right) \tag{36}$$

Here H denotes the HH function

$$H_K^{(l_1 l_2)LM}\left(\Theta, \check{\mathbf{Q}}_1, \check{\mathbf{Q}}_2\right) = \mathcal{N}_K^{(l_1 l_2)LM}\Phi_K^{(l_1 l_2)}(\Theta)\left\{\mathrm{Y}_{l_1}(\check{\mathbf{Q}}_1) \times \mathrm{Y}_{l_2}(\check{\mathbf{Q}}_2)\right\}_{LM}$$
$$\Phi_K^{(l_1 l_2)}(\Theta) = (\cos\Theta)^{l_1}\ (\sin\Theta)^{l_2}\ P_{\frac{K-l_1-l_2}{2}}^{l_2+\frac{1}{2}, l_1+\frac{1}{2}}(\cos 2\Theta) \tag{37}$$

From (36), one easily deduces the procedure for selecting basis functions with fixed quantum numbers $\nu = \{n, K, (l_1 l_2)\, LM\}$. One has to differentiate the generating function $(K+2n)$-times with respect to R and then to set $R = 0$. After that one has to integrate over Θ with the weight $\Phi_K^{(l_1 l_2)}$ to project onto the hypermomentum K; one has to integrate over unit vectors $\check{\mathbf{Q}}_1$ and $\check{\mathbf{Q}}_2$ with weights $\mathrm{Y}_{l_1 m_1}(\check{\mathbf{Q}}_1)$ and $\mathrm{Y}_{l_2 m_2}(\check{\mathbf{Q}}_2)$ to project onto partial angular momenta.

The calculation of matrix elements with the generating function method is thus a two-step process. The first step is the calculation of the generating function for the operator involved. Usually this is accomplished with analytical techniques. The second step is the expansion of the generating function w.r.t. the generator coordinates. Either explicit differentiation or recurrence relations can be used to derive expressions for individual matrix elements [37]. In any case, the work involved is straightforward but extremely tedious; both approaches are best implemented using algebraic manipulation software such as Mathematica or Maple.

A further explicit presentation of the calculation of matrix elements for overlap and Hamiltonian is beyond the scope of the current chapter because of the highly technical details involved. We refer to [13] for explicit details for the three-cluster case.

2.5 The Modified JM Approach

The numerical solution of the JM equations critically depends on an appropriate choice of N, separating the internal from the external region. The determining factor in this is the behavior of the potential energy matrix elements which can be slowly decreasing as a function of n. If this is the case, a sufficiently large value of N has to be chosen.

In the case of three-cluster systems it is known from literature [38] that the potential asymptotically behaves as $1/\rho^3$ in the hyperradius, with a corresponding effect on the matrix elements. It will be shown later that the asymptotic form of the effective potential in our calculation indeed follows this behavior. Potentials with an asymptotic tail $1/\rho^3$ dramatically affect the phase shift in the low energy region [39,40] and special care should be taken to get convergent results. This usually requires more than merely choosing a sufficiently large value of N.

Another factor that aversely affects convergence is the choice of a single basis (i.e. a fixed oscillator length b) for both the cluster and relative motion basis functions. This choice has been made to make the calculation of matrix elements manageable, and within acceptable execution time. The value chosen for b is usually fixed so that the frozen clusters have appropriate physical properties (e.g. ground-state energy). It is indeed important that the individual fragments are well described, as their properties cannot be influenced by the solution of the dynamical equations which only involves their relative behavior. This choice of basis is often far from optimal for a convergent calculation.

To account for these convergence problems we have considered the MJM approach, developed in [1–3]. In this approach we consider a semi-classical approximation for the matrix elements of the potential that is incorporated in the three-term recursion relation. We so introduce an intermediate region between the pure interaction and asymptotic regions. In this intermediate region a modified asymptotic solution holds, obtained from the modified recursion relation, and incorporating potential effects. In particular a long asymptotic tail of the potential can be taken into account in a satisfactory way. The matching point for the boundary condition, originally positioned at the border between interaction and asymptotic regions, is brought down to the border between the near interaction and intermediate regions. This dramatically reduces the dimensions of the remaining set of matrix equations, as well as improves overall convergence by taking into account the asymptotic tail of the potential energy.

3 Resonances in the MJM ^6He and ^6Be in the Three-Cluster Approach

In this section we consider the continuum obtained in a three-cluster MJM calculation for the 6-particle nuclei ^6He and ^6Be, determined by the three-cluster configurations $\alpha + n + n$ and $\alpha + p + p$. Our objective is to highlight the characteristics of

MJM three-cluster calculations, and to produce accurate results for the astrophysically relevant resonances in ^6He and ^6Be.

As the MJM for three-cluster scattering leads to a set of coupled-channel Schrödinger equations (see section 2.2), we will transform the S-matrix to diagonal form. This is usually referred to as the eigenchannel representation of the S-matrix. We will derive the position and width of the resonances eigenphase shifts δ as a function of energy. To be precise, we use the conditions

$$\frac{d^2\delta}{dE^2}|_{E=E_r} = 0, \quad \Gamma = 2\left[\frac{d\delta}{dE}\right]^{-1}|_{E=E_r} \tag{38}$$

We particularly focus on the properties of the low-lying 0^+-and 2^+-resonances in ^6Be and the 2^+-resonance state in ^6He.

Several models and methods have already been applied to investigate the resonance states of ^6He and ^6Be. In the series of chapters [22, 41, 42], the nuclei were considered as three-particle systems, neglecting antisymmetrization. These authors used an effective interaction between the alpha-particle and a nucleon, and a nucleon–nucleon (NN) realistic potential between the two nucleons. The HH's Method was used to describe the bound as well as the three-particle continuum states. In [43] the Complex Scaling Method was used to calculate the characteristic parameters of the resonance states in ^6He and ^6Be. This was done within a three-cluster model, in which full antisymmetrization was taken into account, and the interactions between clusters was obtained from the Minnesota NN-potential. In [44] it was demonstrated that the method of continuation in the coupling constant (CCCM), used for the $\alpha + N + N$ configurations in ^6He and ^6Li, reproduces results, which are very close to those obtained by the CSM. We will compare MJM results to each of these methods. The MJM formulation leads to a good convergence of the computations, even for the (important) Coulomb contribution. An important renormalization effect on the irregular asymptotic solution still remains a restrictive factor in the reduction of the number of basis states in some cases.

3.1 Details of the Calculation

The Volkov potential [45] has been used to describe NN interaction in all calculations of this section. According to [18, 19] this provides an acceptable description for ^6He within the three-cluster model. The effective Volkov potential consists of central forces only, so that total angular momentum L, total spin S and total isospin T are good quantum numbers. We will therefore consider only three-cluster configurations $\alpha + N + N$ with $S = 0$ and $T = 1$. This is known to be the most prominent spin-isospin state for the resonances of interest in the nuclei considered in this work. The Coulomb interaction has also been included, as it is to a great extent responsible for reproducing the 0^+ resonance state in ^6Be.

The oscillator radius b, associated with the basis functions for the nuclear state, is the only free parameter for the MJM calculation. It's value was chosen to optimize the ground-state energy of the α-particle, and equals $b = 1.37$ fm.

As explained before, we have a choice of Jacobi coordinate systems and this can be exploited to simplify the calculations. For a cluster configuration $\alpha + N + N$ one can consider the two Jacobi configurations displayed in Fig. 1. The first (referred to as the "4+2"-configuration) is the most appropriate for the current calculation. Selection rules significantly reduce the number of basis functions. Indeed, quantum numbers $S = 0, T = 1$ coincide for the full six-nucleon system and the two-nucleon subsystem $(N + N)$. Hence the relative motion wave function must be an even function in the coordinate q_1 of this Jacobi system. This means that only even angular momenta l_1 have to be considered. Moreover, for positive parity states only even values of l_2, and for negative parity states only odd values of l_2 should be taken into account. With the second Jacobi configuration from Fig. 1 (the "5+1" one), these constraints are hard to meet and the full oscillator basis would have to be used.

Table 2 shows the number of HH's (N_h) of given hypermomentum K for $L^\pi = 0^+$ and $L^\pi = 2^+$ in the "4+2" Jacobi configuration. This table also indicates the total number of HH's so far (N_c) with $K = K_{\min}, K_{\min} + 2, \ldots, K_{\max}$, i.e. it shows the number of channels for a given $K_{\max} = K$.

3.1.1 Overlap Matrix Elements

In the interaction region we apply full antisymmetrization, i.e. also inter-cluster antisymmetrization. Here we look at its effect on positioning the boundary between the internal (interaction) region and the asymptotic region. No particle exchanges should occur in the asymptotic region. In the MJM the antisymmetrization effects can make themselves felt through the overlap and potential matrix elements. We use the notations of [13] for the overlap matrix elements, and in particular the shorthand notation ν_0 is used for the set $(l_1 l_2)LM$ of quantum numbers:

$$\left\langle n, (K, \nu_0) \left| \widehat{\mathcal{A}} \right| n', \left(K', \nu_0\right) \right\rangle \tag{39}$$

Non-zero matrix elements (39) can be obtained from states within the same many-particle oscillator shell only. As the oscillator shells in the Hyperspherical description are characterized by $N = 2n + K$, the selection rule becomes $2n + K = 2n' + K'$.

Table 2 Number (N_h) and accumulated number (N_c) of Hyperspherical Harmonics up to $K = 10$

$L^\pi = 0^+$	K	0	2	4	6	8	10
	N_h	1	1	2	2	3	3
	N_c	1	2	4	6	9	12
$L^\pi = 2^+$	K	0	2	4	6	8	10
	N_h	–	2	3	5	6	8
	N_c	–	2	5	10	16	24

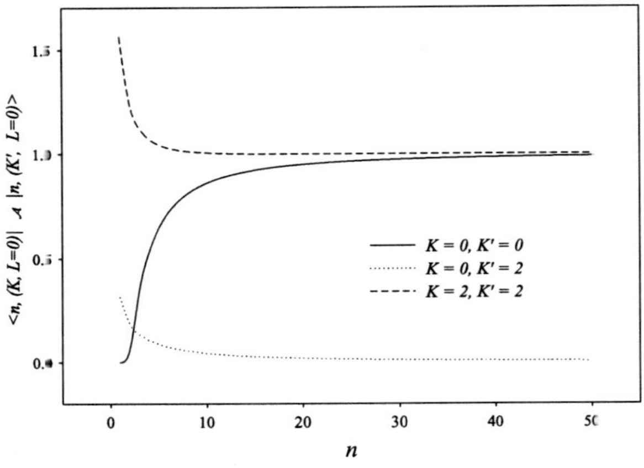

Fig. 2 Matrix elements of the antisymmetrization operator for the 0^+-state in ^6He and ^6Be

In Fig. 2 overlap matrix elements diagonal in n for $L = 0$ and hypermomenta $K = 0$ and $K = 2$ are shown for the six-nucleon three-cluster system. One notices from this figure that the Pauli principle involves oscillator states of at least the 25 lowest shells, for both diagonal and off-diagonal matrix elements in K. The antisymmetrization effects are visible in the deviation from unity for the diagonal matrix elements, and the deviation from zero of the off-diagonal (in K) elements. These effects decrease monotonically with higher n.

In Fig. 3 we compare matrix elements for $L = 0$ diagonal in n and K for some of the K-values, with σ a multiplicity quantum number for states with identical K. Only those matrix elements where the Pauli principle effect is most prominent have been shown. Some states with $K = 4$ and $K = 8$ are affected more strongly by antisymmetrization than others. To understand this we note that the Hyperspherical angles (corresponding to the hyperangular quantum numbers K, l_1, l_2, LM) define the most probable triangular shape and orientation in space of the three-cluster system. The HH's with $K = 4, \sigma = 2$ (characterized by $l_1 = l_2 = 2$) and $K = 8, \sigma = 3$

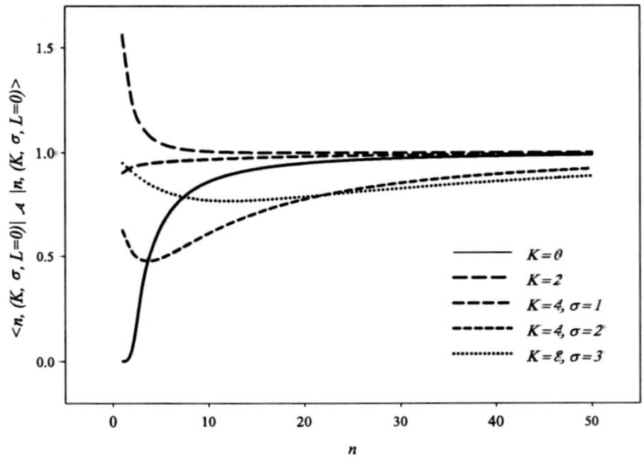

Fig. 3 Matrix elements of the antisymmetrization operator for the 0^+-state in ^6He and ^6Be

(characterized by $l_1 = l_2 = 4$) seem to describe a triangular shape where one of the nucleons is very close to the α-particle.

For larger K-values the probability to find all clusters close to one another within a hypersphere of fixed radius ρ decreases, and one can expect that HH's with large values of K will play a diminishing role in the calculations.

It interesting to compare overlap matrix elements for the three-cluster configuration $\alpha+N+N$ with those of the two-cluster configurations $\alpha+2N$ (such as $\alpha+d$ in ^6Li or $\alpha+2n$ in ^6He). This comparison is shown in Fig. 4, and it indicates that the Pauli principle has a much larger "range" in the three-cluster than in the two-cluster configuration.

3.1.2 Potential Matrix Elements

To investigate the Pauli effect on the potential energy of the three-cluster configuration $\alpha+N+N$ we compare the potential matrix elements with hypermomentum $K = 0$ with full antisymmetrization against those in the folding approximation, where antisymmetrization between clusters is neglected. Figure 5 shows the diagonal potential matrix elements diagonal in n for $K = 0$ and Fig. 6 shows the matrix elements along a fixed row $n = 50$ for $K = 0$. One notes that the folding model results are very close to the fully antisymmetrized ones, especially for larger n.

In Fig. 7 we also display the potential matrix elements between states of the two lowest values of hypermomentum $K = 0$ and $K = 2$. One notices that the $K = 2$ contribution is the largest. The potential energy for $K = 0$ is relatively small, and so is the coupling between $K = 0$ and $K = 2$ states.

The main conclusion is that, in the asymptotic region, the "exact" potential energy can be substituted with the folding approximation. It leads to a considerable reduction in computational effort. We are led to the following setup for three-cluster calculations. In the internal region, consisting of states of the lower oscillator shells and with a large probability to find the clusters close to one another, the fully

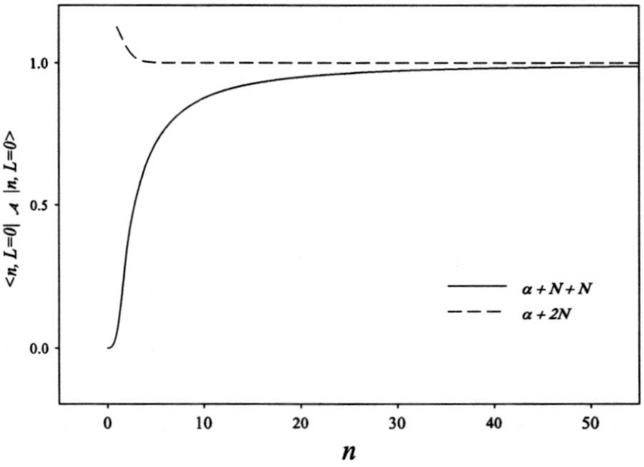

Fig. 4 Matrix elements of the antisymmetrization operator for the 0^+-state of ^6He and ^6Be in the three-cluster $\alpha+N+N$ ($K = 0$) and the two-cluster configuration $\alpha+2N$

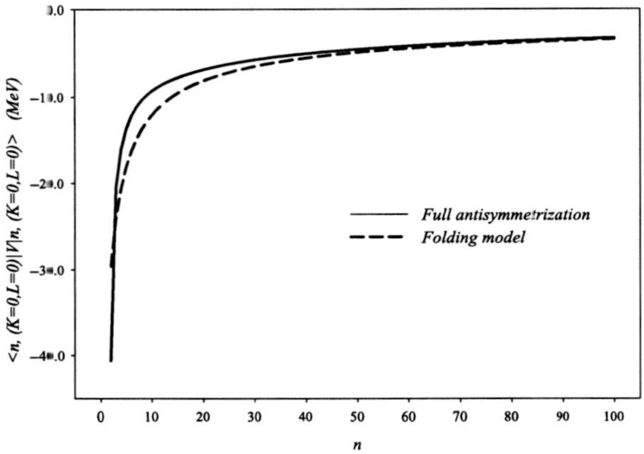

Fig. 5 Diagonal potential matrix elements for the 0^+-state of ^6Be and ^6He ($K = 0$) with full antisymmetrization, and in the folding model

antisymmetrized potential energy is used. In the asymptotic region, where the average distance between clusters is large, we use the folding model potential.

The folding model provides additional insight in the structure of the interaction matrix. It is well known that neither the $\alpha + n$ nor the $n + n$ interaction can create a bound state in the corresponding subsystems of ^6He. Only a full three-cluster configuration $\alpha + n + n$ contains the necessary conditions to create a bound state. In the folding model the total potential energy is a sum of contributions from the three interacting pairs: the α-particle and the first neutron, the α-particle and the second neutron and the neutron–neutron pair. For $K = 0$ the first two contributions are identical, and we only have to consider the $\alpha + n$ and the $n + n$ pairs. Figure 8 shows the diagonal matrix element contribution of both components, and one notices that $\alpha + n$ represents the main contribution.

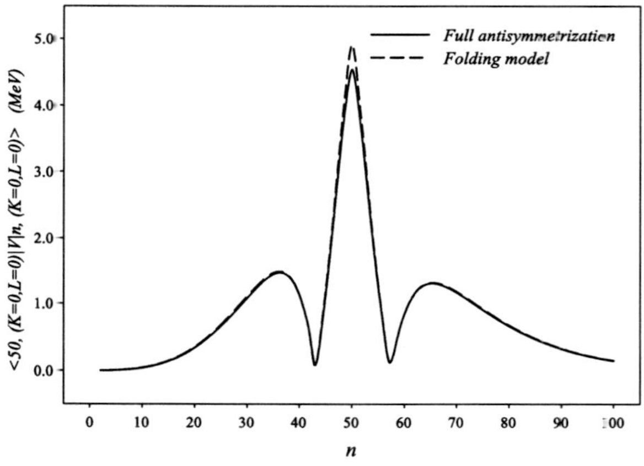

Fig. 6 Off-diagonal potential matrix elements for the 0^+-state of ^6Be and ^6He ($K = 0$) with full antisymmetrization, and in the folding model

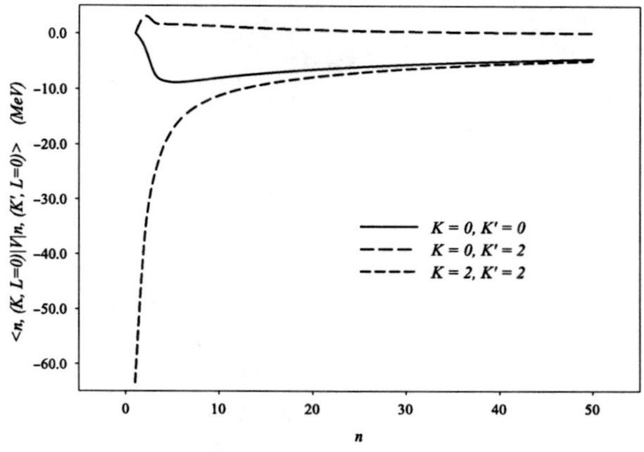

Fig. 7 Diagonal potential matrix elements for the 0^+-state of ^6Be and ^6He, with full antisymmetrization

3.1.3 The Effective Charge

When Coulomb forces are taken into account, one needs to determine the effective charge in order to properly solve the MJM equations. The effective charge unambiguously defines the effective Coulomb interaction in each channel as well as the coupling between different channels. Using the approach suggested in Section 2.1, the effective charges for the 0^+- and 2^+-states of ^6Be were calculated. Part of the corresponding matrices of $\parallel Z_K^{K'} \parallel$ are displayed in Tables 3 and 4 respectively. One notices that the diagonal matrix elements are much larger than the off-diagonal ones. This justifies the approximation [13] of disregarding the coupling of the channels in the asymptotic region.

It is interesting to compare the three-cluster effective charge with the effective charge in the two-cluster configuration. For the latter we can write

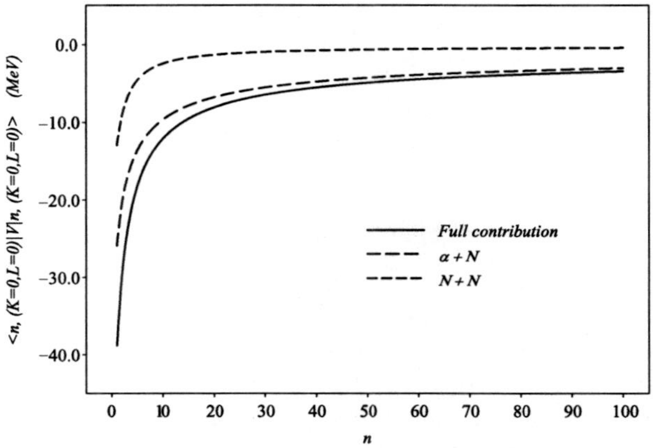

Fig. 8 Folding model contributions in the diagonal potential matrix elements of the energy operator for the 0^+-state of ^6Be and ^6He ($K = 0$)

Table 3 Effective charge matrix for the 0^+-state in ^6Be

$K; l_1, l_2$	0; 0, 0	2; 0, 0	4; 0, 0	4; 2, 2
0; 0, 0	7.274	0.006	−0.129	1.414
2; 0, 0	0.006	7.146	−0.436	0.314
4; 0, 0	−0.129	−0.436	7.428	−0.877
4; 2, 2	1.414	0.314	−0.877	9.098

Table 4 Effective charge matrix for the 2^+-state in ^6Be

$K; l_1, l_2$	2; 2, 0	2; 0, 2	4; 2, 0	4; 0, 2	4; 2, 2
2; 2, 0	7.253	0.400	−0.224	0.546	−0.751
2; 0, 2	0.400	7.244	−0.309	−0.004	−0.601
4; 2, 0	−0.224	−0.309	6.942	−0.186	0.431
4; 0, 2	0.546	−0.004	−0.186	7.694	−0.671
4; 2, 2	−0.751	−0.601	0.431	−0.671	7.345

$$Z = Z_1 Z_2 e^2 \sqrt{\frac{A_1 A_2}{A_1 + A_2}}$$

where A_1 and Z_1 (A_2 and Z_2) are the respective mass and charge of both clusters. For the configuration $\alpha + 2p$ in ^5Be, we then obtain an effective charge

$$Z = \frac{8}{\sqrt{3}} e^2 \simeq 6.65$$

which is independent on the angular momentum of the system. One notice that the two-cluster effective charge is close to the diagonal matrix elements of $\| Z_K^{K'} \|$. We can assume that, if in one of the three-cluster channels the effective charge is very close to the two-cluster one, it could indicate that the two protons move as an aggregate in the asymptotic region. For the 2^+-state we observe at least one channel with this property, carrying the labels $K = 4, l_1 = 2, l_2 = 0$.

3.2 Results

3.2.1 Definition of the Model Space

The model space for the current calculations is primarily determined by the total number of HH's in the internal and external region, and the number of oscillator states. Different sets of HH's can be used in the internal and asymptotic regions. An extensive set of HH's in the internal region will provide a well correlated description of the three-cluster system due to the coupling between states with different hypermomentum. The HH's in the asymptotic region, which are exactly (without Coulomb) or nearly exactly (Coulomb included) decoupled, are responsible for the richness in decay possibilities.

Table 5 0^+-resonance parameters for ^{6}Be, varying $K^{(a)}_{max}$ ($K^{(i)}_{max} = 8$)

$K^{(a)}_{\max}$	0	2	4	6	8
E, MeV	1.434	1.314	1.304	1.298	1.292
Γ, MeV	0.075	0.082	0.084	0.085	0.087

Table 6 0^+-resonance parameters for ^{6}Be, varying $K^{(a)}_{max}$ ($K^{(i)}_{max} = 10$)

$K^{(a)}_{\max}$	0	2	4	6	8	10
E, MeV	1.324	1.204	1.192	1.184	1.176	1.172
Γ, MeV	0.068	0.069	0.071	0.071	0.073	0.072

In the current chapter we restricted both internal and asymptotic HH sets maximal hypermomentum value $K^{(i)}_{\max} = K^{(a)}_{\max} = 10$. By extending the respective subspaces up to these maximal values, we obtain a fair indication of convergence.

We fixed the matching point between internal and asymptotic region at $N = 50$. It is based on the observations of the previous sections, i.e. (1) it is sufficiently large to allow the Pauli principle its full impact, and (2) it is large enough for the semi-classical approximations in the MJM to be valid.

The resonance parameters are obtained from the eigenphase shifts obtained from the eigenchannel representation (diagonalized form) of the S-matrix.

3.2.2 Convergence Study

We first consider the influence of the number of asymptotic channels. We do this by calculating the position and width of the 0^+-state in ^6Be using successively larger asymptotic subspaces. The first calculation includes all HH's up to $K^{(i)}_{\max} = 8$ and extends the asymptotic subspace from $K^{(a)}_{\max} = 0$ to $K^{(a)}_{\max} = 8$. The second calculation includes all HH's up to $K^{(i)}_{\max} = 10$ and extends the asymptotic subspace from $K^{(a)}_{\max} = 0$ to $K^{(a)}_{\max} = 10$. The corresponding results are shown in Tables 5 and 6. One learns from these results that a sufficient rate of convergence has been obtained. Figure 9 displays the first eigenphase shift as a function of energy for $K^{(i)}_{\max} = 10$ and a choice of $K^{(a)}_{\max}$ values from which results in the tables are derived.

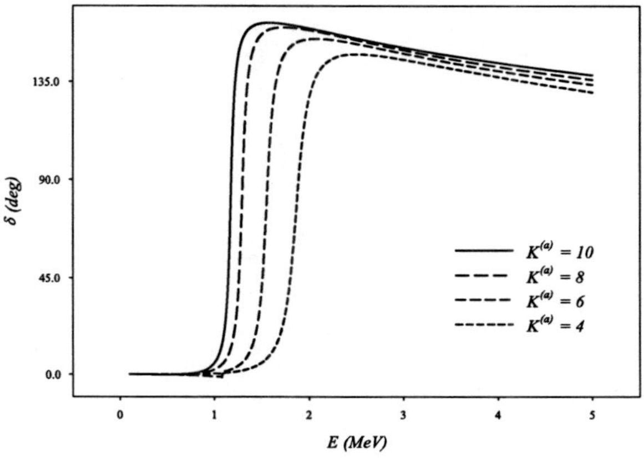

Fig. 9 Eigenphase shifts for the 0^+-state of ^6Be for different $K^{(a)}_{max}$

Table 7 0^+-resonance parameters for ^{6}Be, varying $K^{(i)}_{max}$ ($K^{(a)}_{max} = 0$)

$K^{(i)}_{max}$	0	2	4	6	8	10
E, MeV	–	2.408	2.020	1.688	1.434	1.324
Γ, MeV	–	0.147	0.129	0.097	0.075	0.068

Whereas the inclusion of higher hypermomenta in the asymptotics shows a fast and monotonic convergence, it is also clear from these results that a sufficient number of HH's has to be used for a correct description of the correlations in the internal state. To support this conclusion, we performed a calculation, again for the 0^+-state in ^6Be, in which only one HH with value $K^{(a)} = 0$ was used. In the internal region the number of HH's was varied from $K^{(i)} = 0$ up to $K^{(i)} = 10$. These results appear in Table 7 and corroborate the previous conclusion. In particular they indicate that the effective potential obtained with the $K^{(i)} = 0$ HH only, is unable to produce a resonance. Only after including a $K = 2$ HH does the resonance appear. Further inclusion of higher HH states then lead to an acceptable convergence in a monotonically decreasing fashion for both position and width of the resonance.

3.2.3 Comparisons

In Fig. 10 we display eigenphase shifts for $L^\pi = 0^+$ in ^6Be in the full calculations, i.e. with the maximal number of internal and asymptotic HH's. One notices that the first 0^+-resonance state of ^6Be appears in the first eigenchannel, and that a second (broad) resonance at a higher energy is created in the second eigenchannel (see Table 8 for details on this resonance). The phase shifts in the higher eigenchannels show a smooth behavior as a function of energy without a trace of resonances in the energy range that we consider.

Table 9 compares the results of this work to those obtained in other calculations, in particular by CSM [43], by HHM [41] and by CCCM [44]. In Table 10 our results are compared with the experimental data available from [48, 49]. The agreement with the experimental energy and width of the resonant states is reasonable. The difference

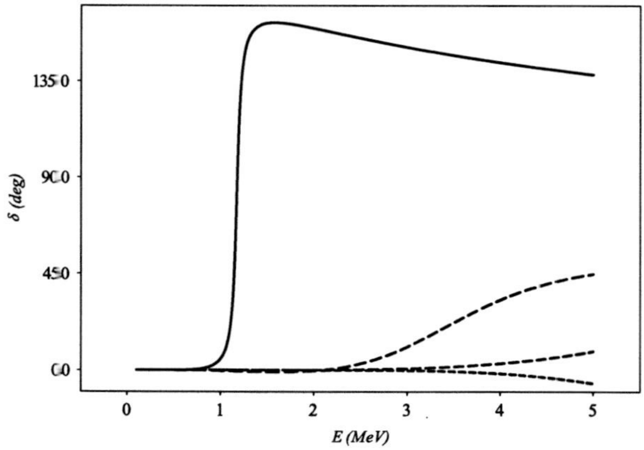

Fig. 10 The eigenphases for the 0^+-state in ^6Be

Table 8 Second resonance state parameters, obtained by MJM, HHM and CSM

Nucleus	L^π	MJM		HH [41]		CSM [46,47]	
		E, MeV	Γ, MeV	E, MeV	Γ, MeV	E, MeV	Γ, MeV
^6He	0_2^+	2.1	4.3	5.0	6.0	3.9	9.4
^6He	2_2^+	3.7	5.0	3.3	1.2	2.5	4.7
^6Be	0_2^+	3.5	6.1				
^6Be	2_2^+	5.2	5.6				

Table 9 Resonance state parameters for ^6He and ^6Be, obtained by MJM, HHM, CSM and CCCM

Method	$^6He; L^\pi = 2^+$		$^6Be; L^\pi = 0^+$		$^6Be; L^\pi = 2^+$	
	E, MeV	Γ, MeV	E, MeV	Γ, MeV	E, MeV	Γ, MeV
MJM	1.490	0.168	1.172	0.072	3.100	0.798
HHM [41]	0.75	0.04				
CSM [43]	0.74	0.06	1.52	0.16	2.81	0.87
CCCM [44]	0.73	0.07				

Table 10 Comparison of resonance state parameter in ^6He and ^6Be between MJM and experiment

	MJM		Experiment [48,49]	
	E, MeV	Γ, MeV	E, MeV	Γ, MeV
$^6He; L^\pi = 2^+$	1.490	0.168	0.822 ± 0.025	0.133 ± 0.020
$^6Be; L^\pi = 0^+$	1.172	0.072	1.371	0.092 ± 0.006
$^6Be; L^\pi = 2^+$	3.100	0.798	3.04 ± 0.05	1.16 ± 0.06

between the experimental and calculated energies of the 2^+-resonance states in both ^6He and ^6Be are probably due to the lack on LS-forces in the present calculations.

It has been pointed out in [41] and [43], that the barrier created by the three-cluster configuration is sufficiently high and wide to accommodate two resonance states, the second one usually being very broad. At first sight, this could be ascribed to an artifact of the HH Model since large values of hypermomentum K create a substantial centrifugal barrier. However the CSM calculations [46, 47], which do not use HH's, also reveal such resonances. Our calculations now also confirm the existence of a second very broad resonance. A comparison of the resonance parameters with those of the HH-method [41] and of the CSM [43] is shown in Table 8. The differences are most probably due to the different descriptions of the system, as well as to the difference in NN-forces.

4 The $^3H(^3H, 2n)^4He$ and $^3He(^3He, 2p)^4He$ Reactions in the MJM Cluster Approach

A second application that shows the strength of the MJM approach is connected to the nuclear reaction $^3He(^3He, 2p)^4He$ which contributes for 89% to the pp-chain of nuclear synthesis, and thus is of particular interest to astrophysics. We will couple the two- and three-cluster descriptions to calculate the reaction properties, by considering a two-cluster entrance, and a three-cluster exit channel

The theoretical analysis of the $^{3}He(^{3}He, 2p)^{4}He$ reaction is usually linked to its mirror companion, $^{3}H(^{3}H, 2n)^{4}He$. A comparison of both leads to a better understanding of the underlying dynamics and of the Coulomb effects of reactions with three-cluster exit channels.

A first microscopic calculation for these reactions was presented in [50]. A two-cluster approach was used for both the entrance and exit channels. The nucleon-nucleon fragment cluster (denoted NN for either pp or nn) carried a simple shell-model description, featuring a pseudo-bound state with positive energy. The experimental cross-section or S-factor at relatively high energy (approximately 1 MeV) was reproduced by adjusting the Majorana exchange parameter of the effective NN-potential. The available experimental data at the (small) energy range relevant in astrophysical reactions were fairly well reproduced. In this model no resonance state appears that would sufficiently amplify the S-factor in the appropriate energy range, and thus would constitute an explanation for the solar neutrino problem.

The previous model was further improved or enhanced in [52–54] by using a more elaborate description of the NN-channel, or by simulating the exit channel description using both the $(^{4}He + N) + N$ and the $^{4}He + (N + N)$ two-cluster configurations. The relative motion of the two clusters was described by a discrete superposition of translated Gaussian functions. In each case the same S-factor shape as a function of energy was obtained.

In this section we study both the $^{3}H\left(^{3}H, 2n\right)^{4}He$ and $^{3}He\left(^{3}He, 2p\right)^{4}He$ reactions and use a correct treatment of the corresponding three-cluster exit channels.

The bound state energy of ^{6}He in the three-cluster description can easily be obtained by diagonalizing the nuclear Hamiltonian. We have compared it to the results obtained for the three-cluster calculation with the Stochastic Variational Method (SVM) [55] [51]. We have used the Minnesota potential without spin-orbit components, and an oscillator parameter $b = 1.285$ fm which minimizes the α-particle

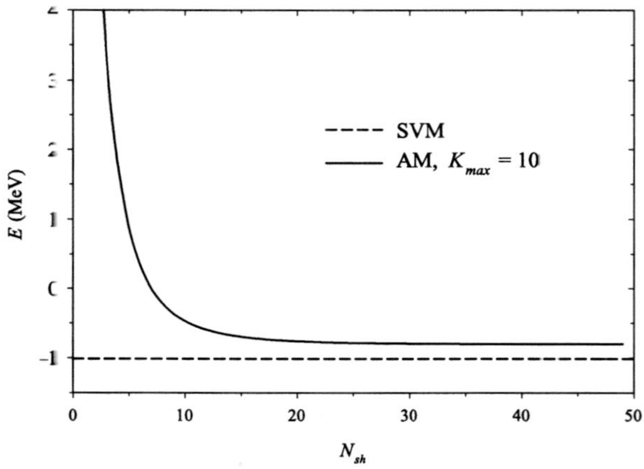

Fig. 11 Ground state of ^{6}He as a function of the number of oscillator shells N_{sh} in the MJM three-cluster model, compared to the results of [51]. The energy is relative to the $\alpha + n + n$ threshold

energy as in [51]. In Fig. 11 we compare our bound-state energy of ^6He as a function of number of oscillator shells N_{sh} to the SVM value. We used all HH's up to $K_{max} = 10$. One notices convergence for $N_{sh} \geq 25$ towards $E = -0.8038$ MeV (relative to the $\alpha + n + n$ sthreshold). This is to be compared to $E = -1.016$ MeV for the full SVM calculation; full convergence towards the SVM result would require additional K values, which is beyond the scope of the current calculation. These results encourage us to combine the two- and three-cluster MJM descriptions to obtain an advanced description of the fusion reactions $^3H(^3H, 2n)^4He$ and $^3He(^3He, 2p)^4He$.

4.1 Model Specifics

We again rely on Section 2 for the details of the microscopic model. The specific cluster configurations used to describe the six-nucleon systems ^6He and ^6Be are analogous to those of Section 3.

4.1.1 A Combined Cluster Model

The six-nucleon wave functions will be built up by using both the two- and three-cluster configurations, each one fully antisymmetrized:

$$\Psi_L = \mathcal{A}\{\Psi_{3N}\,\Psi_{3N}\,f_L(\mathbf{q}_0)\} + \mathcal{A}\{\Psi_\alpha\,\Psi_N\,\Psi_N\,g_L(\mathbf{q}_1, \mathbf{q}_2)\} \tag{40}$$

The Ψ_A (N stands for either nucleon, α for ^4He, and $3N$ for ^3H or ^3He) represents cluster component wave functions, f_L and g_L refer to the wave functions of relative motion for the two-, and three-cluster system. The $\mathbf{q}_i$ are Jacobi coordinates describing the configuration of relative position of the clusters.

All of the reaction dynamics, and in particular the S-matrix elements, is concentrated in the functions describing the relative motion, i.e. f_L and g_L because the internal cluster wave functions are "frozen".

We need to make a choice for the Jacobi coordinates $\mathbf{q}$ and for a classification scheme of the wave-functions to be used in the expansion of f_L and g_L.

For the two-cluster configuration [50] (in $^3H +^3 H$ respectively $^3He +^3 He$) we use the standard spherical coordinates $\mathbf{q}_0 = \{q_0, \widehat{\mathbf{q}}_0\}$ and take the quantum numbers $\mu = \{n, L, M\}$ to classify the basis states. The n is the radial oscillator quantum number. As we have only central components in the NN interaction, the angular momentum L of relative motion will be an integral of motion for the system.

For the three-cluster configurations (in $^4He + p + p$ respectively $^4He + n + n$) we use the Hyperspherical coordinates (33). This choice is consistent with the set of quantum numbers $\nu = \{N, K, (l_1 l_2)LM\} = \left\{n_\rho, K, (l_1 l_2)LM\right\}$, in which $N = 2n_\rho + K$ represents the total number of oscillator quanta, and n_ρ reflects the number of hyperradial excitations. The partial angular momenta l_1 and l_2 are associated with the choice of Jacobi vectors $\mathbf{q}_1$ and $\mathbf{q}_2$. A coupled K-channel calculation with each channel characterized by the set of quantum numbers $\nu_0 = \{K, (l_1 l_2)LM\}$, has to be performed. This type of basis is particularly suitable for the so-called Borromian nuclei, and nuclei with pronounced three-cluster features, when the three-cluster threshold represents the lowest energy decay channel.

4.1.2 The Boundary Conditions

The MJM boundary conditions are expressed in terms of the expansion coefficients of the wave functions of relative motion. They are directly connected to the boundary conditions in coordinate representation. For the two-cluster configurations the asymptotic form of the expansion coefficients in $f_L = \sum c_{n,L}\phi_{n,L}$ can be approximated by

$$c_{n,L} \simeq \sqrt{r_n} f_L(r_n) \tag{41}$$

where the $\{\phi_{n,L}\}$ are the oscillator basis functions, and $r_n = b\sqrt{4n+2L+3}$ is the classical turning point of the three-dimensional oscillator with energy $E_n = \hbar\omega(2n + L + 3/2)$. In three-cluster configurations the expansion is defined by $g_L = \sum d_{n_\rho,L}\phi_{n_\rho,L}$, where the $\{\phi_{n_\rho,L}\}$ now stand for the oscillator basis for three-cluster relative motion. The expansion coefficients behave asymptotically as

$$d_{n_\rho,L} \simeq \sqrt{2}\rho_n^2 g_L(\rho_n) \tag{42}$$

with $\rho_n = b\sqrt{4n_\rho + 2K + 6}$. For clarity we have indicated in the preceding discussion only relevant indices to this section.

We consider both incoming and outgoing waves for the two-cluster configurations (see, for instance, [56])

$$f_L(\mathbf{q}_0) \simeq \left[\psi_L^{(-)}(k_0 q_0) - S_{\{\mu\},\{\mu\}}\,\psi_L^{(+)}(k_0 q_0)\right] Y_{LM}(\widehat{\mathbf{q}}_0) \tag{43}$$

where $S_{\{\mu\},\{\mu\}}$ is a notation to characterize the elastic two-cluster scattering matrix element for the $^3H + ^3H$, resp. $^3He + ^3He$ channels, and $Y_{LM}(\widehat{\mathbf{q}}_0)$ is the spherical harmonic.

Because we are only interested in reactions with a three-cluster exit-channel, the asymptotic wave function can be written as

$$g_L(\mathbf{q}_1, \mathbf{q}_2) = g_L(\rho, \Omega) \simeq \sum_{\nu_0}\left[-S_{\{\mu\},\{\nu_0\}}\,\psi_K^{(+)}(k\rho)\right] H_K^{\nu_0}(\Omega) \tag{44}$$

where $S_{\{\mu\},\{\nu_0\}}$ is the scattering matrix element describing the inelastic coupling between the two- and three-cluster channels, and $H_K^{\nu_0}(\Omega)$ is the HH. One can find in [57] details on the asymptotic behavior of wave function for three-particle systems.

The total cross-section is given by the expression

$$\sigma(E) = \frac{\pi}{k_0^2} \sum_{L,S} \frac{(2L+1)(2S+1)}{4} \sum_{\nu_0} \left| S_{\{\mu\},\{\nu_0\}} \right|^2 \tag{45}$$

with S the total spin of the six-nucleon system.

As discussed in Section 2.1, the asymptotic solutions for incoming and outgoing waves can be written as

$$\psi_{\mathcal{L}}^{(\pm)}(k\rho) = \frac{1}{\sqrt{k}} W_{\pm i\eta,\lambda}(\pm 2ik\rho) / \rho^{\frac{\sigma-1}{2}} \tag{46}$$

where W are the Whittaker functions, and η the Sommerfeld parameter (for the parameters $\mathcal{L}$, λ, σ and η for two- and three-cluster channels: see Table 1).

By using the correspondences (41) and (42) we can now define the boundary conditions for the expansion coefficients

$$\begin{aligned} c_{n,L} &\simeq \sqrt{r_n}\left[\psi_L^{(-)}(k_0 r_n) - S_{\{\mu\},\{\mu\}}\, \psi_L^{(+)}(k_0 r_n)\right] \\ d_{n_\rho,\nu_0} &\simeq \rho_n^2 \left[-S_{\{\mu\},\{\nu_0\}}\, \psi_K^{(+)}(k\rho_n)\right] \end{aligned} \tag{47}$$

or equivalently

$$\begin{aligned} c_{n,L} &\simeq c_{n,L}^{(-)} - S_{\{\mu\},\{\mu\}}\, c_{n,L}^{(+)} \\ d_{n_\rho,\nu_0} &\simeq -S_{\{\mu\},\{\nu_0\}}\, c_{n_\rho,\nu_0}^{(+)} \end{aligned} \tag{48}$$

using the notations

$$\begin{aligned} c_{n,L}^{(\pm)} &\simeq \sqrt{r_n}\psi_L^{(\pm)}(k_0 r_n) \\ d_{n_\rho,\nu_0}^{(\pm)} &\simeq \rho_n^2\, \psi_K^{(\pm)}(k\rho_n) \end{aligned} \tag{49}$$

The matching of internal and asymptotic regions is equivalent to the one in the traditional Resonating Group Method (RGM). The correspondence between the matching point in coordinate space for RGM and in function space for MJM is easily made (see [13]) through the value of the classical oscillator turning point $r_n = b\sqrt{4n+2L+3}$ for 2-cluster systems and $\rho_n = b\sqrt{4n_\rho+2K+6}$ for 3-cluster systems. An appropriate value for the matching point can be obtained by choosing sufficiently large values for the total number of oscillator quanta $N = 2n + L = 2n_\rho + K$ in the internal region.

4.1.3 Shape Analysis

The HH's can reveal information on the spatial distribution of clusters and of the reaction dynamics.

These harmonics define a probability distribution in 5-dimensional coordinate (momentum) space for fixed values of hyperradius:

$$dW^5_{\nu_0}(\Omega) = \left|Y_{\nu_0}(\Omega)\right|^2 d\Omega, \quad dW^5_{\nu_0}(\Omega_k) = \left|Y_{\nu_0}(\Omega_k)\right|^2 d\Omega_k \tag{50}$$

By analyzing the probability distribution, one can retrieve the most probable shape of three-cluster shape or "triangle" of clusters. A full analysis of a function of 5 variables is non-trivial and one usually restricts oneself to some specific variable(s). We integrate the probability distribution $dW^5_{\nu_0}(\Omega)$ over the unit vectors $\widehat{\mathbf{q}}_1, \widehat{\mathbf{q}}_2$ (resp. $\widehat{\mathbf{k}}_1, \widehat{\mathbf{k}}_2$)

$$\begin{aligned} dW_{\nu_0}(\theta) &= \int \left|Y_{\nu_0}(\Omega)\right|^2 \cos^2\theta \sin^2\theta d\theta \, d\widehat{\mathbf{q}}_1 d\widehat{\mathbf{q}}_2 \\ dW_{\nu_0}(\theta_k) &= \int \left|Y_{\nu_0}(\Omega_k)\right|^2 \cos^2\theta_k \sin^2\theta_k d\theta_k \, d\widehat{\mathbf{k}}_1 d\widehat{\mathbf{k}}_2 \end{aligned} \tag{51}$$

and introduce the (new) variable(s)

$$\mathcal{E} = \frac{q_1^2}{\rho^2} = \cos^2\theta, \quad \mathcal{E} = \frac{k_1^2}{k^2} = \cos^2\theta_k$$

In coordinate space these can be interpreted as the squared distance between the pair of clusters associated with coordinate $\mathbf{q}_1$, or, in momentum space, the relative energy of that pair of clusters. We obtain

$$\begin{aligned} W_{\nu_0}(\mathcal{E}) = \frac{dW_{\nu_0}(\theta)}{d\theta} &= \left|N_K^{(l_1,l_2)} \cos^{l_1}\theta \sin^{l_2}\theta P_n^{(l_2+1/2,l_1+1/2)}(\cos 2\theta)\right|^2 \cos^2\theta \sin^2\theta \\ &= \left|N_K^{(l_1,l_2)} (\mathcal{E})^{l_1/2} (1-\mathcal{E})^{l_2/2} P_n^{(l_2+1/2,l_1+1/2)}(2\mathcal{E}-1)\right|^2 \sqrt{\mathcal{E}(1-\mathcal{E})} \end{aligned} \tag{52}$$

This function represents the probability distribution for relative distance between the two clusters, resp. for the energy of relative motion of the two clusters. The kinematical factor $\cos^2\theta \sin^2\theta$ was included to make $W_{\nu_0}(\mathcal{E})$ proportional to the differential cross section in momentum space, provided the exit channel is described by the single HH $Y_{\nu_0}(\Omega)$.

In Fig. 12 we display $W_{\nu_0}(\mathcal{E})$ for some HH's involved in our calculations. These figures show that different HH's account for different shapes of the three-cluster systems. For instance, the HH with $K = 10$ and $l_1 = l_2 = 0$ prefers the two clusters to move with very small or very large relative energy, or, in coordinate space, prefers them to be close to each other, or far apart.

4.2 Results

Again we use the VP as the NN interaction. The Majorana exchange parameter m was set to be 0.54 which is comparable to the one used in [53]. The oscillator radius was set to $b = 1.37$ fm (as in [14, 19]) to optimize the ground state energy of the alpha-particle.

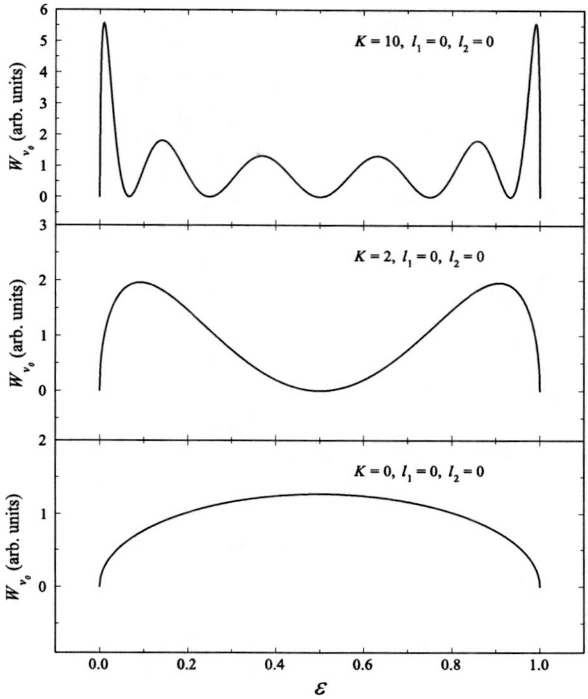

Fig. 12 Function $W_{\nu_0}(\mathcal{E})$ for $K = 0,\ 2$ and 10 and $l_1 = l_2 = 0$

The VP does not contain spin-orbital or tensor components so that total angular momentum L and total spin S are good quantum numbers. Moreover, due to the specific features of the potential, the binary channel is uncoupled from the three-cluster channel when the total spin S equals 1; this means that odd parity states $L^{\pi} = 1^-, 2^-, \ldots$ will not contribute to the reactions.

To describe the continuum of the three-cluster configurations we considered all HH's with $K \leq K_{max} = 10$. In Table 11 we enumerate all contributing K-channels for $L = 0$. For each two- and three-cluster channel we used the same number $n = n_{\rho} = N_{int}$ of basis functions to describe the internal part of the wave function Ψ_L. N_{int} then also defines the matching point between the internal and asymptotic part of the wave function. We used N_{int} as a variational parameter and varied it between 20 and 75, which corresponds to a variation in coordinate space of the RGM matching radius approximately between 14 and 25 fm. This variation showed only small changes in the S-matrix elements, of the order of one percent or less, and do not influence any of the physical conclusions. We have then used $N_{int} = 25$ for the final calculations as a compromise between convergence and computational effort. We also checked the impact of N_{int} on the unitarity conditions of the S-matrix, for instance the relation

$$\left|S_{\{\mu\},\{\mu\}}\right|^2 + \sum_{\nu_0} \left|S_{\{\mu\},\{\nu_0\}}\right|^2 = 1$$

Table 11 Number of Hyperspherical Harmonics for $L = 0$

N_{ch}	1	2	3	4	5	6	7	8	9	10	11	12
K	0	2	4	4	6	6	8	8	8	10	10	10
$l_1 = l_2$	0	0	0	2	0	2	0	2	4	0	2	4

We have established that from $N_{int} = 15$ on this unitarity requirement is satisfied with a precision of one percent or better. In our calculations, with $N_{int} = 25$, unitarity was never a problem. It should be noted that our results concerning the convergence for the three-cluster system with a restricted basis of oscillator functions agree with those of Papp et al [58], where a different type of square-integrable functions was used for three-cluster Coulombic systems.

In Fig. 13 we show the total S-factor for the reaction $^3H\left(^3H,2n\right)^4He$ in the energy range $0 \le E \le 200$ keV. One notices that the theoretical curve is very close to the experimental data. The total S-factor for the reaction $^3He\left(^3He,2p\right)^4He$ is displayed in Fig. 14. It is also close to the available experimental data. The S-factor for both reactions is seen to be a monotonic function of energy, and does not manifest any irregularities to be ascribed to a hidden resonance. Thus no indications are found towards explaining the solar neutrino problem.

The astrophysical S-factor at small energy is usually written as

$$S\left(E\right) = S_0 + S_0' E + S_0'' E^2 \tag{53}$$

We have fitted the calculated S-factor to this formula in the energy range $0 \le E \le 200$ keV. For the reaction $^3H\left(^3H,2n\right)^4He$ we obtain the approximate expression:

$$S\left(E\right) = 206.51 - 0.53\,E + 0.001\,E^2 \quad \text{keV b} \tag{54}$$

and for $^3He\left(^3He,2p\right)^4He$ we find:

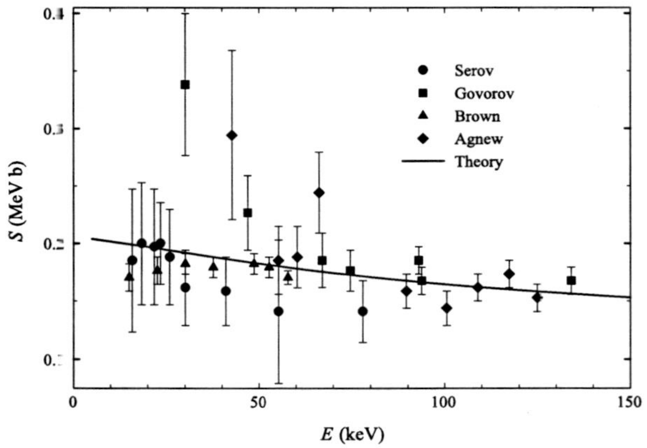

Fig. 13 S-factor of the reaction $^3H\left(^3H,2n\right)^4He$. Experimental data are taken from [59] (Serov), [60] (Govorov), [61] (Brown) and [62] (Agnew)

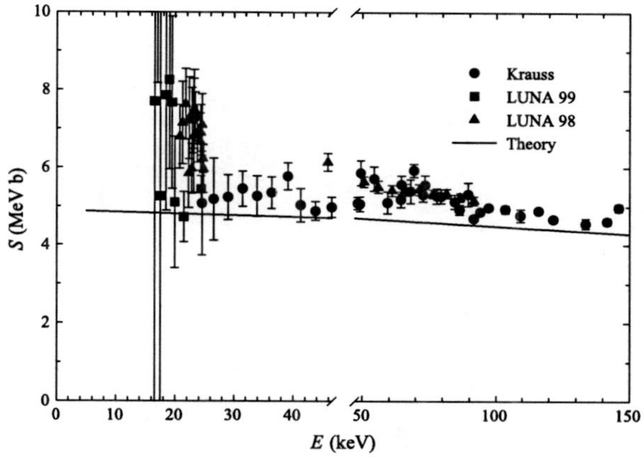

Fig. 14 S-factor of the reaction ${}^3He\left({}^3He,2p\right)^4He$. Experimental data are from [63] (Krauss), [64] (LUNA 99) and [65] (LUNA 98)

$$S(E) = 4.89 - 3.99\ E + 2.3\ \ 10^{-4}\ E^2 \quad \text{MeV b} \tag{55}$$

One notices significant differences in the S-factor for the ${}^6He$ and ${}^6Be$ systems. The NN-interaction induces the same coupling between the clusters of entrance and exit channels for both ${}^6He$ and ${}^6Be$. It is the Coulomb interaction that distinguishes both systems, and accounts for the pronounced differences in the cross-sections and S-factors.

We compare the calculated S-factor to fits of experimental results for the reaction ${}^3He\left({}^3He,2p\right)^4He$:

$$\begin{aligned} S(E) &= 5.2 - 2.8\ E + 1.2\ E^2 \quad \text{MeV b [66]} \\ S(E) &= (5.40 \pm 0.05) - (4.1 \pm 0.5)\,E + (2.3 \pm 0.5)\,E^2 \quad \text{MeV b [67]} \\ S(E) &= (5.32 \pm 0.08) - (3.7 \pm 0.6)\,E + (1.95 \pm 0.5)\,E^2 \quad \text{MeV b [68]} \end{aligned} \tag{56}$$

The constant and linear terms of the fit display a good agreement. The difference in energy ranges between the calculated ($0 \le E \le 200$ keV) and experimental ($0 \le E \le 1000$ keV) fits make it difficult to attribute any significant interpretation to the discrepancy in the quadratic term.

The HH's method now allows to study some details of the dynamics of the reactions considered. In Figs. 15 and 16 we show the different three-cluster K-channel contributions (W_{ν_0}) to the total S-factor of the reactions. In Fig. 15 these contributions (in % with respect to the total S-factor) are displayed for some fixed energy (1 keV), while Fig. 16 shows the dependency of W_{ν_0} (in absolute value) on the energy of the entrance channel. One notices that three HH's dominate the full result, namely the $\{K = 0; l_1 = l_2 = 0\}$, $\{K = 2; l_1 = l_2 = 0\}$ and $\{K = 4; l_1 = l_2 = 2\}$, and this is true in both reactions. The contribution of these states to the S-factor is more then 95%. There also is a small difference between the reactions ${}^3H\left({}^3H,2n\right)^4He$ and ${}^3He({}^3He,2p)^4He$, which is completely due to the Coulomb interaction.

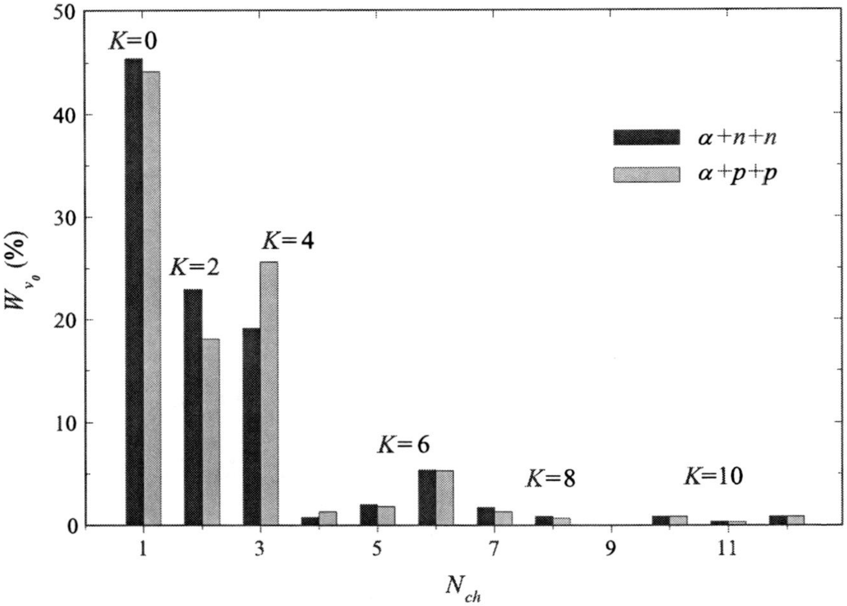

Fig. 15 Three-cluster channel contributions to the total S-factor for the reactions $^3H\left(^3H, 2n\right)^4He$ and $^3He\left(^3He, 2p\right)^4He$ in a full calculation with $K_{max} = 10$

The Figs. 15 and 16 yield an impression of the convergence of the results. We notice that the contribution of the HH's with $K > 6$ is small compared to the dominant ones. This is corroborated in Fig. 17 where we show the rate of convergence of the S-factor in calculations with K_{max} ranging from 0 up to 10. Our full $K_{max} = 10$ basis is seen to be sufficiently extensive to account for the proper rearrangement of

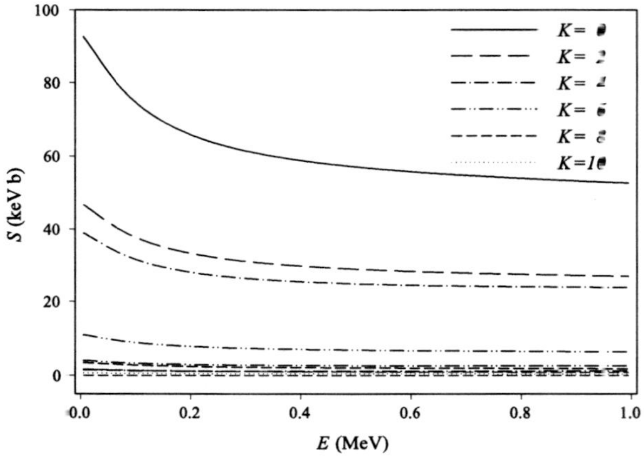

Fig. 16 Three-cluster channel contributions to the total S-factor of the reactions $^3H(^3H, 2n)^4He$ in a full calculation with $K_{max} = 10$, in the energy range $0 \leq E \leq 1000\,\text{keV}$

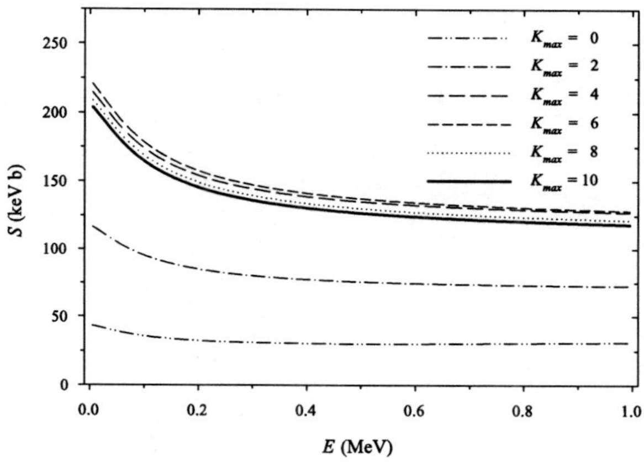

Fig. 17 Convergence of the S-factor of the reaction $^3H(^3H, 2n)^4He$ for K_{max} ranging from 0 to 10

two-cluster configurations into a three-cluster one, as the differences between results becomes increasingly smaller.

To emphasize the importance for a correct three-cluster exit-channel description, we compare the present calculations to those in [50] , where only two-cluster configurations $^4He + 2n$ resp. $^4He + 2p$ were used to model the exit channels. In both calculations we used the same interaction and value for the oscillator radius. In Fig. 18 we compare both results for $^3H(^3H, 2n)^4He$. An analogous picture is obtained for the reaction $^3He(^3He, 2p)^4He$.

4.2.1 Cross Sections

Having calculated the S-matrix elements, we can now easily obtain the total and differential cross sections. In this section we will calculate and analyze one-fold differential cross sections, which define the probability for a selected pair of clusters to be detected with a fixed energy E_{12}. To do so we shall consider a specific choice

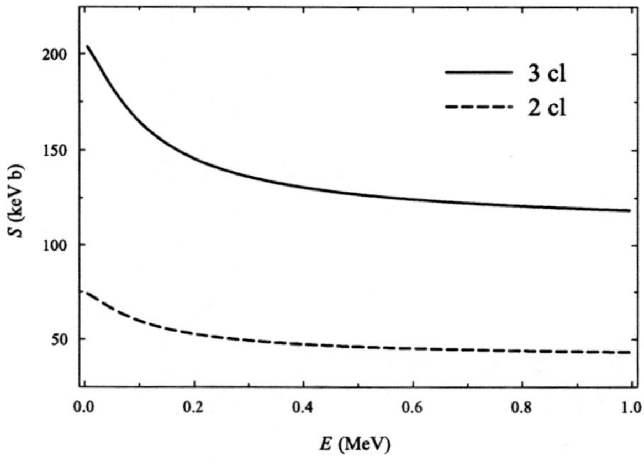

Fig. 18 Comparison of the S-factor of the reaction $^3H(^3H, 2n)^4He$ in a calculation with a three-cluster exit-channel and a pure two-cluster model

of Jacobi coordinates in which the first Jacobi vector $\mathbf{q}_1$ is connected to the distance between these clusters, and the modulus of vector $\mathbf{k}_1$ is the square root of relative energy E_{12}. With this definition of variables, the cross section is

$$d\sigma\left(E_{12}\right) \sim \frac{1}{E} \int d\widehat{\mathbf{k}}_1 d\widehat{\mathbf{k}}_2 \left| \sum_{\nu_0} S_{\{\mu\}\{\nu_0\}} Y_{\nu_0}\left(\Omega_k\right) \right|^2 \sin^2\theta_k \cos^2\theta_k d\theta_k \tag{57}$$

After integration over the unit vectors and substitution of $\sin\theta_k$, $\cos\theta_k$, $d\theta_k$ with

$$\begin{aligned} \cos\theta_k &= \sqrt{\frac{E_{12}}{E}}; \quad \sin\theta_k = \sqrt{\frac{E-E_{12}}{E}} \\ d\theta_k &= \frac{1}{2}\frac{1}{\sqrt{\left(E-E_{12}\right)E_{12}}} dE_{12} \end{aligned} \tag{58}$$

one can easily obtains $d\sigma\left(E_{12}\right)/dE_{12}$.

In Fig. 19 we display the partial differential cross sections of the reactions ${}^3H\left({}^3H,2n\right)^4He$ and ${}^3He\left({}^3He,2p\right)^4He$ for the energy $E = 10$ keV in the entrance channel. The solid lines correspond to the case of two neutrons (protons)

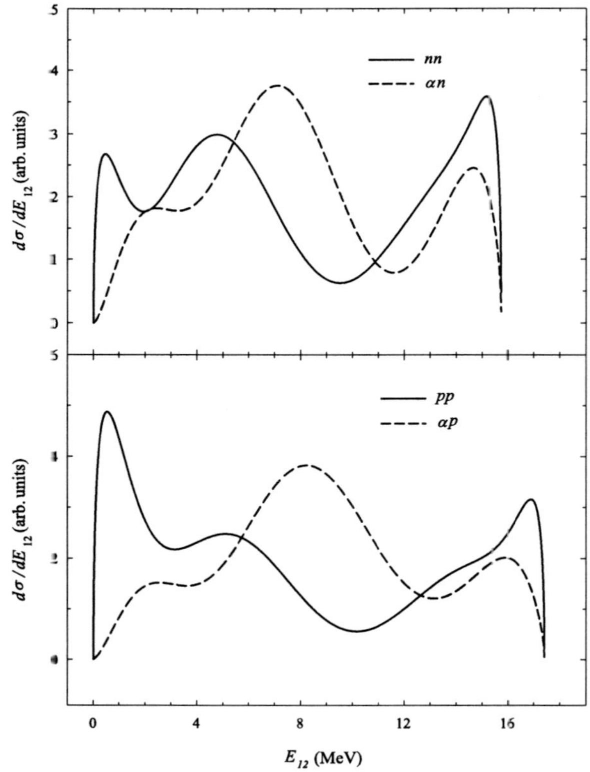

Fig. 19 Partial differential cross sections of the reactions ${}^3H({}^3H,2n)^4He$ and ${}^3He({}^3He,2p)^4He$

with relative energy E_{12}, while the dashed lines represent the cross sections of the α-particle and one of the neutrons (protons) with relative energy E_{12}.

We wish to emphasize the cross section in which two neutrons or two protons are simultaneously detected. One notices a pronounced peak in the cross section around $E_{12} \simeq 0.5$ MeV. This peak is even more pronounced for the reaction $^3He\left(^3He, 2p\right)^4 He$. It means that at such energy two neutrons or two protons could be detected simultaneously with large probability. We believe that this peak can explain the relative success of a two-cluster description for the exit channels at that energy. The pseudo-bound states of nn- or pp-subsystems used in this type of calculation then allows for a reasonable approximation of the astrophysical S-factor.

Special attention should be paid to the energy range 1-3 MeV in the $^4He + n$ and $^4He + p$ subsystems. This region includes $3/2^-$ and $1/2^-$ resonance states of these subsystems with the Volkov potential. In Fig. 19 (dashed lines) we see that it yields a small contribution to the cross sections of the reactions $^3H\left(^3H, 2n\right)^4 He$ and $^3He\left(^3He, 2p\right)^4 He$. This contradicts the conclusions of [53] and [54] where the $1/2^-$ state of the $^4He + N$ subsystem played a dominant role. We suspect this dominance to be due to the interplay of two factors: the weak coupling between incoming and outgoing channels, and the spin-orbit interaction.

In Fig. 20 we compare our results for the total proton yield (reaction $^3He\left(^3He, 2p\right)^4 He$) to the experimental data from [66]. The latter were obtained for incident energy $E\left(^3He\right) = 0.19$ MeV. One notices a qualitative agreement between the calculated and experimental data.

The cross sections, displayed in Figs. 19 and 20, were obtained with the maximal number of HH's ($K \leq 10$). These figures should now be compared to the Fig. 12,

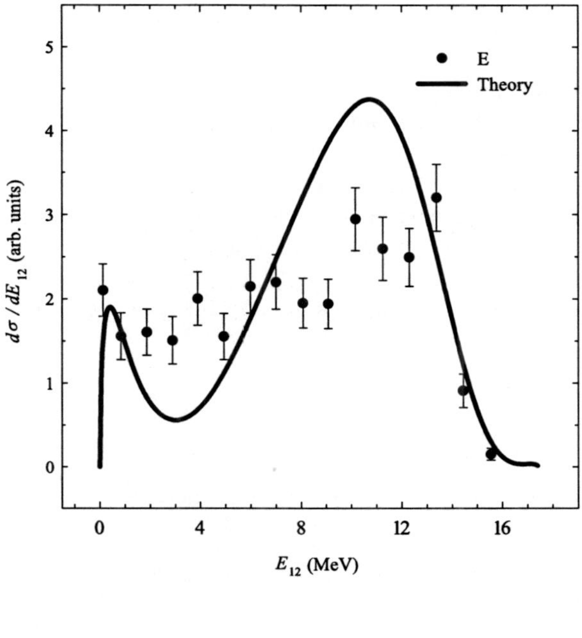

Fig. 20 Calculated and experimental differential cross section for the reaction $^3He(^3He, 2p)^4He$. Experimental data (E) is taken from [66]

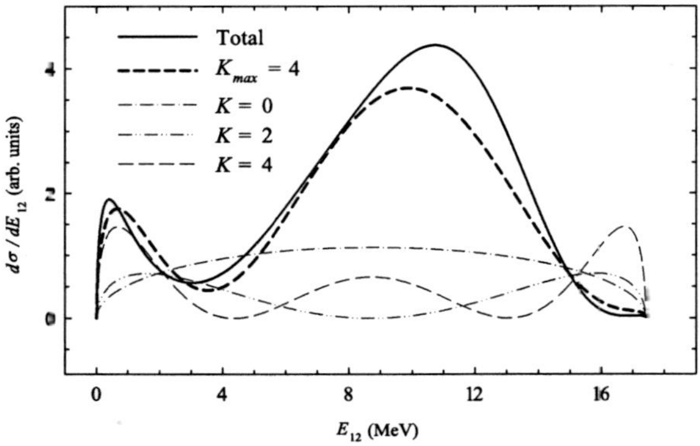

Fig. 21 Partial cross sections for the reaction $^3He(^3He, 2p)^4He$ obtained for individual $K = 0, 2$ and 4 components, compared to the coupled calculation with $K_{max} = 4$ and the full calculations with $K_{max} = 10$

which displays partial differential cross sections for a single K-channel. The cross sections, displayed in Figs. 19 and 20, differ considerably from those in Figs. 12 and comparable ones, even for those HH's which dominate the wave functions of the exit channel. An analysis of the cross section shows that the interference between the most dominant HH's strongly influences the cross-section behavior. To support this statement we display the proton cross sections obtained with hypermomenta $K = 0$, $K = 2$, $K = 4$ to those obtained with the full set of most important components $K_{max} \leq 4$ in Fig. 21. One observes a huge bump around 10 MeV which is entirely due the interference of the different HH components. We also included the full calculation ($K_{max} \leq 10$) to indicate the rate of convergence for this cross-section.

5 Conclusion

In this chapter we have presented a three-cluster description of light nuclei on the basis of the Modified J-Matrix method (MJM). Key steps in the MJM calculation of phase shifts and cross sections have been analyzed, in particular the issue of convergence. Results have been reported for ^6He and ^6Be. They compare favorably to available experimental data. We have also reported results for coupled two- and three-cluster MJM calculations for the $^3He\left(^3He, 2p\right)^4 He$ and $^3H\left(^3H, 2n\right)^4 He$ reactions with three-way disintegration. Again comparison indicates good agreement with other calculations and with experimental data.

References

1. W. Vanroose, J. Broeckhove, and F. Arickx, "Modified J-matrix method for scattering," *Phys. Rev. Lett.*, vol. **88**, p. 10404, Jan. 2002.
2. V. S. Vasilevsky and F. Arickx, "Algebraic model for quantum scattering: Reformulation, analysis, and numerical strategies," *Phys. Rev.*, vol. **A55**, pp. 265–286, 1997.

3. J. Broeckhove, F. Arickx, W. Vanroose, and V. S. Vasilevsky, "The modified J-matrix method for short range potentials," *J. Phys. A Math. Gen.*, vol. **37**, pp. 7769–7781, Aug. 2004.
4. G. F. Filippov and I. P. Okhrimenko, "Use of an oscillator basis for solving continuum problems," *Sov. J. Nucl. Phys.*, vol. **32**, pp. 480–484, 1981.
5. G. F. Filippov, "On taking into account correct asymptotic behavior in oscillator-basis expansions," *Sov. J. Nucl. Phys.*, vol. **33**, pp. 488–489, 1981.
6. G. F. Filippov, V. S. Vasilevsky, , and L. L. Chopovsky, "Generalized coherent states in nuclear-physics problems," *Sov. J. Part. Nucl.*, vol. **15**, pp. 600–619, 1984.
7. G. F. Filippov, V. S. Vasilevsky, and L. L. Chopovsky, "Solution of problems in the microscopic theory of the nucleus using the technique of generalized coherent states," *Sov. J. Part. Nucl.*, vol. **16**, pp. 153–177, 1985.
8. G. Filippov and Y. Lashko, "Structure of light neutron-rich nuclei and nuclear reactions involving these nuclei," *El. Chast. Atom. Yadra*, vol. **36**, no. 6, pp. 1373–1424, 2005.
9. G. F. Filippov, Y. A. Lashko, S. V. Korennov, and K. Katō, "$^6He +^6 He$ clustering of ^{12}Be in a microscopic algebraic approach," *Few-Body Syst.*, vol. **34**, pp. 209–235, 2004.
10. V. Vasilevsky, G. Filippov, F. Arickx, J. Broeckhove, and P. V. Leuven, "Coupling of collective states in the continuum: an application to ^{4}He," *J. Phys. G: Nucl. Phys.*, vol. **G18**, pp. 1227–1242, 1992.
11. A. Sytcheva, F. Arickx, J. Broeckhove, and V. S. Vasilevsky, "Monopole and quadrupole polarization effects on the α-particle description of ^8Be," *Phys. Rev. C*, vol. **71**, p. 044322, Apr. 2005.
12. A. Sytcheva, J. Broeckhove, F. Arickx, and V. S. Vasilevsky, "Influence of monopole and quadrupole channels on the cluster continuum of the lightest p-shell nuclei," *J. Phys. G: Nucl. Phys.*, vol. **32**, pp. 2137–2155, Nov. 2006.
13. V. S. Vasilevsky, A. V. Nesterov, F. Arickx, and J. Broeckhove, "Algebraic model for scattering in three-s-cluster systems. I. Theoretical background," *Phys. Rev.*, vol. **C63**, p. 034606, 2001.
14. V. S. Vasilevsky, A. V. Nesterov, F. Arickx, and J. Broeckhove, "Algebraic model for scattering in three-s-cluster systems. II. Resonances in three-cluster continuum of ^6He and ^6Be," *Phys. Rev.*, vol. **C63**, p. 034607, 2001.
15. V. Vasilevsky, F. Arickx, J. Broeckhove, and V. Romanov, "Theoretical analysis of resonance states in ^4H, ^4He and ^4Li above three-cluster threshold," *Ukr. J. Phys.*, vol. **49**, no. 11, pp. 1053–1059, 2004.
16. V. Vasilevsky, F. Arickx, J. Broeckhove, and T. Kovalenko, "A microscopic model for cluster polarization, applied to the resonances of ^7Be and the reaction $^6Li\left(p,^3He\right)^4He$," in *Proceedings of the 24 International Workshop on Nuclear Theory, Rila Mountains, Bulgaria, June 20–25, 2005* (S. Dimitrova, ed.), pp. 232–246, Sofia, Bulgaria: Heron Press, 2005.
17. F. Arickx, J. Broeckhove, P. Hellinckx, V. Vasilevsky, and A. Nesterov, "A three-cluster microscopic model for the ^5H nucleus," in *Proceedings of the 24 International Workshop on Nuclear Theory, Rila Mountains, Bulgaria, June 20–25, 2005* (S. Dimitrova, ed.), pp. 217–231, Sofia, Bulgaria: Heron Press, 2005.
18. V. S. Vasilevsky, A. V. Nesterov, F. Arickx, and P. V. Leuven, "Dynamics of $\alpha + N + N$ channel in ^{6}He and ^{6}Li," *Preprint ITP-96-3E*, p. 19, 1996.
19. V. S. Vasilevsky, A. V. Nesterov, F. Arickx, and P. V. Leuven, "Three-cluster model of six-nucleon system," *Phys. Atomic Nucl.*, vol. **60**, pp. 343–349, 1997.
20. Y. A. Simonov, *Sov. J. Nucl. Phys.*, vol. **7**, p. 722, 1968.
21. M. Fabre de la Ripelle, "Green function and scattering amplitudes in many-dimensional space," *Few-Body Syst.*, vol. **14**, pp. 1–24, 1993.
22. M. V. Zhukov, B.V. Danilin, D.V. Fedorov, J.M. Bang, J.I. Thompson, and J.S. Vaagen, "Bound state properties of borromean halo nuclei: ^{6}He and ^{11}Li," *Phys. Rep.*, vol. **231**, pp. 151–199, 1993.
23. F. Zernike and H. C. Brinkman, *Proc. Kon. Acad. Wetensch.*, vol. **33**, p. 3, 1935.
24. M. V. Zhukov, B. V. Danilin, D. V. Fedorov, J. S. Vaagen, F. A. Gareev, and J. Bang, "Calculation of ^{11}Li in the framework of a three-body model with simple central potentials," *Phys. Lett. B*, vol. **265**, pp. 19–22, Aug. 1991.

25. L. V. Grigorenko, B. V. Danilin, V. D. Efros, N. B. Shul'gina, and M. V. Zhukov, "Structure of the ^{8}Li and ^{8}B nuclei in an extended three-body model and astrophysical S_{17} factor," *Phys. Rev. C*, vol. **57**, p. 2099, May 1998.
26. L. V. Grigorenko, R. C. Johnson, I. G. Mukha, I. J. Thompson, and M. V. Zhukov, "Three-body decays of light nuclei: ^{6}Be, ^{8}Li, ^{9}Be, ^{12}O, ^{16}Ne, and ^{17}Ne," *Eur. Phys. J. A*, vol. **15**, pp. 125–129, 2002.
27. M. V. Zhukov, D. V. Fedorov, B. V. Danilin, J. S. Vaagen, and J. M. Bang, "^{9}Li and neutron momentum distributions in ^{11}Li in a simplified three-body model," *Phys. Rev. C*, vol. **44**, pp. 12–14, July 1991.
28. D. V. Fedorov, A. S. Jensen, and K. Riisager, "Three-body halos: gross properties," *Phys. Rev. C*, vol. **49**, pp. 201–212, Jan. 1994.
29. E. Garrido, D. V. Fedorov, and A. S. Jensen, "Breakup reactions of ^{11}Li within a three-body model," *Phys. Rev. C*, vol. **59**, pp. 1272–1289, Mar. 1999.
30. I. F. Gutich, A. V. Nesterov, and I. P. Okhrimenko, "Study of tetraneutron continuum states," *Yad. Fiz.*, vol. **50**, p. 19, 1989.
31. T.Ya. Mikhelashvili, Y. F. Smirnov, and A. M. Shirokov, "The continuous spectrum effect on monopole excitations of the ^{12}C nucleus considered as a system of a particles," *Sov. J. Nucl. Phys.*, vol. **48**, p. 969, 1988.
32. E. J. Heller and H. A. Yamani, "New L^2 approach to quantum scattering: theory," *Phys. Rev.*, vol. **A9**, pp. 1201–1208, 1974.
33. H. A. Yamani and L. Fishman, "J-matrix method: extensions to arbitrary angular momentum and to Coulomb scattering," *J. Math. Phys.*, vol. **16**, pp. 410–420, 1975.
34. Y. I. Nechaev and Y. F. Smirnov, "Solution of the scattering problem in the oscillator representation," *Sov. J. Nucl. Phys.*, vol. **35**, pp. 808–811, 1982.
35. A. Perelomov, "Coherent states for arbitrary Lie group," *Comment Math. Phys.*, vol. **26**, pp. 222–236, 1972.
36. A. M. Perelomov, *Generalized Coherent States and Their Applications*. Berlin: Springer, 1987.
37. F. Arickx, J. Broeckhove, P. V. Leuven, V. Vasilevsky, and G. Filippov, "The algebraic method for the quantum theory of scattering," *Am. J. Phys.*, vol. **62**, pp. 362–370, 1994.
38. D. V. Fedorov, A. S. Jensen, and K. Riisager, "Three-body halos: gross properties," *Phys. Rev.*, vol. **C49**, pp. 201–212, 1994.
39. F. Calogero, *Variable Phase Approach to Potential Scattering*. New-York and London: Academic Press, 1967.
40. V. V. Babikov, *Phase Function Method in Quantum Mechanics*. Moscow: Nauka, 1976.
41. B. V. Danilin, M. V. Zhukov, S. N. Ershov, F. A. Gareev, R. S. Kurmanov, J. S. Vaagen, and J. M. Bang, "Dynamical multicluster model for electroweak and charge-exchange reactions," *Phys. Rev.*, vol. **C43**, pp. 2835–2843, 1991.
42. B. V. Danilin, T. Rogde, S. N. Ershov, H. Heiberg-Andersen, J. S. Vaagen, I. J. Thompson, and M. V. Zhukov, "New modes of halo excitation in ^{6}He nucleus," *Phys. Rev.*, vol. **C55**, pp. R577–R581, 1997.
43. A. Csoto, "Three-body resonances in ^{6}He, ^{6}Li, and ^{6}Be, and the soft dipole mode problem of neutron halo nuclei.," *Phys. Rev.*, vol. **C49**, pp. 3035–3041, 1994.
44. N. Tanaka, Y. Suzuki, and K. Varga, "Exploration of resonances by analytical continuation in the coupling constant," *Phys. Rev.* vol. **C56**, pp. 562–565, 1997.
45. A. B. Volkov, "Equilibrum deformation calculation of the ground state energies of 1p shell nuclei," *Nucl. Phys.*, vol. **74**, pp. 33–58, 1965.
46. S. Aoyama, S. Mukai, K. Kato, and K. Ikeda, "Binding mechanism of a neutron-rich nucleus ^{6}He and its excited states," *Prog. Theor. Phys.*, vol. **93**, pp. 99–114, Jan. 1995.
47. S. Aoyama, S. Mukai, K. Kato, and K. Ikeda, "Theoretical predictions of low-lying three-body resonance states in ^{6}He," *Prog. Theor. Phys.*, vol. **94**, pp. 343–352, Sept. 1995.
48. F. Ajzenberg-Selove, "Energy levels of light nuclei A = 5–10," *Nucl. Phys.*, vol. **A490**, p. 1, 1988.

49. D. R. Tilley, C. M. Cheves, J. L. Godwin, G. M. Hale, H. M. Hofmann, J. H. Kelley, C. G. Sheu, and H. R. Weller, "Energy levels of light nuclei A=5, 6, 7," *Nucl. Phys. A*, vol. **708**, pp. 3–163, Sept. 2002.
50. V.S. Vasilevsky and I.Yu. Rybkin, "Astrophysical S factor of the reactions $t(t, 2n)\alpha$ and $^3He(^3He, 2p)\alpha$," *Sov. J. Nucl. Phys.*, vol. **50**, pp. 411–415, 1989.
51. Y. S. K. Varga and R. G. Lovas, "Microscopic multicluster description of neutron-halo nuclei with a stochastic variational method," *Nucl. Phys*, vol. A571, pp. 447–466, 1994.
52. S. Typel, G. Bluge, K. Langanke, and W. A. Fowler, "Microscopic study of the low-energy ^{3}He(^{3}He,2p)^{4}He and ^{3}H(^{3}H,2n)^{4}He fusion cross sections," *Z. Phys.*, vol. **A339**, p. 249, 1991.
53. P. Descouvemont, "Microscopic analysis of the $^3He(^3He, 2p)^4He$ and $^3H(^3H, 2n)^4He$ reactions in a three-cluster model," *Phys. Rev.*, vol. **C50**, pp. 2635–2638, 1994.
54. A. Csoto and K. Langanke, "Large-space cluster model calculations for the $^3He(^3He, 2p)^4He$ and $^3H(^3H, 2n)^4He$ reactions," *Nucl. Phys.*, vol. **A646**, p. 387, 1999.
55. K. Varga and Y. Suzuki, "Precise solution of few body problems with stochastic variational method on correlated gaussian basis," *Phys. Rev.*, vol. **C52**, pp. 2885–2905, 1995.
56. R. F. Barrett, B. A. Robson, and W. Tobocman, "Calculable methods for many-body scattering," *Rev. Mod. Phys.*, vol. **55**, pp. 155–243, Jan. 1983.
57. L. D. Faddeev and S. P. Merkuriev, *Quantum Scattering Theory for Several Particle Systems*. Dordrecht, Boston, London: Kluwer Academic Publishers, 1993.
58. Z. Papp, I. N. Filikhin, and S. L. Yakovlev, "Integral equations for three-body Coulombic resonances," *Few Body Syst. Suppl.*, vol. **99**, p. 1, 1999.
59. V. I. Serov, S. N. Abramovich, and L. A. Morkin, "Total cross section measurement for the reaction $T(t, 2n)^4He$," *Sov. J. At. Energy*, vol. **42**, p. 66, 1977.
60. A.M. Govorov, L. Ka-Yeng, G. M. Osetinskii, V. I. Salatskii, and I. V. Sizov *Sov. Phys. JETP*, vol. **15**, p. 266, 1962.
61. R. E. Brown and N. Jarmie, "Hydrogen fusion-energy reactions," *Radiat. Eff.*, vol. **92**, p. 45, 1986.
62. H. M. Agnew, W. T. Leland, H. V. Argo, R. W. Crews, A. H. Hemmendinger, W. E. Scott, and R. F. Taschek, "Measurement of the cross section for the reaction $T + T \to He^4 + 2n + 11.4$ MeV," *Phys. Rev.*, vol. **84**, pp. 862–863, 1951.
63. A. Krauss, H. W. Becker, H. P. Trautvetter, and C. Rolfs, "Astrophysical $S(E)$ factor of $^3He(^3He, 2p)^4He$ at solar energies," *Nucl. Phys.*, vol. **A467**, pp. 273–290, 1987.
64. R. Bonetti, C. Broggini, L. Campajola, P. Corvisiero, A. D'Alessandro, M. Dessalvi, A. D'Onofrio, A. Fubini, G. Gervino, L. Gialanella, U. Greife, A. Guglielmetti, C. Gustavino, G. Imbriani, M. Junker, P. Prati, V. Roca, C. Rolfs, M. Romano, F. Schuemann, F. Strieder, F. Terrasi, H. P. Trautvetter, and S. Zavatarelli, "First measurement of the $^3He\left(^3He, 2p\right)^4 He$ cross section down to the lower edge of the solar gamow peak," *Phys. Rev. Lett.*, vol. **82**, pp. 5205–5208, June 1999.
65. M. J. The LUNA C̃ollaboration, A. D'Alessandro, S. Zavatarelli, C. Arpesella, E. Bellotti, C. Broggini, P. Corvisiero, G. Fiorentini, A. Fubini, and G. Gervino, "The cross section of $^3He\left(^3He, 2p\right)^4 He$ measured at solar energies," *Phys. Rev.*, vol. **C57**, pp. 2700–2710, 1998.
66. M. R. Dwarakanath and H. Winkler, "$^3He(^3He, 2p)^4He$ total cross-section measurements below the Coulomb barrier," *Phys. Rev.*, vol. **C4**, pp. 1532–1540, 1971.
67. C. Arpesella et al., "The cross section of $^3He(^3He, 2p)^4He$ measured at solar energies," *Phys. Rev.*, vol. **C57**, pp. 2700–2710, 1998.
68. R. Bonetti et al., "First measurement of the $^3He +^3 He \to^4 He + 2p$ cross section down to the lower edge of the solar gamow peak," *Phys. Rev. Lett.*, vol. **82**, pp. 5205–5208, 1999.

Part V
Other Related Methods: Chemical Physics Application

A Generalized Formulation of Density Functional Theory with Auxiliary Basis Sets

Benny G. Johnson and Dale A. Holder

Abstract We present a generalized formulation of Kohn–Sham Density Functional Theory (DFT) using auxiliary basis sets for fitting of the electron density that significantly extends the range of applicability of this method by removing the current computational bottleneck of the exchange-correlation integrals. This generalization opens the door to the development of a new fitted DFT method that is directly analogous to a J-matrix method, allowing the exchange-correlation energy and potential of atomic and molecular systems to be calculated with an order of magnitude reduction in computational cost and no loss in accuracy. However, in contrast with prior approximate exchange-correlation methods, this computational advantage is realized within a rigorous theoretical framework as with other J-matrix methods. Generalized equations are presented for the self-consistent field energy, and example applications are discussed. In particular, it is shown that the stationary condition of the energy with respect to the fitting coefficients can be removed without penalty in complexity of the derivative theory, a characteristic drawback of most fitted exchange-correlation treatments. Results on accuracy and efficiency from an implementation of the new theory are presented and discussed.

1 Introduction

Quantum chemical methods, which explore chemical phenomena directly using rigorous quantum physics, have established themselves as valuable tools in the arsenals of many organic, inorganic, and physical chemists. From first principles, quantum chemistry can calculate the total energy and wavefunction of molecular systems, from which a prediction of virtually any experimentally observable chemical property can be obtained.

B.G. Johnson
Quantum Simulations, Inc., 5275 Sardis Road, Murrysville, PA 15668, USA
e-mail: johnson@quantumsimulations.com

A.D. Alhaidari et al. (eds.), *The J-Matrix Method*, 311–352.

All quantum chemistry methods involve computing the following electronic total energy expression:

$$E = E_1 + E_J + E_X + E_C \tag{1}$$

where the individual contributions are the one-electron, Coulomb, exchange and correlation energies, respectively. The last term, arising from the correlation of the motions of the electrons to each other, is the most difficult to treat, and most often is by far the most expensive part of a quantum chemistry calculation.

In recent years, Density Functional Theory [1–3] (DFT) has emerged as an accurate alternative first-principles quantum mechanical simulation approach in chemistry, which is very cost-effective compared with conventional correlated methods. Once practiced in chemistry only by a small group of specialists, the last decade has witnessed an explosion in growth in its usage, and DFT has become firmly established in mainstream chemistry research. Perhaps the most compelling testament to this fact is that the Nobel Prize in Chemistry in 1998 was awarded for work in DFT.

In several systematic validation studies DFT has exhibited good performance, and has often given results of quality comparable to or better than second-order perturbation theory but at much lesser cost. These encouraging results have provided incentive for the development of enhanced functionality within DFT programs. The computational attractiveness of DFT stems from its treatment of electronic exchange and correlation (XC) at the self-consistent field (SCF) level via a functional of the one-electron charge density (and sometimes its derivatives), rather than requiring a post-SCF calculation of correlation (*e.g.* perturbation theory, configuration interaction), which is very expensive relative to the initial SCF procedure.

The inclusion of correlation effects in an accurate fashion at the SCF level generally implies that the approximate density functional used in practice has a mathematical form that is quite complicated. Specifically, this has required computer implementations of DFT for practical molecular calculations to resort to numerical quadrature to evaluate the exchange-correlation integrals involving the density functional. Sophisticated and accurate techniques have been developed for this purpose [4].

The numerical calculation of the XC integrals must be approached mindfully, as there are potential difficulties with grid-based methods (associated with translational and rotational invariance) that do not arise in methods where all the requisite integrals are evaluated analytically, for example, as in Hartree–Fock (HF) theory. However, with care these can be rigorously handled, as we have shown [5,6]. Given this, the single major drawback of the numerical integration scheme is its large computational cost.

One of the most important areas of research in modern quantum chemistry is the continual search for ways to improve computational efficiency. There is a compelling need to broaden the spectrum of applicability of these methods, in order to bring their powerful advantages to bear on as wide a range of chemical problems as possible, maximizing their potential impact in research. Before proceeding, it is important to note that the computational challenges facing quantum chemistry today fall into two important categories:

- The need for quantum mechanical calculations on molecules that are as large as possible, given up to the maximum practical amount of computing resources: The exciting recent burst of progress in linear-scaling methods has come about in response to this problem.
- The need for quantum mechanical calculations in "real time": This involves computing energies and properties for small and medium-sized molecules in a matter of seconds or even less. Examples of the need for real-time simulation include applications in which calculations that are individually small must be repeated a very large number of times, such that the total time required becomes impractical. This includes mapping of molecular potential surfaces and dynamics calculations using quantum mechanical forces.

Each of these areas has it own unique challenges, but in all cases the old adage "Time is money" applies. It is the second category which is the primary concern motivating this work, though it will be seen that the solution proposed will benefit the first category substantially also. Usually it is the case that advances in the first area offer no benefit in the second, on the other hand.

To put this problem in further perspective, we will briefly look at the current status of advanced techniques for SCF methods, which include HF theory and the Kohn–Sham (KS) formulation of DFT [2]. Each term in equation (1) is usually computed separately (except in DFT where exchange and correlation are often combined into a single functional). For practical purposes, the one-electron term has a negligible cost and thus does not warrant further consideration. For many years, the challenge of the two-electron repulsion integrals, which comprise the Coulomb term in DFT (and the Coulomb and exchange terms in HF theory), was the main obstacle. Early on, and particularly relevant to the current work, this problem was addressed in DFT by projecting the charge density onto an auxiliary density basis set [7], resolving the four-center integrals into combinations of three-center and two-center integrals, thus avoiding the more costly four-center integrals altogether. More recently, much progress on four-center integrals methods has followed, while in recent years there have been many exciting breakthroughs in evaluation of the Coulomb and exchange energy in an amount of work that grows only linearly in the size of the system.

Besides evaluating the individual energy terms in equation (1), the only other computationally significant step is the energy and wavefunction optimization, required due to the iterative nature of solution of the SCF equations. This is often based on matrix diagonalization and is formally cubic in system size. Much progress has recently been made on reduced-scaling methods here also, speeding up this aspect of the calculation significantly for large molecules.

This systematic attack on the costliest computational terms has produced remarkable progress in methodology. Because of this, the XC contribution, already dominant for small and medium-sized systems, has greater significance than ever, often dominating for large molecules as well. Large production calculations on parallel computers have XC as the single most expensive component [8], and even sometimes as the dominant component. Practical experience with the NWChem program [9, 10], a state-of-the-art package designed specifically for massively parallel quantum chemistry, has shown that for large calculations the XC contribution can regularly account for a whopping 90% of the total job time [11].

This mandates that XC efficiency be the next area of intensive focus. The calculation of the numerical integrals to sufficient accuracy is now often the rate-limiting step, especially for vibrational frequencies (second derivatives) [12]. This is unfortunate, since the XC problem now restricts the scope of application of DFT methods in both areas of research mentioned above. This is unlike the two-electron integrals and diagonalization problems, since the latest advances there give some relief at the larger end, while at the smaller end these already go relatively fast and do not pose a serious problem to begin with.

What is the current state of the art in XC technology? Work has already been done that has successfully reduced the XC cost to asymptotically linear in system size [13, 14]. While an important definite advancement, it is not a solution to the problem at hand, for two reasons. First, as with all methods relying on some type of spatial cutoffs, it offers no improvement for small molecules, where the relevant length scale is simply not large enough, and in fact in this case can sometimes be a detriment over less complicated schemes. Second, for large molecules, the XC method is often competing with linear-scaling Coulomb methods that are *analytic*, rather than numerical. On the face of it, the notion that a linear-scaling numerical method might be slower than a linear-scaling analytic method is not surprising, even when comparing two different but both computationally substantial energy contributions. In fact, this is indeed the case, as we have observed in practical work on very large systems [9]. Therefore, it must be particularly noted that the above-discussed difficulties with XC cost represent the state of affairs at present, *after* and in spite of these linear-scaling XC breakthroughs.

The XC contribution presents its own unique computational challenges. The need to speed up this part of the calculation has long been recognized as important, and much effort has been focused with several approaches proposed. However, all have drawbacks of varying types that go along with them, which make the computational advantages they offer much less appealing, such that most have not gained widespread acceptance or use. These are briefly described in the following section.

It will become clear that a new, more powerful approach is needed. There is a significant opportunity for considerable advancement beyond current technology in the evaluation of density functional integrals. It is the development of an improved technique for the treatment of the exchange-correlation terms in DFT, one that significantly improves the computation time for systems of *all* sizes, which is the subject of this work. The innovation is in creating a new formulation of XC with auxiliary basis sets that is on a rigorous theoretical foundation and specifically that is variational, reducing the XC cost by considerably more than an order of magnitude and the total cost by close to an order of magnitude, while also overcoming the drawbacks from which other treatments of XC suffer. These are attributes that no other XC method has been able to achieve simultaneously.

To establish context for the present research, we begin with a brief survey of existing approaches to the XC computational problem, outlining the strengths and weaknesses of each. Then, the underlying theory for the present research is presented, which opens the way for a completely new approach that removes the XC bottleneck entirely, while possessing all the desirable aspects of present methods with none of their drawbacks.

2 Current XC Methodologies

2.1 Fitting of the XC Potential

One of the most significant early breakthroughs in computational DFT came for the Coulomb (J) term. As mentioned above, the four-center integral problem was addressed by fitting the charge density using a one-center auxiliary density basis [7]. This was carried out by minimizing the Coulomb self-energy of the difference density, subject to a charge conservation constraint, resolving the costly four-center integrals into combinations of cheaper three-center and two-center integrals. Though it should be noted this is an approximation to the original "exact" (four-center) J expression, experience with the resulting method has shown that this is acceptable in the majority of practical cases, as the results differ little, and the quality of predicted chemical properties does not suffer [15, 16]. One particularly attractive advantage of J fitting, in contrast with other methods that depend on spatial cutoffs, is that it speeds up *all* calculations substantially, not just for large molecules.

Salahub et al. recognized the advantages of J fitting and sought to carry them over to the XC problem by developing an auxiliary basis set method for XC [17]. While this motivation is an important one, their particular approach did not possess some of the key attributes of the J fitting method. Developing a robust fitting method for J is substantially simpler than for XC. The J integrals can be evaluated analytically, and the form of the J functional easily yields linear least-squares fitting equations, neither of which is the case for XC. Perhaps of greatest importance, however, as discussed later on, is that by back-substituting the optimized Coulomb energy into the energy expression, the variational principle is recovered.

For XC fitting, the problem is much more difficult due to the complicated non-linear forms of XC functionals. In Salahub's method, the XC operator is expanded directly in the auxiliary basis set, also resulting in a set of linear fitting equations. However, to do this, the dependence of the XC fit coefficients on the density had to be neglected, and the resulting method hence was *not* variational. This is serious, since loss of the variational principle creates serious problems when calculating energy derivatives. Unlike in other DFT methods, coupled-perturbed equations must now be solved at *first* order to obtain the correct result, highly undesirable for an SCF method. Since the reason for introducing XC fitting was to speed up the calculation, for derivatives this very expensive consequence was dealt with by simply leaving out the orbital gradient contribution altogether, introducing substantial errors into the calculated derivatives in order to avoid the extra computation.

Though this XC fitting method still has some proponents, it is not generally recognized as satisfactory. The goal of introducing fitting for the purpose of solving the XC cost problem is a laudable one, however. Furthermore, for J, this approach has proven quite successful in practice. Salahub's method does indeed achieve the goal of delivering substantial computational improvement, but it comes along with other attributes that are simply not acceptable. However, it remains that the concept of XC fitting has substantial potential, and this work investigates the prospect of developing it via an entirely new approach.

2.2 Matrix Representation of the Density

Zheng and Almlöf [18] have introduced a method in which the density and density gradient are represented as matrix elements in an auxiliary basis instead of on a grid. The matrix representations of these fundamental quantities are then directly used to construct the matrix representation of the XC potential, by diagonalization of the matrix representation of the density (not to be confused with the density matrix), application of the appropriate XC functional form, and back-transformation. The "auxiliary" basis initially used was the orbital basis itself.

This projection-based approach is intriguing and succeeds in removing the numerical integration aspect of the calculation, which has the potential for improving the calculation time. This approach does require the evaluation of four-center overlap integrals. Even though these are "only" overlap integrals, all other accepted XC methodologies effectively involve no more than two-center contributions. It has been recommended that four-index contributions to the XC potential be avoided, especially for second derivative calculations [19]. In any case, the need for four-index integrals will be seen to be unnecessarily complicated compared to the new approach developed here.

2.3 Quadrature Cutoff Schemes

Numerical integration for molecules is generally based on multi-center quadrature weights that in principle involve consideration of each nuclear center for each quadrature point. Together with a normalization requirement, this leads to a cubic formal cost dependence on molecule size. The most commonly used weights are those of Becke [4], which have shown to be reliable and stable, but are costly. Efforts to improve the cost of the weights rely on the weight function's property that nuclei far away from a given quadrature point generally do not make a significant contribution to the weight. This can safely be done [13] with the Becke scheme, since Becke's weights have been proven numerically stable.

However, Scuseria et al. recently proposed a different approach [20,21], in which Becke's inter-center weight function is replaced with a new function, which is flat for a greater distance from the nuclei than Becke's and steeper between them. Hence, by construction a greater fraction of the weight contributions are made insignificant and can be neglected. This approach to attempted computational improvement is particularly dangerous. Becke pointed out that careful attention was required when optimizing the steepness of the weight function [4], as computational expediency and numerical accuracy are forces working in opposite directions in numerical integration. Replacing Becke's function with a significantly more precipitous one decreases numerical stability, potentially to the point of failure, since it is a direct move towards a "step" function as the partitioning between pairs of nuclei.

In practical molecular applications, independent tests by other researchers [22] have not confirmed that the Scuseria scheme is capable of delivering the improvements promised without seriously degrading the results obtained. In one respect, therefore, this method is reminiscent of what was perhaps the earliest attempt to negotiate the XC bottleneck, by simply reducing the number of quadrature points until the cost becomes manageable. Though Scuseria's approach is more sophisticated, it fails for the same reason, namely that computational gains are made by directly eroding the effectiveness of the grid as a numerical integrator. This leads to results that are at best uncertain and at worst nonsensical, especially for derivatives, as has previously been noted [5].

2.4 Analytic Integration

Analytic integration or "gridless" DFT has recently been investigated as a possible means of substantially reducing the XC cost. However, solving this problem has so far been easier said than done; the extremely complicated forms of most density functionals are what prompted the move to numerical integration in the first place. Though an obvious area for investigation, progress has been much slower than desired. Limited progress has been made in integrating the Dirac uniform electron gas functional [23], the simplest exchange functional. Even in this case, though, to achieve an analytic method an approximate functional form had to be introduced, so the objective of analytically integrating the original target was not entirely achieved.

A more significant drawback, however, is that the technique developed is specific to the Dirac functional only. Other useful XC functionals have far more complicated mathematical forms than this, and so far no real insight has been gained on how to integrate these, as had originally been hoped. The lack of a general analytic procedure is a significant impediment, since in searching for an analytic method one must start from scratch with each functional. This makes analytic integration not as attractive as it first seems, since not only has analytic integration so far proven intractable, but even if it *were* available the development costs for building *ad hoc* energy, derivative, and property programs for *each* individual density functional would be truly staggering.

In this respect, numerical integration has a decided advantage in that it *is* a fully general procedure, capable of integrating all functionals currently used, and easily applied to new functionals as they become available. All functionals can make use of the large body of sophisticated computational machinery that has been developed and optimized over the years for a wide range of calculations, and further advancements in numerically based XC techniques benefit *all* functionals immediately.

At the present time, no general procedure for analytic integration appears forthcoming on the horizon. Though one may someday be discovered, the current focus of research on this topic has actually shifted away from trying to integrate established functionals analytically, to a different goal of developing entirely new functionals having a form which *can* be analytically integrated [24]. This places yet another significant constraint into the process of trying to develop chemically useful functionals. From the results in this area so far [23], the level of approximation needed to obtain an analytically integrable functional will likely be even greater that that of using an auxiliary basis set, and certainly far larger in magnitude than the error in the numerical integration schemes already being used.

None of the above methods has gained widespread acceptance or use because of various drawbacks that prevent them from becoming a truly complete solution to the XC cost problem. Particularly, most of the "fast" methods derive their advantage from exploiting the simplification of electronic interactions at large distances, and thus cannot offer any benefit until the molecule in question reaches a certain, usually rather sizeable threshold. This does not help with the goal of real-time calculations on smaller molecules.

In fact, the method discussed so far that has been by far the most successful is not an XC method at all, but rather the J fitting method. By now, this method is well established and proven. The reason J fitting "caught on" whereas XC fitting did not is because J fitting is based on a solid theoretical foundation. Building a rigorous framework for XC fitting is a harder problem than for J fitting, but judging by the success of the latter, it would suggest that there are substantial gains to be had by finding a proper extension to XC. One specific aspect of J fitting we would like to achieve for XC is that it benefits *all* calculations, not just large ones.

We believe XC fitting warrants serious further examination. The research reported here is the development and implementation of a *rigorous* theory of XC with auxiliary basis sets. By solving the XC cost problem, the last remaining hurdle in making numerical integration a fully general and completely acceptable procedure would be overcome, and the applications of computational DFT would be broadened considerably.

3 Theory

For the purpose of achieving our goal, we will develop a generalization of KS DFT with the inclusion of fitted densities, as presented in this section. The notation below is mostly standard, but with some changes to the fitting equations to better illustrate the connection of these with the orbital equations. To avoid cluttering the notation, alpha and beta spin labels are not written explicitly, as these are not central to the points we are investigating. They can easily be added to the equations as appropriate by direct analogy with standard unrestricted HF theory.

The opportunity to improve the XC computational cost can be seen from a few general yet important observations about the DFT SCF equations. Throughout the following analysis we deliberately refrain from specifying a definite form for the energy expression, so that the results will hold for all energy functionals meeting the assumed criteria. It is given only that the electronic total energy, denoted E, is a functional of the molecular orbitals (MO's) and the fitted density. The former is constructed from a set of atomic orbital (AO) basis functions ϕ_μ and MO coefficients $C_{\mu i}$, while the latter is constructed from density basis functions ϕ_m and fitting coefficients p_m.

The MO's are determined by minimizing the energy with respect to them, subject to certain constraints. For our purposes, we may take the constraints to be that the same-spin MO's be orthonormal. In matrix form, this is

$$\mathbf{C}^T\mathbf{S}\mathbf{C} = \mathbf{1} \tag{2}$$

where $\mathbf{S}$ is the AO basis overlap matrix. For our purposes it is usually more convenient to write the orbital dependence of the energy in terms of the density matrix instead of the MO coefficients. The density matrix $\mathbf{P}$ is

$$P_{\mu\nu} = \sum_{i}^{occupied} C_{\mu i} C_{\nu i} \tag{3}$$

which is related to the one-particle charge density as

$$\rho(\mathbf{r}) = \sum_{\mu\nu} P_{\mu\nu}\phi_\mu(\mathbf{r})\phi_\nu(\mathbf{r}) \tag{4}$$

As for the fitted density, though currently in practice the same particular set of fitting equations is always used, at this point we will avoid specifying how the fitting coefficients are determined. It suffices to say only that they are found by solving an additional set of equations, of sufficient number to determine them uniquely. In the scope of this analysis, the determining equations could be in principle whatever we choose, but historically and practically these have been obtained by minimization of some function, which we will denote as Z, again possibly subject to some constraints. Though any suitable function could be used, in practice Z is always a least-squares function of some type, *e.g*. squared error in the density or electric field, with the only common constraint on the fit coefficients being charge conservation. Note particularly that Z does not have to be the electronic energy E. We will return to this point later.

Suppose we view the fitting coefficients p_m as independent parameters that are on equal footing with the density matrix elements, rather than as auxiliary intermediate quantities which help generate the Fock matrix, as is usually the case. Then, determining the best MO's and fit coefficients, and hence the energy, involves solving two sets of minimization equations that are coupled, the results of which are substituted into the functional E. Thus, E and Z are both functions of $\mathbf{P}$ and $\mathbf{p}$, with E optimized with respect to $\mathbf{P}$ and Z optimized with respect to $\mathbf{p}$.

What do these optimization equations look like? Though E and Z are still deliberately unspecified, investigating the consequences of the constrained minimization requirements is nonetheless quite revealing. For the orbital equations, if we define the Fock matrix **F** as the energy derivative

$$\mathbf{F} = \frac{\partial E}{\partial \mathbf{P}} \tag{5}$$

then it is straightforward to show that minimization of E subject to equation (2) leads to

$$\mathbf{FC} = \mathbf{SC}\boldsymbol{\varepsilon} \tag{6}$$

where $\boldsymbol{\varepsilon}$ is a diagonal matrix of orbital eigenvalues (using canonical orbitals for convenience). The derivation is omitted for brevity, but it is easily seen by a simple generalization of the well-known derivation of the Roothaan–Hall equations for HF theory in a finite basis set. Of course, this is the same familiar form of the SCF equations.

This is a key result, since it shows that with the general definition of the Fock matrix in equation (5), the form of equation (6) is a consequence solely of minimization of the energy subject to the orthonormality constraints, and is *independent* of the actual form of the energy functional E. Furthermore, and particularly relevant here, this result holds whether a fitted density is employed or not. This is because the orbital constraint does not explicitly depend on the fit coefficients, and thus the form of the orbital equations is unchanged upon the introduction of fitting. All explicit dependence on the fitted density and on the particular form of E is contained entirely within **F**.

Minimization of the fit functional gives rise to another set of equations, which are coupled to the orbital equations. Introducing the partial derivatives of the fit functional

$$\mathbf{f} = \frac{\partial Z}{\partial \mathbf{p}} \tag{7}$$

then we must solve

$$\mathbf{f} = \lambda \mathbf{s} \tag{8}$$

$$\mathbf{p} \cdot \mathbf{s} = N_e \tag{9}$$

where N_e is the number of electrons the fitted density is constrained to hold, **s** is the vector of integrals of the density basis functions (*i.e.* the amounts of charge they contain), and λ is a Lagrange multiplier. Similarly to the orbital equations, all explicit dependence on Z is contained in **f**. The usual SCF equations are no longer self-contained, since they now depend (through **F**) on **p**, but are simply augmented with the fit equations and solved simultaneously with them.

What are the implications of these results, and why are they important? The property of invariance of the SCF equations to E and Z is not new; however it has apparently not been noted that this property contains an opportunity to improve the computational efficiency of DFT considerably. The focus of this research is to determine how this property can be exploited to computational advantage in the XC problem, which has not been done before.

The key is that these equations form a rigorous theoretical framework of a variationally optimized theory for *any* choice of E and Z. It is useful to compare this formulation to the previous DFT theories discussed. In all previous cases, pains were taken to ensure the fitting equations are linear in the fit coefficients. While the motivation for doing so, namely that the resulting equations are easily solved, is a reasonable one, it is in fact overly restrictive, and can be so to the point of detriment of the overall theory. For Coulomb fitting, it is relatively straightforward to find a fit functional Z that yields linear least-squares fitting equations (the self-energy of the residual density), and to construct a variational theory using it. It is therefore not surprising that this fitted method has gained the widest use.

For XC fitting, the problem is much more difficult due to the complicated nonlinear forms of XC functionals. Achieving linear XC fit equations in prior methods [17] required that the dependence of the fit coefficients on the MO's be neglected, and hence the resulting method was not variational. This is a heavy price to pay, and we argue that it is an unacceptable trade. The computational advantage afforded is overwhelmed by the loss of ability to calculate derivative quantities that correspond to the computed energy (without resorting to costly coupled-perturbed equations at first order, which is not done since it would defeat the original purpose of fitting).

How can the situation be rectified? Within the present formulation, these concerns do not apply. In particular, as has been shown, it is not necessary to restrict the fit equations to a linear form, as in Coulomb fitting, in order for the variational principle to be retained. This fact has possibly not been previously noted, and has certainly not been used to advantage. Continuing the given derivation leads to an analytic derivative theory that is consistent with the energy, for which coupled-perturbed equations can be avoided at the first derivative level, and which has a cost scaling of the same order as the corresponding SCF calculation. This is very attractive, as these are extremely important properties of HF and regular KS theories. The only question for a particular choice of E and Z is whether the resulting equations can be solved in a computationally expedient manner. However, since E and Z are as yet still unspecified, we will use this degree of freedom to advantage to search for an improved DFT energy expression that removes the XC computational bottleneck without sacrificing the accuracy of the original theory.

Let us assume for the moment that this goal can in fact be achieved. Why would it be better than existing approaches? The benefits are numerous. The new proposed theory will:

- Substantially improve cost for all molecular sizes
- Obey the variational principle
- Possess numerical stability

- Allow large grids to be used when necessary
- Immediately apply to all commonly used functionals
- Ensure that development of analytic derivative theory poses no problems

As is evident from the survey of existing approaches, no currently used method possesses all of these advantages. Importantly, the proposed theory will achieve all of these without suffering any of the limitations of the other methods reviewed.

4 Method Development

The next goal is to use this new formulation to find a modified energy expression that significantly improves the cost of the XC calculation. On its own the formulation does or does not say whether a particular choice of energy functional will lead to equations that are useful or even equations that can be solved. This is not a detriment; to the contrary, the new formulation accommodates a great deal of flexibility for experimenting with development of improved energy expressions. For example, the J fitting equations were carefully set up as linear equations in such a way that the variational principle is retained. Linearity is attractive since the fit equations can be solved exactly at each SCF iteration (for the current set of MO's, that is); however it is encouraging to note that this is in fact not a specific requirement in order to have a variational fitted theory that is also suitable for practical computations. In that event, the fit equations may need to be solved self-consistently as well, but on its own, this is not particular cause for concern. After all, this is successfully dealt with routinely in traditional SCF methods.

Note particularly that the new formulation also encompasses "regular" HF and non-fitted KS theory as special cases, as is easily seen by considering the case where E has no dependence on $\mathbf{p}$. Variational Coulomb fitting is encompassed by the new formulation as well. This is a bonus, since it suggests that established computational machinery for solving the SCF equations can be used across different formulations of the energy expression. Programs that have long used the exact same extrapolation code to accelerate the convergence of both HF and KS calculations have already evidenced this, for example. Some modifications will be required when introducing a new method, but starting with a good framework these can likely be minimized.

Having set the stage, let us attempt to find a modified energy expression that achieves our objective. First, it is logical that any attempt to re-introduce fitting in the XC term should be done in conjunction with J fitting, since otherwise the two-electron integral problem will recur, especially for smaller calculations to be done in real time. Actually specifying an appropriate improved form for XC fitting is more difficult than for J fitting, however. The first obvious direct analogies to J fitting fail. For example, one might first consider substituting the difference density into the XC functional and optimizing, as was done for J. This will not work, since the difference density is not a true probability density (it can, and with normalization, will have negative regions), and hence the XC "energy" of the difference density is undefined. The quadratic form of the Coulomb energy functional, on the other hand, is perfectly appropriate in this regard.

Directly minimizing the squared deviation from the usual XC energy of the XC energy obtained with the fitted density will not work either. Though the equations involved are well defined this time, the two-center and one-center XC contributions will not separate and allow the costliest terms to be eliminated, as happened in J fitting. Thus, the reason this particular approach is not useful is because it will not improve the computational cost.

Given that the derivation shows that the orbital and fit equations are on more of an equal footing than is usually considered, perhaps the most appropriate way to approach XC fitting is simply to substitute the fitted density into the XC functional instead of the orbital density, and variationally optimize the resulting expression. That is, the first new theory proposed for investigation involves the following modification of the XC energy expression:

$$E_{XC} = \int f_{XC}[\rho(\mathbf{r})]d\mathbf{r} \rightarrow \int f_{XC}[\tilde{\rho}(\mathbf{r})]d\mathbf{r} \tag{10}$$

where f_{XC} is the XC functional and the fitted density is given by

$$\tilde{\rho}(\mathbf{r}) = \sum_m p_m \phi_m(\mathbf{r}) \tag{11}$$

Evaluation of the density on the grid is one of the two computationally significant XC steps in an SCF calculation (the other is XC matrix element evaluation). This transition allows the *fitted* density in equation (11) to take the place of the orbital-based density in equation (4) on the grid. Likewise, derivation of the proper SCF equations as indicated by the generalized theory shows for this expression that the role of the usual two-center XC matrix integral involving the orbital basis functions is replaced by a simpler one-center integral involving the density basis functions. Using a local functional for convenience of illustration, this transition is

$$\int \phi_\mu(\mathbf{r})\phi_\nu(\mathbf{r})\frac{\partial f_{XC}[\rho]}{\partial \rho}d\mathbf{r} \rightarrow \int \phi_m(\mathbf{r})\frac{\partial f_{XC}[\tilde{\rho}]}{\partial \tilde{\rho}}d\mathbf{r} \tag{12}$$

This substantial simplification in number and complexity of the XC integrals is a direct illustration of how the generality of the theory presented can be exploited to computational advantage, as was our stated intent. Clearly, this simplification benefits the entire spectrum of molecular sizes.

The treatment of the fitting equations now must be specified. Perhaps the most obvious and aesthetic choice would be to take $Z = E$, *i.e.* optimize the energy with respect to the fit coefficients as well as the orbitals. However, an even simpler procedure is possible that meets all of the stated requirements. The original Coulomb fitted density [7], defined by taking

$$Z = \frac{1}{2}\iint [\rho - \tilde{\rho}](\mathbf{r}_1)\frac{1}{|\mathbf{r}_1 - \mathbf{r}_2|}[\rho - \tilde{\rho}](\mathbf{r}_2)d\mathbf{r}_1 d\mathbf{r}_2 \tag{13}$$

was implemented in equation (10). It was found that this fitting functional is itself capable of satisfying the goals of this research. Note that this procedure is not the same as the original Coulomb fitting scheme [7], which uses the original two-center density of equation (4) in the XC functional. Rather, the density determined by the Coulomb fit is substituted into the XC functional, but still with variational optimization of the energy according to the new SCF equations that follow from the generalized formulation given this particular choice of the energy expression.

Provided that this expression performs acceptably in other required respects, the use of equation (13) has considerable advantages over taking $Z = E$. Retaining the Coulomb fit form preserves the linearity of the fit equations, ensuring that they can be solved exactly on each SCF iteration. This is a substantial benefit, as it avoids the need to solve the fit equations self-consistently as well.

In principle, the only potential drawback to using a fit functional other than E is that that this gives rise to additional terms in the SCF equations and subsequent analytic derivative expressions. For example, the Fock matrix elements may be written as

$$F_{\mu\nu} = \frac{\partial E}{\partial P_{\mu\nu}} + \sum_m \frac{\partial E}{\partial p_m}\frac{\partial p_m}{\partial P_{\mu\nu}} \tag{14}$$

If E is also made stationary with respect to $\mathbf{p}$, the second term can be avoided. However, because of the particularly simple form of the Coulomb fit equations, and because these do not contain any explicit XC dependence, this term does not pose a problem. All the computational simplification for the XC terms afforded by $Z = E$ is retained, since it turns out that the additional terms arising because the energy is non-stationary with respect to the fit coefficients are "off the grid." That is, with the choice of equation (13) as the fit functional, the computation of the numerical XC terms is exactly the same; the difference lies only in how they are manipulated after the numerical integration, which is the most costly step, has been carried out. In practice the cost of these extra terms is insignificant compared to the grid-based work.

In the new formulation, the construction of the XC potential matrix becomes

$$F_{\mu\nu}^{XC} = \sum_m (\mu\nu|m)c_m^{XC} \tag{15}$$

The vector of XC potential coefficients in the fit basis is

$$\mathbf{c}^{XC} = \mathbf{A}^{-1}\frac{\partial E_{XC}[\tilde{\rho}]}{\partial \mathbf{p}} \tag{16}$$

where $\mathbf{A}$ is the matrix of two-center Coulomb integrals $(m|n)$. Note that equation (16) is of the exact same form as the usual fitted J matrix, but with a different right-hand side, namely the projections of the XC potential onto the fitting basis functions, as given on the right side of equation (12). This is a direct consequence of using the Coulomb fit functional variationally. Obtaining these coefficients merely involves solving an additional linear system and, again, is not significant in practice.

In summary, the new fitted XC method is most similar to the J-matrix method of Dunlap, et al. [7] discussed above, extended properly to the XC energy and potential as outlined here. In the initial implementation, the solution is not partitioned into separate spaces for treatment of the short-range and long-range interactions; all interactions are treated directly. Such partitioning can be an effective technique for improving the efficiency of larger calculations, and can be implemented in the future to enhance the new method even further. In this work, the method is implemented and tested for atomic and molecular systems; no inherent difficulty exists in implementing the method for systems with periodic boundary conditions. The criteria for choice of basis set are currently the same as with Dunlap's method [7] and other common methods in which the XC terms are not included in the expansion. Optimization of the expansion basis set specifically for J-matrix-like XC calculations is a topic for future research that can potentially yield further improvement. As discussed in the *Results and Discussion* section below, here we have begun by taking a sample of representative expansion basis sets already commonly used in J-matrix calculations to begin gaining insight into the effect of the choice of basis on the method.

5 Results and Discussion

In the bottom line, establishing feasibility of the new method comes down to satisfying criteria in three separate areas simultaneously: speed, accuracy, and convergence. In order to have the desired impact, the proposed method must be substantially faster than its predecessor, must be at least as accurate, *and* must be sufficiently reliable and robust with regard to achieving SCF convergence. To summarize the findings detailed below, the new method met and often exceeded expectations in these areas for all cases. It was shown that the XC cost can be reduced by more than an order of magnitude with no loss in accuracy.

5.1 Preliminary Evaluation

The first logical step in implementing the new method was to modify an existing DFT program to allow the new energy expression to be evaluated with densities already produced by the program. This allowed useful experimentation with the new energy expression before a self-consistent implementation was made. Though such densities do not optimize the new energy expression, experience has shown that such preliminary calculations have predictive value as to the quality of the fully self-consistent theory. This has been the case, for example, with using the HF density to test various DFT XC functionals [25, 26].

The implementation work was carried out within a development version of the NWChem quantum chemistry program from Pacific Northwest National Laboratory [9, 10]. NWChem provided the best starting point because of its strong framework for fitted DFT, including the standard fitted Coulomb method, as well as the existing non-variational fitted XC method for comparison purposes. The DFT package in NWChem was extended to take the standard Coulomb fit density already produced, evaluate it numerically on the grid and substitute it into an XC functional to obtain an XC energy. The resulting XC energy was used in place of the original XC energy to obtain the approximation to the energy of the new method as

$$E_{new}^{approx} = E_{SCF} - E_{XC}[\rho] + E_{XC}[\tilde{\rho}] \tag{17}$$

This simple expression allowed preliminary assessment of the quality of the new energy expression to reveal any serious deficiencies before proceeding further. In the initial implementation, symmetry was not exploited to facilitate the computation. As with other possible efficiency enhancements previously mentioned, this can readily be added as a future further improvement. The purpose of the initial investigation is to characterize the fundamental qualitative performance of the method, and as seen below, performance of the prototype is already quite good.

Pople's G2 neutral test set [27] of 148 molecules, which has emerged as a *de facto* standard benchmark set for DFT, was used to conduct a preliminary comparison of molecular XC energies and atomization energies using the *same* charge densities, *i.e.* from the Coulomb fit SCF method. The local S-VWN and gradient-corrected B-LYP functionals were used with the standard 6-31G* orbital basis set and Ahlrichs density basis set. Single points were carried out at the MP2/6-31G* reference geometries used in G2 theory. Zero-point corrections are not included in the calculated atomization energies. The uncorrected values are directly comparable, and the zero-point effect on the *difference* between the standard and new method is negligible. The results are summarized in Table 1, which shows the changes in the XC and atomization energies upon going to the new fit method.

These results are very encouraging. Use of the fitted density in place of the usual density for evaluation of the XC energy resulted in a change of less than 0.1% in all cases, with the mean change in XC energy on the order of only 0.01%. The absolute magnitudes of the energy changes are of course dependent on the size of the system; however, these are considerably smaller than those usually corresponding to a change in the orbital basis set, for example. Such small changes in the energy itself are a promising indication that chemical properties, which correspond to differences or derivatives of the total energy, will not be significantly changed.

This is indeed the case for the first test of atomization energies, calculated using the approximation to the new energy expression. The results, summarized in the second half of Table 1, show that for this test set the mean changes in atomization energy using the new method are less than 1 kcal/mol. This is considerably smaller than the inherent error in these DFT methods, and is also well under the threshold

Table 1 Changes in XC energies and atomization energies on G2 neutral test set resulting from use of fitted density in XC functional

	S-VWN/6-31G*/Ahlrichs	B-LYP/6-31G*/Ahlrichs
XC energies		
Mean % change	−0.0024	−0.0098
Mean % absolute change	0.0065	0.011
Range (h)	−0.0084−0.012	−0.011−0.0079
Range (%)	−0.062−0.015	−0.093−0.024
Atomization energies (kcal/mol)		
Mean change	0.73	0.32
Mean absolute change	0.84	0.48
Range	−1.20−5.22	−1.23−5.47
Mean % change	0.12	0.11
Mean % absolute change	0.23	0.24
Range (%)	−5.96−1.26	−7.03−1.71

of 2 kcal/mol generally considered "chemical accuracy" for atomization energies. The largest absolute changes in atomization energy of 5.22 and 5.47 kcal/mol occur for disilane, and still represent a less than 1% relative change for this molecule. The largest relative change occurs for Na_2, which has the smallest atomization energy in the test set. Here there is a change of 1.2 kcal/mol in a quantity that is less than 20 kcal/mol, still small in absolute terms.

For the remainder of the validation study a smaller molecular test set was used for practical manageability. This set is the G2 subset of 32 neutral molecules from a prior well-known DFT systematic validation study [28] that established the use of the G2 set as a standard benchmark for DFT. A methodology patterned after that study was used in this work. If the validation results on this smaller standard set continue to be as clearly and unambiguously successful as those in Table 1, the overall assessment of feasibility of the new method will be just as clear.

5.2 SCF Convergence

Given the success of initial testing of the new energy expression, a self-consistent implementation was carried out. Generalization of NWChem to implement the new method starting from its current capabilities was relatively straightforward. The routine evaluating the density in the fit basis developed in the prior objective was now included in the SCF procedure itself, rather than as a single point at the end. This is directly responsible for the computational improvement in one of the two dominant components of the XC cost. For spin-unrestricted calculations the Coulomb fit routine was generalized to fit the alpha and beta densities separately, since these are needed separately by the XC functional (only the total density is needed for the Coulomb fit).

An additional routine was written to evaluate the XC potential vector in the fit basis (the fitted analogue of numerical XC matrix integrals). This one-center routine was substituted for the usual two-center numerical integration for the XC matrix. This accounts for the computational savings for the other major component of the

XC cost. The resulting vector must be then transformed to the AO basis according to equations (15) and (16). Again, in direct analogy with the fitted J matrix method, the XC matrix in equation (15) uses the same three-center Coulomb integrals as comprise the fitted Coulomb matrix itself, and thus if desired the Coulomb and XC matrices can even be formed together.

In order for the method to be feasible, the introduction of the XC fit must not degrade the SCF convergence. The convergence behavior must be similar in nature to that of non-fitted XC methods in terms of number of orbital SCF iterations required and overall convergence robustness.

To assess the convergence behavior, the number of cycles required to achieve convergence was compiled for the molecular test set using four different DFT methods and two combinations of basis sets. The first method ("No fit") does not use fitting for the Coulomb or XC terms (*i.e.* four-center two-electron integrals and equation (4) in the XC terms), and is the method used in reference [28]. The second method ("J fit") is the standard Coulomb fit method [7], using three-center integrals for the Coulomb term but no approximation for XC. The third method ("Old J+XC fit") is the method of Salahub et al. [17], which additionally expands the XC potential in a second auxiliary basis. As previously discussed, this method was developed to address the cost of the XC terms but is unsatisfactory because the approximation introduced in the XC fit sacrifices the variational principle and leads to several undesirable consequences. The fourth method ("New J+XC fit") is the method of this work, which suffers from none of the theoretical and practical drawbacks of the previous XC fit method, and is most directly comparable in its approach to the Coulomb-only fit method.

The 6-31G*/Ahlrichs basis set combination was used, as well as the minimal STO-3G orbital basis set combined with the familiar DGauss A1 fitting basis, which is perhaps the most widely used fit basis and is considerably smaller than the Ahlrichs basis. A comment on this latter combination is warranted. The speed of the calculation is increased by using a smaller fitting basis; however, this also lessens the quality of the fit and hence possibly the results produced. It is therefore of particular interest to observe the performance of the method in the limit of small orbital and fitting basis sets, both for SCF convergence and for chemical properties in the next objective.

It is reiterated that the objective of the new formulation is to obtain equivalent DFT results at significantly reduced cost. Specifically, improving the agreement with experiment is not the purpose of the fit. If the XC fit were perfect, such as with a complete fitting basis, the new method would yield exactly the same results as without the fit. Thus, a large change in results due to the fit (one way or the other) would be viewed unfavorably. In practice, fit basis incompleteness does produce a change, but the important issue is whether the magnitude of the change is insignificant for practical purposes, such as in comparison to the error in the original method. It is specifically recognized that some of the methods used in the study, for example S-VWN/STO-3G, produce results that are in poor agreement with experiment and as such are not generally recommended. The point in this work is not whether these results agree with experiment, though, but whether the XC fit maintains the integrity of the method with these small basis sets. For establishing feasibility, if the performance standards can be met even at the smaller end of the spectrum of basis set size, as well as for the larger basis sets used, it is reasonable to expect this to continue throughout.

Table 2 presents the average number of SCF cycles required on a set of 30 of the 32 test molecules. The CN and NO molecules were omitted because of a well-known difficulty in obtaining convergence for π radicals with DFT. A solution to this problem is known [13], but is not implemented in NWChem. Manual assistance was required to obtain convergence for these systems, and thus the number of SCF cycles was not meaningful. This problem affects convergence for all DFT methods, and is no way inherent or specific to the new method.

These results clearly show no significant difference in the convergence behavior of the method from when there is no fit or only a Coulomb fit used. Convergence is obtained equally well whether the local or gradient-corrected functional is used. In fact, all four methods, including the non-variational old XC fit method perform similarly here, especially for the larger basis sets. This is most probably due to the larger fitting basis, as the deviation from variationality of the old XC fit method is proportional to the incompleteness of the fit basis. For the smaller basis sets, the only result that appears out of line with the others is for the old XC fit method with B-LYP, which is more than one cycle greater than all the others. Furthermore, convergence failed on CH_3 and NH_2 (not expected to be inherently problematic) for the old method without assistance, while no difficulty was observed with the other methods. It is well established that the old XC method has convergence problems in general, and these are apparently being brought out here with the smaller basis sets. The new XC fit method does not suffer from these problems, as it is completely variational.

The complete individual B-LYP results are included in the Appendix in Tables 12 and 13 for inspection. Comparing individual cases, no real pattern of discrepancy was observed between the new method and the Coulomb-only fit. The number of cycles required by each was most commonly the same, and the most common difference observed was only one cycle in either direction. There were also several

Table 2 Average number of SCF cycles to convergence for various DFT methods

	No fit	J fit	Old J+XC fit	New J+XC fit
S-VWN	7.8	7.8	8.1	8.0
B-LYP	8.5	8.5	8.1	8.4
Orbital basis = 6-31G*				
Density basis = Ahlrichs				
XC potential basis[1] = Ahlrichs				
S-VWN	8.3	8.3	8.7	7.6
B-LYP	8.8	8.5	10.3	8.9
Orbital basis = STO-3G				
Density basis = DGauss A1 Coulomb				
XC potential basis[1] = DGauss A1 Exchange				

[1]Used by Old J+XC fit method only.

instances, particularly with the smaller bases, of a difference of several cycles, but this was as likely to favor one method as the other, such that the resulting averages were essentially the same as seen above. Most likely this is not significant and is thought to arise from numerical stability issues in the NWChem SCF algorithm for calculations very near the convergence threshold. In any case, the results do not indicate any systematic disadvantage (or advantage) of the new method with respect to convergence.

5.3 Chemical Accuracy

The accuracy of the new method for chemical properties should be verified before the computational efficiency is assessed, since if accuracy is compromised it does not matter how fast the calculations go. Atomization energies, molecular geometries and dipole moments were calculated on the test set and compared with results from no fitting, Coulomb fitting only, and the old method of XC fitting, using the same functionals and basis sets as in the SCF convergence study. The results are statistically summarized in Tables 3–5 and the B-LYP data are presented in the Appendix.

To be considered chemically successful, the introduction of XC fitting should not perturb the results too far from the non-fitted XC case. Specifically, the method can be considered successful if the differences between the fitted and non-fitted results are small compared with the deviation of the original results (Coulomb fit only) from experiment, *i.e.* if the perturbation introduced is small compared to the original error in the method. A reasonable criterion is that the change in the error should ideally be no more than 10%. A change that is an order of magnitude smaller than the error in the result is effectively negligible since the accuracy of the result is unchanged for practical purposes. Achieving essentially equivalent results through considerably faster calculations is therefore a definite success.

Table 3 Mean errors in atomization energies (kcal/mol) for various DFT methods

	No fit	J fit	Old J+XC fit	New J+XC fit
S-VWN/6-31G*/Ahlrichs				
Mean error	46.9	46.9	46.9	47.1
Mean absolute error	46.9	47.0	47.0	47.1
Change				0.1 (0.3%)
B-LYP/6-31G*/Ahlrichs				
Mean error	12.3	12.4	12.3	12.4
Mean absolute error	12.9	13.0	12.9	12.9
Change				−0.1 (−0.7%)
S-VWN/STO-3G/A1				
Mean error	78.7	79.5	79.5	81.9
Mean absolute error	79.1	79.9	79.9	82.3
Change				2.4 (3.0%)
B-LYP/STO-3G/A1				
Mean error	34.8	35.6	35.6	38.2
Mean absolute error	40.4	41.1	41.1	43.7
Change				2.6 (6.4%)

Table 4 Mean errors in dipole moments (debye) for various DFT methods

	No fit	J fit	Old J+XC fit	New J+XC fit
S-VWN/6-31G*/Ahlrichs				
Mean error	−0.02	−0.02	−0.02	−0.01
Mean absolute error	0.26	0.26	0.26	0.25
Change				−0.01 (−3.2%)
B-LYP/6-31G*/Ahlrichs				
Mean error	−0.10	−0.10	−0.10	−0.09
Mean absolute error	0.25	0.25	0.26	0.24
Change				−0.01 (−2.5%)
S-VWN/STO-3G/A1				
Mean error	−0.53	−0.53	−0.52	−0.52
Mean absolute error	0.75	0.75	0.75	0.74
Change				−0.01 (−1.3%)
B-LYP/STO-3G/A1				
Mean error	−0.58	−0.58	−0.57	−0.57
Mean absolute error	0.77	0.77	0.76	0.76
Change				−0.003 (−0.4%)

The atomization energy results are summarized in Table 3, with the B-LYP individual results presented in Tables 14 and 15. These results were obtained by fully self-consistent calculation (as opposed to the atomization energies summarized in Table 1). Note that the "No fit" values are slightly different than those reported in reference [28] due to the use of the G2 reference geometries and omission of zero-point corrections for simplicity. First of all, we observe that all four methods produce similar results on this data set; the variation among the methods is small compared to their deviations from experiment. When this is the case, the "best" method is simply the one producing the results the quickest. The most relevant comparison is, again, the new method with the Coulomb-only results, *i.e.* before and after the new XC fit. The change in absolute error upon introduction of the new fit is reported as "Change" below. For the larger basis sets, the change in atomization energy caused by the fit is practically nonexistent – a change in the error of less than 1%. For the smaller basis sets, the change in error is greater, as expected, but is still well below the stated goal of less than 10% change in the error. Furthermore, the atomization energies by the new XC fit appear stable and well behaved, with no significant individual deviations observed. The new fit is a clear success here.

The performance on dipole moments is summarized in Table 4 with the individual B-LYP results in Tables 16 and 17 in the Appendix. As with the atomization energies, the performance of the new method is excellent. The magnitude of the change in mean absolute error is no more than one-hundredth of a debye unit for all combinations of functionals and basis sets. The maximum relative change in the error is only −3.2%, occurring for S-VWN with the 6-31G* and Ahlrichs basis sets.

The results for equilibrium geometries are summarized in Table 5 with the individual B-LYP results again presented in the Appendix (Tables 18 and 19). The geometry optimizations for the new method were carried out with forces obtained by

Table 5 Mean errors in equilibrium bond lengths (angstroms) and bond angles (degrees) for various DFT methods

Bond lengths	No fit	J fit	Old J+XC fit	New J+XC fit
S-VWN/6-31G*/Ahlrichs				
Mean error	0.015	0.015	0.022	0.017
Mean absolute error	0.022	0.022	0.022	0.023
Change				0.0003 (1.3%)
B-LYP/6-31G*/Ahlrichs				
Mean error	0.021	0.021	0.022	0.022
Mean absolute error	0.021	0.021	0.022	0.022
Change				0.001 (3.8%)
S-VWN/STO-3G/A1				
Mean error	0.036	0.037	0.037	0.041
Mean absolute error	0.048	0.049	0.049	0.053
Change				0.005 (9.4%)
B-LYP/STO-3G/A1				
Mean error	0.051	0.052	0.052	0.050
Mean absolute error	0.060	0.060	0.060	0.063
Change				0.002 (3.6%)
Bond angles				
S-VWN/6-31G*/Ahlrichs				
Mean error	−0.3	−0.5	−0.6	−0.4
Mean absolute error	2.4	2.3	2.9	2.2
Change				−0.1 (−4.5%)
B-LYP/6-31G*/Ahlrichs				
Mean error	−0.9	−1.0	−0.6	−1.0
Mean absolute error	2.6	2.6	2.9	2.7
Change				0.1 (4.9%)
S-VWN/STO-3G/A1				
Mean error	−2.3	−2.3	−2.3	−2.7
Mean absolute error	4.7	4.7	4.7	5.2
Change				0.5 (11%)
B-LYP/STO-3G/A1				
Mean error	−4.3	−4.3	−5.1	−5.6
Mean absolute error	6.3	6.3	6.9	5.6
Change				−0.7 (−12%)

finite difference of the energy (since analytic gradients are not yet available), while all other methods used analytic gradients as already implemented in NWChem.

The changes introduced by the new fit are slightly greater here than for the other properties. The change in mean absolute error is less than 10% for bond lengths by all methods and for bond angles when using the larger basis sets. For bond angles with the small basis sets the 10% threshold is slightly exceeded (11% and −12%). This is not problematic, as bond angles are known to be particularly sensitive to the theoretical method [28], and it still represents a small change on the order of half a degree.

Of more concern is that geometry optimization failed to converge for the new method for several molecules in the test set. There were four failures with the larger basis sets in Table 18 and eight failures with the smaller basis sets in Table 19. By itself, this might potentially indicate a problem with the method, but considered in the context of excellent performance elsewhere on SCF convergence and the other chemical properties studied, this does not seem plausible.

The reasons for these problems are not yet entirely confirmed, but all indications point to a problem with numerical error in the finite difference gradients adversely affecting the geometry optimization algorithm in NWChem. For the failed cases, optimization typically proceeded smoothly at first but then "stalled" at the end once the RMS force approached the convergence tolerance. No SCF convergence difficulties were observed throughout the optimization process, either with obtaining convergence or with the number of cycles required, at perturbed or unperturbed geometries; given this, it is unlikely that the problem with the gradient originates with the method itself. Many of the failures occurred for high-symmetry molecules, which would tend to exacerbate a problem with finite difference forces calculated by Cartesian perturbations. And finally, when geometry convergence was achieved the results are good, without problematic deviations. Given that SCF convergence is solid, it is doubtful that a problem will remain once the gradient is calculated analytically.

We note also that there was one convergence failure for the old XC fit method in Table 19, for H_2O_2 with the small basis sets. This is likely due to the non-rigorous treatment of the gradient by this method. The "analytic" forces computed neglect terms that account for non-stationary wavefunction, and hence the forces are not consistent with the energy. The problem is greater with smaller basis sets.

5.4 Computational Efficiency

Now that we have firmly established that the new energy expression easily meets its requirements in terms of SCF convergence and chemical accuracy, it remains only to assess the quantitative improvement in computational cost. Using the notation

$G =$ total number of molecular grid points
$N =$ number of orbital basis functions
$M =$ number density basis functions
$A =$ number of atoms
$N_G =$ average number of orbital basis functions having a significant value at a single grid point
$M_G =$ average number of density basis functions having a significant value at a single grid point
$A_G =$ average number of atoms making a significant contribution to the weight of a single grid point

the cost scaling of the various components of the XC calculation are as given in Table 6.

The rate-limiting steps are the evaluation of the charge density and the XC matrix elements on the grid, both of which scale without fitting as $O(GN^2)$ (*i.e.* cubic in system size) for small molecules, before spatial cutoffs can achieve any significant

Table 6 Cost scaling in XC calculations

	Without XC fit		With XC fit	
	Formal	Asymptotic	Formal	Asymptotic
Quadrature weights	GA^2	$G{A_G}^2$	GA^2	$G{A_G}^2$
Basis functions	GN	GN_G	GM	GM_G
Charge density	GN^2	$G{N_G}^2$	GM	GM_G
XC functional	G	G	G	G
XC matrix	GN^2	$G{N_G}^2$	GM	GM_G

benefit, while in the limit of large molecules the cost goes as $O(G{N_G}^2)$, which is linear in system size since N_G is a constant. For the new approach, the small-molecule cost instead goes as $O(GM)$, which is only quadratic in system size, reducing to $O(GM_G)$ for large molecules, which is also linear. The expected speedup in the rate-limiting steps starts out proportional to N^2/M, increasing with molecular size until approaching a limiting value of ${N_G}^2/M_G$, which is estimated in the range of 15–30 assuming a balanced combination of orbital and density basis sets.

The computational improvement was investigated as a function of molecule size, molecule shape, basis set, and whether a local or gradient-corrected functional is used. Again, exhaustive presentation of the results is not possible (or necessary); the relevant trends are summarized and discussed here. The bottom line is, once again, that the new method proved to be an overwhelming success.

Table 7 gives XC timing results for the non-fitted and fitted cases for a series of straight-chain alkanes, for a single representative SCF cycle at the B-LYP/6-31G*/Ahlrichs level of theory. This particular method was chosen for a detailed discussion here since it is likely the most similar to practical research work. The performance with gradient-corrected functionals is of interest since these yield better results and are also more expensive than local functionals. The 6-31G* orbital basis is very common in practical work, while the pairing with the Ahlrichs fit basis, on the larger end of the spectrum of density basis sets, is perhaps even slightly conservative in the ratio of fit functions to orbital functions. Furthermore, the one-dimensional molecular shape is the worst case for the new method (as is seen later); the asymptotic limit is reached sooner, and the maximum achievable speedup is not as great, since the amount of significant work per grid point is smallest for linear molecules. The speedups obtained with this combination are therefore likely to indicate the minimum improvement that can be expected in practice.

All timing calculations were performed in serial on a Sun Ultra 60 Model 2360 workstation (2 processors, 1 GB total memory) at Pacific Northwest National Laboratory. For all calculations the quadrature grid used was an Euler–Maclaurin–Lebedev grid with 50 radial points and 194 angular points per atom (before cutoffs). This grid is considerably smaller than the default grid in NWChem, but is more representative of a grid typically used for production work while still achieving perfectly acceptable accuracy. Had the default NWChem grid been used, the calculations would have been dominated by XC even more and the absolute improvement offered by the new method even greater.

Table 7 XC CPU timings (s) for a single SCF cycle on straight-chain alkanes, B-LYP/6-31G*/Ahlrichs

Molecule	CH_4		C_5H_{12}		$C_{10}H_{22}$		$C_{15}H_{32}$		$C_{20}H_{42}$		$C_{25}H_{52}$	
Orbital fns	23		99		194		289		384		479	
Fit fns	104		400		770		1140		1510		1880	
CPU time:	No fit	Fit	No fit	Fit	No fit	Fit	No fit	Fit	No fit	Fit	No fit	Fit
Weights	0.07	0.09	1.99	2.06	10.75	10.90	21.51	21.65	32.82	33.12	44.20	44.48
Basis fns	0.34	0.82	2.80	7.66	6.56	28.98	10.36	43.00	14.29	58.09	18.41	74.05
Density	0.59	0.24	15.82	1.61	47.00	3.61	78.41	5.41	112.00	7.38	139.10	9.29
XC functional	0.09	0.10	0.29	0.35	0.57	0.67	0.81	0.96	1.06	1.26	1.34	1.67
XC matrix	0.70	0.10	15.97	0.73	45.97	1.69	75.70	2.51	107.17	3.41	137.53	4.23
Total	1.80	1.36	36.89	12.43	110.89	45.89	186.85	73.60	267.42	103.35	340.68	133.83
Speedup:												
Density		2.5		9.8		13.0		14.5		15.2		15.0
XC matrix		7.0		22		27.2		30.2		31.4		32.5
Rate-limiting steps		3.8		13.6		17.5		19.5		20.3		20.5

As expected, the speedup of the rate-limiting steps, evaluation of the charge density and the XC matrix, is considerable. The first thing to note is that significant improvement is obtained immediately even for the smallest molecule, methane. This illustrates one important breakthrough of the new formulation, namely that unlike most "fast" XC methods, it is *not* dependent on neglect of electronic interactions at large distances (spatial cutoffs) to derive its computational advantage. Cutoff-based methods offer zero improvement for methane, for example. The new method does not replace the use of cutoffs, either, as evidenced by the fact that asymptotically linear scaling is indeed observed in the cost of the dominant XC terms here (effectively reached by $C_{15}H_{32}$); spatial cutoffs are complementary and benefit the new fit significantly as well. The asymptotically limiting values for improvement of the dominant XC terms are factors of 15 and 30, respectively, for an overall improvement of a factor of 20 on the rate-limiting steps in this case.

Of course, the best measure of benefit is the impact on the calculation time *as a whole*, which was also predicted to be significant given that the XC component often dominates the DFT SCF calculation. From the characterization of the gains on the XC rate-limiting steps in the current prototype implementation, we can at this stage reliably estimate the final achievable improvement to overall calculation by the production implementation to be developed. The remaining refinements of the prototype necessary to realize the full computational power of the new method are perhaps surprisingly simple. These concern the need for more efficient treatment of the XC quadrature weights and basis function evaluation on the grid.

Without the XC fit, the quadrature weights and basis functions were not particularly significant computationally, together comprising less than 20% of the total XC time. However, upon effectively eliminating the previously dominant steps through the fit, these contributions become the new dominant steps, jumping up towards 90% of the total. In order to leverage the fitting gains fully, improvement is needed on these new rate-limiting terms.

Having the weights and basis function values as the new rate-limiting steps is a fortunate point to reach, as the necessary optimizations are straightforward to carry out. First, unlike the other XC terms in Table 6, the weights and basis functions are independent of the density, do not change from cycle to cycle during the SCF, and thus can be computed once and reused. NWChem currently recalculates both of these on each cycle. Storing the quadrature weights in-core is entirely practical, as only one value needs to be stored per grid point, and has been implemented successfully in other quantum chemistry programs in the past [8].

The basis function values can also be pre-computed and stored, but in this case the storage requirements are more serious, scaling as M_G values per grid point, which on most current single-CPU systems eventually becomes impractical. When sufficient storage is available, however, this is the way to go, and an in-core approach is tailor-made for large-scale calculations on distributed memory parallel systems, which will be developed in future work. It is perfectly plausible to expect superlinear parallel speedups to be attained with this approach. Note that without the new XC fit this is not possible. Though high parallel efficiencies are well known for

XC, the prior rate-limiting steps of the density and XC matrix must be re-evaluated at each SCF cycle, and thus re-using the basis functions cannot make a significant impact without the new fit.

There will also be many situations where the basis function will have to be re-evaluated on each cycle, but there is also considerable improvement possible in that case. Not surprisingly, due to its previous low importance, the current NWChem implementation of the basis functions in particular was found to be extremely inefficient, performing many redundant computations, and the existing NWChem implementation is far from indicative of the attainable performance. What can be expected from an efficient implementation? There is every reason to expect that the cost of the basis functions can be reduced to the same level as say, the charge density. A simple argument for this is that the angular factors can be pre-computed on the grid, which depends only on the highest angular momentum in the basis set and not on the total number of basis functions. The radial functions require evaluation of an exponential (once for each primitive for contracted functions, though most fitting functions are uncontracted); however, storage and reuse of only the radial components requires considerably less storage and is a much more manageable proposition. Together with the efficient evaluation of the angular components, then on each cycle the basis functions can be formed with a single multiplication per function (per grid point). This is on the same order of work required for the charge density, which takes a multiplication and an addition per function. The need to focus serious attention on optimization of basis function evaluation is another compelling testament to the power of the new XC formulation.

Therefore, in a production implementation, the quadrature weights are evaluated only once during the SCF, and the basis function cost can be brought down to the level of the charge density, either incurred each cycle or only once depending on available storage. This together with the performance measurements from Table 7 was used to create estimates of the impact of the production implementation on the *total* job time in Table 8. While expected to be fairly accurate, these estimates are conservative in that they assume no further improvement can be made in the density and XC matrix steps over the prototype implementation, which is unlikely given that no such optimization was yet attempted.

The results show that the factor of 20 improvement on the rate-limiting steps projects to a factor of 16 on the total XC calculation, retaining the majority of that benefit. This reduces the total XC cost to only a few percent of its prior stature, decisively removing it as the computational bottleneck. For this test set, the XC cost comprised 80%–95% of the total job time when the two-electron integrals could be stored, and 30% when the integrals were calculated in direct mode for the largest job. This translates to a projected speedup on the total job of a factor of 3–7 in the former case and 1.4 in the latter. Note that the maximum gain possible for the entire job depends on the extent to which the XC part dominates, but by effectively eliminating the XC cost (reducing it by more than an order of magnitude), the new method is capable of closely approaching this maximum gain in practice. It is expected that total speedups of an order of magnitude can be routinely realized in the production version.

Table 8 Projected speedups for entire SCF calculation with production implementation of new XC fit, B-LYP/6-31G*/Ahlrichs

Molecule	CH_4	C_5H_{12}	$C_{10}H_{22}$	$C_{15}H_{32}$	$C_{20}H_{42}$	$C_{25}H_{52}$
% XC time (before fit)	94.7	92.5	88.4	84.6	80.7	29.9
Projected total XC speedup						
Efficient basis function evaluation	2.6	8.2	10.4	11.4	11.8	11.8
Reuse of basis function values	3.8	12.1	14.9	16.1	16.6	16.6
Projected overall job speedup						
Efficient basis function evaluation	2.4	5.3	5.0	4.4	3.8	1.4
Reuse of basis function values	3.3	6.6	5.7	4.8	4.1	1.4

Given that the quadrature weight and basis function issues can be straightforwardly and satisfactorily handled, we will henceforth continue to focus on the improvement to the charge density and XC matrix.

In addition to the one-dimensional straight-chain alkanes studied, calculations were performed on a series of two-dimensional (graphite sheets), and three-dimensional molecules (diamond-like carbon clusters). Representative results showing how the XC fit speedup is affected by molecular shape are given in Table 9. As expected, for the more compact diamond-like structures the advantage obtained is greater than for the graphite structure having the same number of carbons, because the amount of significant grid work per point is greater. In practice, most large organic molecules are effectively approximately two-dimensional for the purpose of spatial cutoffs, and so speedups can often be expected in practice that are greater than these alkane results.

Table 10 investigates the effect on the speedup of using a local or gradient-corrected functional. Overall, it appears that the speedup is effectively the same, with the local functional (S-VWN) having a slight advantage in density evaluation and the gradient-corrected functional having more of an advantage on the XC matrix.

Table 9 Effect of molecular shape on new XC fit speedups, B-LYP/6-31G*/Ahlrichs

Number of carbons	5	10	15	20	25
Number of orbital basis functions	75	150	225	300	375
Number of density basis functions	220	440	660	880	1100
	Speedup				
Graphite sheet					
Density	5.2	9.04	11.7	13.9	13.9
XC matrix	15	22	26.3	29.8	24.9
Rate-limiting steps	8.2	13.3	16.5	19.1	17.9
Diamond cluster					
Density	7.4	13.2	17.9	21.2	24.7
XC matrix	16	23.9	29.7	33.3	34.5
Rate-limiting steps	10.	16.7	21.8	25.3	28.2

Table 10 Effect of functional type on new XC fit speedups

Molecule	CH_4	C_5H_{12}	$C_{10}H_{22}$	$C_{15}H_{32}$	$C_{20}H_{42}$	$C_{25}H_{52}$
S-VWN/6-31G*/Ahlrichs						
Density	2.3	11.8	15.8	17.5	14.1	14.0
XC matrix	5.4	19.1	24.9	26.0	17.1	18.0
Rate-limiting steps	3.2	14.1	18.7	20.3	15.2	15.5
B-LYP/6-31G*/Ahlrichs						
Density	2.5	9.8	13.0	14.5	15.2	15.0
XC matrix	7.0	22	27.2	30.2	31.4	32.5
Rate-limiting steps	3.8	13.6	17.5	19.5	20.3	20.5

The S-VWN speedups for $C_{20}H_{42}$ and $C_{25}H_{52}$ seem slightly out of line from other results, even though they are quite impressive. The reason for this deviation is not entirely clear. The absolute timings for the density and XC matrix by the new fit are small here (only a few seconds), and thus a small fluctuation in timing (such as with system load) can cause a noticeable change in the observed speedup.

Table 11 shows the computational advantage on the alkane series of different pairings of orbital and fit basis sets. In addition to the 6-31G*/Ahlrichs combination used to this point (speedups reported in Table 7), the standard A1 Coulomb fitting density basis set, which is smaller than Ahlrichs but is perhaps the most commonly used in practical work, was paired with 6-31G* as well as the smaller 3-21G orbital basis. In addition, to observe the effect with a large orbital basis, the standard 6-311++G** orbital basis was also paired with the Ahlrichs fit basis. The other two possible pairings among these basis sets (3-21G with Ahlrichs and 6-311++G** with A1) were not investigated because they are likely too unbalanced for practical work.

With 3-21G/A1 the smallest basis sets, the speedups are similar to but slightly less than with 6-31G*/Ahlrichs, yielding a limiting value of a factor of 16. The ratio of fit basis to orbital basis size is similar for these combinations, about 3.2 and 3.9 respectively. Gaining more than an order of magnitude for 3-21G/A1 is

Table 11 Effect of orbital and density basis set pairing on new XC fit speedups

Molecule	CH_4	C_5H_{12}	$C_{10}H_{22}$	$C_{15}H_{32}$	$C_{20}H_{42}$	$C_{25}H_{52}$
B-LYP/3-21G/A1						
Density	2.5	9.4	11.2	11.7	12.3	12.2
XC matrix	8.6	20.0	22.8	24.4	24.9	25.5
Rate-limiting steps	4.0	12.5	14.7	15.6	16.1	16.2
B-LYP/6-31G*/A1						
Density	3.9	15.1	21.3	22.6	22.4	22.5
XC matrix	14	38.1	48.2	48.7	51.5	52.1
Rate-limiting steps	6.4	21.9	29.3	30.6	31.2	31.3
B-LYP/6-311++G**/Ahlrichs						
Density	14.2	67.6	120.	153	155	167
XC matrix	46.2	163	251	322	241	266
Rate-limiting steps	23.1	97.8	163	208	183	198

particularly impressive given that this is attainable for such a small orbital basis. For 6-31G*/A1, another commonly used combination, the speedups are greater than for 6-31G*/Ahlrichs as expected due to the smaller fitting basis. The speedup on the rate-limiting steps rises from a factor of 20 to a factor of 30.

The final combination, involving the largest orbital and density basis sets, yields speedups that are particularly spectacular – over two orders of magnitude. The ratio of fit basis to orbital basis size is low here, around 2.0, and it is not yet known if this pairing is reliable in terms of chemical accuracy. This is certainly an interesting question for follow-up. However, there is reason to be optimistic. Even if the accuracy is not satisfactory as is, some of the extra orbital basis size comes from additional polarization functions, which are less important in DFT (*i.e.* only as they contribute to the charge density) than in conventional correlated methods. Furthermore, the Ahlrichs basis contains a large proportion of high angular momentum functions, and there is a question of whether the emphasis in the basis might usefully be shifted from the costly towards more capability for valence density modeling, giving more fitting power with the same basis set size. In any case, several techniques for optimizing the fit basis set for maximum speedup are planned, and the maximum speedups in Table 11 are by no means out of the question for practical work.

In summary, the computational advantage afforded by the new XC fit is as expected or better than expected in all cases. The observed speedups on the rate-limiting XC steps are consistently greater than a factor of 10, typically significantly so, with speedups over a factor of 200 obtained in the best case. These results were obtained before any attempt to optimize the prototype implementation. The production implementation will make the entire XC cost essentially nonexistent relative to its present size and thus yield a significant speedup to the total SCF calculation time.

Finally, though explicit timings were not presented for the old fitted XC method [17], a relevant comparison can easily be drawn. In addition to its lack of theoretical rigor which makes it unacceptable as a general solution to the XC cost problem (as evidenced by its failure to gain widespread use), recall that the old method evaluates the orbital-based density in equation (4) on the grid, not the fitted density, and uses an auxiliary basis for expanding the XC potential only. That is, the old method can only offer potential improvement on one of the two rate-limiting steps, the XC matrix, and does not change the cost of the charge density step. This means that the maximum improvement possible by the old method is only a factor of two in the XC part, which is minuscule compared to the gains obtained here.

5.5 Negative Fitted Densities

As established in the preceding results, the energy expression chosen performed more than adequately, and hence no alteration or refinement was necessary. There is one related issue, however, that warrants further discussion. Since the fitted density is not constructed from a sum of squares as the orbital-based density is, the possibility exists that the fitted density used in the XC functional can take on negative values. In previous fitting experience, this has only been observed as slightly negative values at large distances from the molecule that make a negligible contribution to the result [29].

Negative densities were also observed in the present work, though so far this has not presented a serious problem. The negative values were monitored by computing the numerical integral of the region of negative density. The result is the number of "electrons" contained in the total region of negative density. In the calculations performed in this study, the negative densities have all been small in magnitude. With the Ahlrichs basis set most commonly no negative density was observed, and the maximum negative content observed over all calculations was 0.0033 electrons. Negative density values were more common, not surprisingly, with the smaller A1 basis set, with the largest negative content observed being 0.017 electrons. Though undesirable, put in perspective these values are the same order of magnitude than the change that would result from omitting the fit charge conservation constraint. That effect is typically on the order of 10^{-3}–10^{-2} electrons, yet unconstrained fits have been shown generally to be perfectly acceptable in most practical calculations.

We point out that the origin of the negative density is not inherent to the new energy expression. Negative densities can and do occur with Coulomb-only fits. The difference, however, is that since the Fock matrix is linear in the fit coefficients for a Coulomb fit, no special treatment of negative densities is needed. Though unphysical, any negative contributions will be exactly canceled (in a charge-constrained fit) by excess positive contributions elsewhere. Things are not this simple for an XC fit however. Most XC functionals are (rightly) undefined for negative values of the density, and currently the negative density values are handled adequately simply by setting them to zero. Doing so does not in fact introduce any problems with computational robustness as might first be thought, as it is equivalent to defining the XC functional to be identically zero for negative values of the density. This maintains continuity in the XC functional and its first derivatives, and thus there will be no problems with SCF convergence caused by this treatment, as verified in practice. Also, as long as the magnitude of the negative density remains small, its neglect will not adversely affect the calculated energy and properties, as also empirically verified by the present results.

However, it would be desirable if the unphysical negative densities could be eliminated altogether, and this will be a point of further investigation. For one thing, a relevant question is whether the size of the fitting basis sets might even be further reduced to improve computational efficiency. In doing so the problem of negative densities becomes of greater concern, such that simple neglect of negative contributions may no longer be adequate.

What reason is there to expect that negative densities can in fact be eliminated or significantly reduced while keeping an atom-centered fit basis of manageable size? One important note is that we have observed that the magnitude of negative density is not increased in going from Coulomb-only to the new XC fit. The new fit does not exacerbate (or alleviate) this problem. This is encouraging, because it suggests that the method itself does not have a strong influence on the negative density and thus this is likely a basis set issue entirely. With the Ahlrichs basis, negative density occurs sometimes, but rarely, and when it does the magnitude is small. The Ahlrichs basis is considerably larger than A1, but nonetheless it is small enough to be used to advantage in practical calculations. This suggests there may be room for improvement by keeping the basis the same size and making a simple adjustment of the basis parameters to lessen or remove negative density without sacrificing quality otherwise. This question has not previously been investigated in fit basis set development since it has not been a relevant issue before. Plus, the average magnitude of negative density is already small with currently used basis sets, so not much improvement is needed. Investigation of this issue may give insight into how to use even smaller fit basis sets and improve the speedup even more.

6 Conclusion

In summary, the generalized formulation of fitted DFT presented has led to a new XC fitting method that removes the computational bottleneck of the XC integrals. The new method delivers the same accuracy as Coulomb-fitted DFT, but in dramatically reduced computation time. The new method is directly analogous to a fitted J-matrix method and is completely rigorous theoretically, suffering from none of the drawbacks plaguing prior XC fitting methods. All these attributes have never before been simultaneously achieved in prior attempts to improve the XC cost. Therefore, the viability of the new method has been clearly established and the groundwork has been laid for exciting future development work.

Acknowledgments This work was supported by the U. S. National Science Foundation through Small Business Innovation Research award number DMI-9960903 and by Pacific Northwest National Laboratory. We are grateful to the staff of the William R. Wiley Environmental Molecular Sciences Laboratory at Pacific Northwest National Laboratory for their assistance in this work. We are particularly indebted to Carlos A. Gonzalez for many stimulating discussions on this topic.

References

1. P. Hohenberg and W. Kohn, *Phys. Rev. B* **136**, 864 (1964).
2. W. Kohn and L. J. Sham, *Phys. Rev. A* **140**, 1133 (1965).
3. R. G. Parr and W. Yang, *Density-Functional Theory of Atoms and Molecules* (Oxford Univ. Press, New York, 1989).
4. A. D. Becke, *J. Chem. Phys.* **88**, 2547 (1988).
5. B. G. Johnson and M. J. Frisch, *Chem. Phys. Lett.* **216**, 133 (1993).
6. B. G. Johnson, P. M. W. Gill and J. A. Pople, *Chem. Phys. Lett.* **220**, 377 (1994).
7. B. I. Dunlap, J. W. D. Connolly and J. R. Sabin, *J. Chem. Phys.* **71**, 3396 (1979).
8. B. G. Johnson, unpublished work.
9. E. J. Bylaska, W. A. de Jong, K. Kowalski, T. P. Straatsma, M. Valiev, D. Wang, E. Aprà, T. L. Windus, S. Hirata, M. T. Hackler, Y. Zhao, P.-D. Fan, R. J. Harrison, M. Dupuis, D. M. A. Smith, J. Nieplocha, V. Tipparaju, M. Krishnan, A. A. Auer, M. Nooijen,

E. Brown, G. Cisneros, G. I. Fann, H. Früchtl, J. Garza, K. Hirao, R. Kendall, J. A. Nichols, K. Tsemekhman, K. Wolinski, J. Anchell, D. Bernholdt, P. Borowski, T. Clark, D. Clerc, H. Dachsel, M. Deegan, K. Dyall, D. Elwood, E. Glendening, M. Gutowski, A. Hess, J. Jaffe, B. Johnson, J. Ju, R. Kobayashi, R. Kutteh, Z. Lin, R. Littlefield, X. Long, B. Meng, T. Nakajima, S. Niu, L. Pollack, M. Rosing, G. Sandrone, M. Stave, H. Taylor, G. Thomas, J. van Lenthe, A. Wong and Z. Zhang, "NWChem, A Computational Chemistry Package for Parallel Computers, Version 5.0" (2006), Pacific Northwest National Laboratory, Richland, Washington 99352-0999, USA. A modified version.
10. R. A. Kendall, E. Aprà, D. E. Bernholdt, E. J. Bylaska, M. Dupuis, G. I. Fann, R. J. Harrison, J. Ju, J. A. Nichols, J. Nieplocha, T. P. Straatsma, T. L. Windus and A. T. Wong, *Computer Phys. Comm.* **128**, 260 (2000).
11. J. A. Nichols, private communication.
12. B. G. Johnson and M. J. Frisch, *J. Chem. Phys.* **100**, 7429 (1994).
13. B. G. Johnson, Ph.D. thesis, Carnegie Mellon University (1993).
14. J. M. Perez-Jorda and W. Yang, *Chem. Phys. Lett.* **241**, 469 (1995).
15. J. Andzelm and E. Wimmer, *J. Chem. Phys.* **96**, 1280 (1992).
16. Y. Jung, A. Sodt, P. M. W. Gill and M. Head-Gordon, *Proc. Nat. Acad. USA* **102**, 6692 (2005).
17. D. R. Salahub, In *Density Functional Methods in Chemistry*, J. K. Labanowski and J. W. Andzelm, Eds. (Springer-Verlag, New York, 1991).
18. Y.C. Zheng and J. Almlöf, *Chem. Phys. Lett.* **214**, 397 (1993).
19. A. Komornicki and G. Fitzgerald, *J. Chem. Phys.* **98**, 1398 (1993).
20. R. E. Stratmann, G. E. Scuseria and M. J. Frisch, *Chem. Phys. Lett.* **257**, 213 (1996).
21. J. Heyd and G. E. Scuseria, *J. Chem. Phys.* **120**, 7274 (2004).
22. E. Apra and J. A. Nichols, private communication.
23. P. M. W. Gill, *Mol. Phys.* **89**, 433 (1996).
24. S. B. Andrews, O. Treutler, S. W. Taylor and P. M. W. Gill, to be published.
25. N. Oliphant and R. J. Bartlett, *J. Chem. Phys.* **100**, 6550 (1994).
26. M. Tobita, S. Hirata and R. J. Bartlett, *J. Chem. Phys.* **118**, 5776 (2003).
27. L. A. Curtiss, K. Raghavachari, G. W. Trucks and J. A. Pople, *J. Chem. Phys.* **94**, 7221 (1991).
28. B. G. Johnson, P. M. W. Gill and J. A. Pople, *J. Chem. Phys.* **98**, 5612 (1993).
29. B. I. Dunlap, private communication.

Appendix

Table 12 Number of SCF cycles to convergence for the B-LYP functional using the 6-31G* orbital basis set and the Ahlrichs fitting basis set

Molecule	No fit	J fit	Old J+XC fit	New J+XC fit
BeH	9	9	13	16
C2H2	6	6	6	7
CH.	9	9	9	9
CH2-1	14	14	14	14
CH2-3	8	8	7	8
CH2CH2	6	6	6	7
CH3	11	11	10	11
CH3CH3	6	6	6	7
CH3OH	7	7	8	7
CH4	6	6	6	7
CO	8	8	7	8
CO2	7	7	8	7
F2	6	6	6	6
H2	5	5	5	5
H2CO	8	8	8	8
H2NNH2	7	7	7	7
H2O	7	7	8	7
H2O2	7	7	8	7
HCN	7	7	7	7
HCO	8	8	8	8
HF	7	7	7	7
Li2	21	21	6	7
LiF	13	13	13	13
LiH	8	8	9	8
N2	6	7	7	6
NH	15	13	10	12
NH2	9	9	10	10
NH3	6	6	6	6
O2	8	8	8	8
OH	11	11	11	11

Table 13 Number of SCF cycles to convergence for the B-LYP functional using the STO-3G orbital basis set and the DGauss A1 Coulomb and exchange fitting basis sets

Molecule	No fit	J fit	Old J+XC fit	New J+XC fit
BeH	16	10	17	17
C2H2	7	7	7	7
CH.	18	7	17	8
CH2-1	11	11	11	11
CH2-3	17	13	11	23
CH2CH2	7	7	8	6
CH3	11	12	*	9
CH3CH3	7	7	10	7
CH3OH	7	7	10	7
CH4	7	7	9	6
CO	7	7	6	7
CO2	7	7	6	7
F2	6	6	6	6
H2	3	3	3	3
H2CO	7	7	9	7
H2NNH2	7	7	7	6
H2O	7	7	13	6
H2O2	7	7	10	6
HCN	7	13	10	7
HCO	13	13	13	9
HF	7	7	11	6
Li2	5	5	6	7
LiF	14	15	17	18
LiH	7	8	11	7
N2	5	5	8	5
NH	8	10	17	15
NH2	13	12	*	14
NH3	7	7	12	6
O2	7	7	7	9
OH	13	13	16	16

* SCF convergence failure

Table 14 Density functional and experimental atomization energies (kcal/mol). DFT values obtained with the B-LYP functional using the 6-31G* orbital basis set and the Ahlrichs fitting basis set

Molecule	No fit	J fit	Old J+XC fit	New J+XC fit	Experiment
BeH	56.8	56.9	56.9	57.3	46.9
C2H2	399.3	399.3	399.3	399.1	388.9
CH	83.5	83.5	83.5	83.4	79.9
CH2-1	175.8	175.8	175.8	175.6	170.6
CH2-3	188.2	188.2	188.2	188.2	179.6
CH2CH2	559.1	559.2	559.2	558.9	531.9
CH3	305.5	305.5	305.5	305.5	289.2
CH3CH3	706.5	706.6	706.7	706.3	666.3
CH3OH	506.3	506.5	506.4	506.1	480.8
CH4	417.3	417.4	417.4	417.2	392.5
CN	185.3	185.3	185.3	185.1	176.6
CO	260.4	260.5	260.5	260.1	256.2
CO2	399.8	399.9	399.8	399.5	381.9
F2	55.7	55.8	55.8	56.5	36.9
H2	109.4	109.4	109.4	109.4	103.3
H2CO	377.8	377.9	377.9	377.8	357.2
H2NNH2	434.3	434.4	434.4	434.7	405.4
H2O	219.9	220.0	219.8	220.0	219.3
H2O2	268.2	268.4	268.0	268.5	252.3
HCN	316.1	316.1	316.1	315.9	301.8
HCO	286.0	286.1	286.1	285.7	270.3
HF	129.8	129.8	129.8	129.9	135.2
Li2	20.2	20.2	20.2	20.5	24.0
LiF	136.9	136.9	136.9	137.6	137.6
LiH	56.8	56.8	56.8	57.4	56.0
N2	234.4	234.5	234.5	234.2	225.1
NH	86.0	86.0	86.0	86.1	79.0
NH2	182.0	182.0	182.0	182.1	170.0
NH3	290.8	290.9	290.9	291.1	276.7
NO	164.1	164.3	164.2	163.8	150.1
O2	139.0	139.2	138.8	139.1	118.0
OH	103.5	103.5	103.4	103.5	101.3

Table 15 Density functional and experimental atomization energies (kcal/mol). DFT values obtained with the B-LYP functional using the STO-3G orbital basis set and the DGauss A1 Coulomb and exchange fitting basis sets

Molecule	No fit	J fit	Old J+XC fit	New J+XC fit	Experiment
BeH	56.8	56.9	56.9	57.3	46.9
C2H2	399.3	399.3	399.3	399.1	388.9
CH	83.5	83.5	83.5	83.4	79.9
CH2-1	175.8	175.8	175.8	175.6	170.6
CH2-3	188.2	188.2	188.2	188.2	179.6
CH2CH2	559.1	559.2	559.2	558.9	531.9
CH3	305.5	305.5	305.5	305.5	289.2
CH3CH3	706.5	706.6	706.7	706.3	666.3
CH3OH	506.3	506.5	506.4	506.1	480.8
CH4	417.3	417.4	417.4	417.2	392.5
CN	185.3	185.3	185.3	185.1	176.6
CO	260.4	260.5	260.5	260.1	256.2
CO2	399.8	399.9	399.8	399.5	381.9
F2	55.7	55.8	55.8	56.5	36.9
H2	109.4	109.4	109.4	109.4	103.3
H2CO	377.8	377.9	377.9	377.8	357.2
H2NNH2	434.3	434.4	434.4	434.7	405.4
H2O	219.9	220.0	219.8	220.0	219.3
H2O2	268.2	268.4	268.0	268.5	252.3
HCN	316.1	316.1	316.1	315.9	301.8
HCO	286.0	286.1	286.1	285.7	270.3
HF	129.8	129.8	129.8	129.9	135.2
Li2	20.2	20.2	20.2	20.5	24.0
LiF	136.9	136.9	136.9	137.6	137.6
LiH	56.8	56.8	56.8	57.4	56.0
N2	234.4	234.5	234.5	234.2	225.1
NH	86.0	86.0	86.0	86.1	79.0
NH2	182.0	182.0	182.0	182.1	170.0
NH3	290.8	290.9	290.9	291.1	276.7
NO	164.1	164.3	164.2	163.8	150.1
O2	139.0	139.2	138.8	139.1	118.0
OH	103.5	103.5	103.4	103.5	101.3

Table 16 Density functional and experimental dipole moments (debye). DFT values obtained with the B-LYP functional using the 6-31G* orbital basis set and the Ahlrichs fitting basis set

Molecule	No fit	J fit	Old J+XC fit	New J+XC fit	Experiment
CH	1.350	1.349	1.348	1.354	1.46
CH3OH	1.613	1.611	1.623	1.618	1.70
CN	0.933	0.960	0.913	0.948	1.15
CO	0.145	0.145	0.152	0.147	0.11
H2CO	2.053	2.053	2.044	2.058	2.33
H2NNH2	2.143	2.141	2.139	2.150	1.75
H2O	2.019	2.017	2.040	2.024	1.85
H2O2	1.662	1.661	1.687	1.667	2.20
HCN	2.786	2.785	2.787	2.788	2.98
HF	1.795	1.794	1.793	1.797	1.82
LiF	5.375	5.375	5.370	5.387	6.33
LiH	5.519	5.520	5.455	5.617	5.88
NH	1.534	1.533	1.538	1.536	1.39
NH3	1.835	1.833	1.831	1.841	1.47
NO	0.207	0.239	0.232	0.235	0.15
OH	1.693	1.692	1.714	1.696	1.66

Table 17 Density functional and experimental dipole moments (debye). DFT values obtained with the B-LYP functional using the STO-3G orbital basis set and the DGauss A1 Coulomb and exchange fitting basis sets

Molecule	No fit	J fit	Old J+XC fit	New J+XC fit	Experiment
CH	1.020	1.023	1.022	1.020	1.46
CH3OH	1.304	1.307	1.313	1.309	1.70
CN	0.775	0.850	0.851	0.770	1.15
CO	0.550	0.551	0.546	0.552	0.11
H2CO	1.209	1.217	1.220	1.205	2.33
H2NNH2	2.144	2.146	2.152	2.154	1.75
H2O	1.621	1.622	1.628	1.625	1.85
H2O2	1.287	1.288	1.293	1.290	2.20
HCN	2.229	2.228	2.233	2.221	2.98
HF	1.138	1.138	1.146	1.130	1.82
LiF	2.247	2.244	2.269	2.236	6.33
LiH	4.704	4.721	4.708	4.905	5.88
NH	1.222	1.215	1.221	1.215	1.39
NH3	1.812	1.812	1.815	1.817	1.47
NO	0.491	0.495	0.490	0.503	0.15
OH	1.159	1.160	1.166	1.157	1.66

Table 18 Density functional and experimental equilibrium geometries. Bond lengths are in angstroms and bond angles are in degrees. DFT values obtained with the B-LYP functional using the 6-31G* orbital basis set and the Ahlrichs fitting basis set

Molecule	No fit	J fit	Old J+XC fit	New J+XC fit	Experiment
BeH					
Be–H	1.356	1.356	1.357	1.356	1.343
C2H2					
C–C	1.215	1.215	1.215	1.215	1.203
C–H	1.073	1.073	1.073	1.073	1.061
CH					
C–H	1.146	1.146	1.147	1.147	1.120
CH2-1					
C–H	1.132	1.132	1.133	1.132	1.111
H–C–H	99.1	99.1	99.1	99.2	102.4
CH2-3					
C–H	1.089	1.089	1.090	1.089	1.078
H–C–H	133.5	133.4	133.7	133.4	136.0
CH2CH2					
C–C	1.341	1.341	1.341	1.341	1.339
C–H	1.095	1.095	1.096	1.095	1.085
H–C–H	116.2	116.2	116.1	116.2	117.8
CH3					
C–H	1.090	1.090	1.090	*	1.079
CH3CH3					
C–C	1.540	1.540	1.540	*	1.526
C–H	1.104	1.104	1.104		1.088
H–C–H	107.5	107.5	107.4		107.4
CH3OH					
C–O	1.435	1.435	1.436	1.434	1.421
C–Ha	1.101	1.101	1.101	1.101	1.093
C–Hb	1.110	1.110	1.110	1.110	1.093
O–H	0.980	0.980	0.980	0.980	0.963
O–C–Ha	106.4	106.4	106.7	106.5	107.0
C–O–H	106.8	106.8	106.6	107.0	108.0
Hb–C–Hb	108.4	108.4	108.3	108.3	108.5
CH4					
C–H	1.1	1.100	1.101	*	1.086
CN					
C–N	1.2	1.187	1.187	1.187	1.172
CO					
C–O	1.2	1.151	1.151	1.151	1.128
CO2					
C–O	1.2	1.183	1.183	1.184	1.162
F2					
F–F	1.4	1.434	1.433	1.433	1.417
H2					
H–H	0.7	0.748	0.750	0.748	0.741
H2CO					
C–O	1.2	1.218	1.218	1.218	1.208
C–H	1.1	1.121	1.122	1.121	1.116
H–C–H	114.9	114.9	114.8	114.9	116.5

Table 18 (continued)

Molecule	No fit	J fit	Old J+XC fit	New J+XC fit	Experiment
H2NNH2					
N–N	1.5	1.464	1.468	1.462	1.447
N–Hb	1.0	1.033	1.034	1.033	1.008
N–Ha	1.0	1.028	1.028	1.027	1.008
N–N–Hb	111.1	111.0	110.9	111.2	109.2
N–N–Ha	105.4	105.4	105.2	105.5	109.2
Ha–N–Hb	105.6	105.6	105.6	105.8	113.3
H2O					
O–H	0.980	0.980	0.980	0.980	0.909
H–O–H	102.7	102.7	103.0	102.7	103.9
H2O2					
O–O	1.494	1.493	1.505	1.493	1.475
O–H	0.985	0.986	0.985	0.985	0.950
O–O–H	98.5	98.5	97.8	98.5	94.8
H–O–O–H	120.7	120.7	127.1	120.7	120.0
HCN					
C–N	1.169	1.169	1.169	1.169	1.153
C–H	1.078	1.078	1.078	1.078	1.065
HCO					
C–O	1.196	1.196	1.196	1.196	1.117
C–H	1.141	1.141	1.142	1.141	1.110
H–C–O	122.9	122.9	123.0	123.0	127.4
HF					
H–F	0.945	0.945	0.946	0.945	0.917
Li2					
Li–Li	2.728	2.728	2.731	2.730	2.670
LiF					
Li–F	1.582	1.582	1.583	*	1.564
LiH					
Li–H	1.629	1.629	1.625	1.629	1.595
N2					
N–N	1.118	1.118	1.118	1.118	1.098
NH					
N–H	1.060	1.060	1.061	1.061	1.045
NH2					
N–H	1.046	1.046	1.047	1.046	1.024
H–N–H	101.1	101.1	101.2	101.1	103.4
NH3					
N–H	1.030	1.030	1.031	1.029	1.012
H–N–H	113.8	113.8	113.9	113.7	106.0
NO					
N–O	1.176	1.176	1.177	1.177	1.151
O2					
O–O	1.239	1.240	1.242	1.241	1.207
OH					
O–H	0.995	0.995	0.995	0.995	0.971

* Geometry optimization convergence failure

Table 19 Density functional and experimental equilibrium geometries. Bond lengths are in angstroms and bond angles are in degrees. DFT values obtained with the B-LYP functional using the STO-3G orbital basis set and the DGauss A1 Coulomb and exchange fitting basis sets

Molecule	No fit	J fit	Old J+XC fit	New J+XC fit	Experiment
BeH					
Be–H	1.327	1.328	1.332	1.330	1.343
C2H2					
C–C	1.213	1.213	1.213	1.214	1.203
C–H	1.088	1.089	1.089	1.091	1.061
CH					
C–H	1.193	1.194	1.193	*	1.120
CH2-1					
C–H	1.169	1.170	1.169	1.174	1.111
H–C–H	96.0	95.9	95.9	95.9	102.4
CH2-3					
C–H	1.107	1.109	1.109	1.113	1.078
H–C–H	126.2	125.9	126.2	124.8	136.0
CH2CH2					
C–C	1.345	1.344	1.345	1.343	1.339
C–H	1.105	1.107	1.106	1.110	1.085
H–C–H	116.0	116.0	116.0	116.2	117.8
CH3					
C–H	1.099	1.100	1.100	*	1.079
CH3CH3					
C–C	1.564	1.564	1.563	*	1.526
C–H	1.109	1.110	1.110		1.088
H–C–H	108.2	108.2	108.1		107.4
CH3OH					
C–O	1.487	1.487	1.487	1.488	1.421
C–Ha	1.114	1.116	1.116	1.119	1.093
C–Hb	1.123	1.124	1.124	1.127	1.093
O–H	1.046	1.046	1.046	1.045	0.963
O–C–Ha	106.4	106.5	106.5	106.7	107.0
C–O–H	100.5	100.5	100.6	100.6	108.0
Hb–C–Hb	107.6	107.6	107.7	107.5	108.5
CH4					
C–H	1.104	1.106	1.106	*	1.086
CN					
C–N	1.230	1.230	1.230	1.230	1.172
CO					
C–O	1.205	1.206	1.206	1.205	1.128
CO2					
C–O	1.245	1.245	1.245	1.245	1.162
F2					
F–F	1.398	1.398	1.398	1.392	1.417
H2					
H–H	0.734	0.739	0.738	0.749	0.741
H2CO					
C–O	1.264	1.265	1.265	*	1.208
C–H	1.137	1.138	1.138		1.116
H–C–H	113.0	113.1	113.1		116.5

Table 19 (continued)

Molecule	No fit	J fit	Old J+XC fit	New J+XC fit	Experiment
H2NNH2					
N–N	1.590	1.589	1.589	*	1.447
N–Hb	1.090	1.090	1.091		1.008
N–Ha	1.090	1.090	1.091		1.008
N–N–Hb	98.3	98.4	98.4		109.2
N–N–Ha	98.3	98.4	98.4		109.2
Ha–N–Hb	97.7	97.7	97.7		113.3
H2O					
O–H	1.044	1.045	1.045	1.045	0.909
H–O–H	96.2	96.2	96.2	96.3	103.9
H2O2					
O–O	1.496	1.496	*	*	1.475
O–H	1.059	1.059			0.950
O–O–H	98.0	98.1			94.8
H–O–O–H	121.3	121.3			120.0
HCN					
C–N	1.209	1.209	1.209	1.208	1.153
C–H	1.097	1.098	1.098	1.100	1.065
HCO					
C–O	1.251	1.251	1.252	1.252	1.117
C–H	1.157	1.158	1.158	1.163	1.110
H–C–O	120.4	120.4	120.4	120.3	127.4
HF					
H–F	1.012	1.013	1.014	1.009	0.917
Li2					
Li–Li	2.709	2.709	2.707	2.707	2.670
LiF					
Li–F	1.469	1.468	1.470	1.480	1.564
LiH					
Li–H	1.538	1.539	1.541	1.530	1.595
N2					
N–N	1.199	1.199	1.200	1.200	1.098
NH					
N–H	1.131	1.132	1.133	1.137	1.045
NH2					
N–H	1.108	1.109	1.110	1.112	1.024
H–N–H	95.9	95.9	95.8	95.8	103.4
NH3					
N–H	1.079	1.080	1.080	*	1.012
H–N–H	118.3	118.3	118.3		106.0
NO					
N–O	1.265	1.264	1.266	1.267	1.151
O2					
O–O	1.314	1.314	1.315	1.317	1.207
OH					
O–H	1.068	1.068	1.069	1.070	0.971

* Geometry optimization convergence failure

Index

Printed in the United States
112010LV00002BC/14/P

9 781402 060724

d in the United States